BLOODY CRIMES

ALSO BY JAMES L. SWANSON

Manhunt: The 12-Day Chase for Lincoln's Killer

BLOODY CRIMES

THE CHASE FOR JEFFERSON DAVIS AND THE DEATH PAGEANT FOR LINCOLN'S CORPSE

JAMES L. SWANSON

wm
WILLIAM MORROW
An Imprint of HarperCollins*Publishers*

The maps on pages 275, 300, and 380 were created by Kieran McAuliffe.

All interior artworks, unless otherwise indicated, are from the author's private collection. Grateful acknowledgment is made to the Library of Congress for photographs that appear on the front and back endpapers. Additionally, the author wishes to thank the following for use of the photographs that appear throughout the text: Library of Congress (pp. 4, 112, 137, 191, 211, 226, 264, 334, 335, 343, 346, 362, 377, 399, 403); Ed Steers, Jr. (p. 104); Ford's Theatre, National Park Service (p. 129); National Museum of Health and Medicine, Walter Reed Army Medical Center (p. 135); Stack's (p. 142); Indiana Historical Society (pp. 203, 260); Terrell Library, Washington State University (p. 207); Abraham Lincoln Presidential Library and Museum (p. 230); U.S. Army Center of Military History (p. 360); The Jefferson Davis Home and Presidential Library, Beauvoir (p. 364); New Orleans Public Library (pp. 379, 384); North Carolina State Archives (p. 385); Wes Cowan (p. 395).

HarperCollins books may be purchased for educational, business, or sales promotional use. For information please write: Special Markets Department, HarperCollins Publishers, 10 East 53rd Street, New York, NY 10022.

FIRST EDITION

Designed by Richard Oriolo

Library of Congress Cataloging-in-Publication Data

Swanson, James L.

Bloody crimes : the chase for Jefferson Davis and the death pageant for Lincoln's corpse / James L. Swanson. — 1st ed.

p. cm.

Includes bibliographical references and index.

ISBN 978-0-06-123378-4

1. Davis, Jefferson, 1808–1889—Captivity, 1865–1867. 2. Lincoln, Abraham, 1809–1865—Death and burial. 3. Fugitives from justice—United States—Case studies. 4. Political prisoners—United States—Case studies. 5. United States—History—Civil War, 1861–1865—Prisoners and prisons. I. Title.

E477.98.S93 2010

973.7'7092—dc22

2010029404

10 11 12 13 14 OV/RRD 10 9 8 7 6 5 4 3 2 1

In memory of my mother, Dianne M. Swanson (1931–2008),
who looked forward to this book but had no chance to read it.

In remembrance of John Hope Franklin (1915–2009),
with gratitude for three decades of teaching, counsel, and friendship,
and with fond memories of University of Chicago days.

CONTENTS

LIST OF ILLUSTRATIONS ix

INTRODUCTION xi

Prologue 1

1: "Flitting Shadows" 3

2: "In the Days of Our Youth" 42

3: "Unconquerable Hearts" 66

4: "Borne by Loving Hands" 98

5: "The Body of the President Embalmed!" 131

6: "We Shall See and Know Our Friends in Heaven" 160

7: "The Cause Is Not Yet Dead" 194

8: "He Is Named for You" 232

9: "Coffin That Slowly Passes" 268

10: "By God, You Are the Men We Are Looking For" 304

11: "Living in a Tomb" 333

12: "The Shadow of the Confederacy" 359

Epilogue 388

ACKNOWLEDGMENTS 404

BIBLIOGRAPHY 408

NOTES 433

INDEX 449

LIST OF ILLUSTRATIONS

1. "Bloody Crimes" carte de visite of Columbia and her eagle xiii
2. Senator Jefferson Davis on the eve of the Civil War 4
3. Fall of Richmond paper flag 35
4. Currier & Ives print of Richmond in flames 40
5. Abraham Lincoln oil portrait, as he appeared in 1865 43
6. The Petersen House 104
7. Sketch of Lincoln on his deathbed 112
8. The empty bed, just after Lincoln died 128
9. Bloody pillow 129
10. "The President Is Dead" broadside 132
11. Diagram of the bullet's path through Lincoln's brain 134
12. The bullet that killed Lincoln 135
13. Allegorical print of Booth trapped inside the bullet 137
14. Portrait engraving of George Harrington 142
15. Invitation to Lincoln's funeral 187
16. "Post Office Department" silk ribbon, April 19 funeral 190
17. Lincoln's hearse, Washington, D.C. 191
18. Photograph of General E. D. Townsend 203
19. War Department pass for Lincoln funeral train 207
20. Lincoln's funeral car 211
21. Silk mourning ribbon of the U.S. Military Railroad 213
22. President Lincoln's hearse, Philadelphia 221
23. The New York funeral procession 226
24. Lincoln in coffin, New York City 230

25. Memorial arch, Sing Sing, New York 233
26. Viewing pavilion, Cleveland, Ohio 253
27. Terre Haute & Richmond Railroad timetable 260
28. Photograph of memorial arch, Chicago 264
29. Lincoln's old law office; Springfield, May, 1865 272
30. A map of the Abraham Lincoln funeral train route 275
31. *Harper's Weekly* woodcut of burial in Springfield, Illinois 283
32. The first reward poster for Jefferson Davis 297
33. A map of Jefferson Davis's escape route 300
34. Photograph of Davis in the suit he wore at capture 310
35. $360,000 reward poster for Davis 319
36. Three caricatures depicting Davis in a dress 323
37. The raglan, shawl, and spurs Davis wore on the day of capture 328
38. Print of Davis ridiculed in prison 334
39. Sketch of Davis in his cell 335
40. Lincoln's home draped in bunting, May 24, 1865 339
41. Davis as a caged hyena wearing a ladies' bonnet 343
42. "The True Story . . ." print ridiculing Davis 346
43. Oil portrait of Jefferson Davis, ca. 1870s 360
44. Davis and family on their porch at Beauvoir, Mississippi 362
45. Oscar Wilde–inscribed photograph 364
46. Jefferson Davis late in life at Beauvoir 377
47. Davis lying in state, New Orleans, 1889 379
48. A map of the Davis funeral train route 380
49. Davis's New Orleans funeral procession, 1889 384
50. Raleigh, North Carolina, floral display and procession, 1893 385
51. The ghosts of Willie and Abraham haunting Mary Lincoln 389
52. Photographs of porcelain Lincoln memorial obelisk 395
53. The site of Jefferson Davis's capture, near Irwinville, Georgia 399
54. Jefferson Davis's library at Beauvoir, Mississippi 403

INTRODUCTION

My book *Manhunt: The 12-Day Chase for Lincoln's Killer* told the story of John Wilkes Booth's incredible escape from the scene of his great crime at Ford's Theatre and his run to ambush, death, and infamy at a Virginia tobacco barn. But the chase for Lincoln's killer was not the only thrilling journey under way as the Civil War drew to a close in April 1865. While the hunt for Lincoln's murderer transfixed the nation, two other men embarked on their own, no less dramatic, final journeys. One, Jefferson Davis, president of the Confederate States of America, was on the run, desperate to save his family, his country, and his cause. The other, Abraham Lincoln, the recently assassinated president of the United States, was bound for a different destination: home, the grave, and everlasting glory.

The title of this book has three origins—as a prophecy, a promise, and an elegy.

In October 1859, abolitionist John Brown launched his doomed raid on the U.S. arsenal at Harpers Ferry, Virginia, as a way of inciting a slave uprising. This daring but foolhardy attack, viewed as an affront to the institution of slavery, enraged the South and brought the United States closer to irrepressible conflict and civil war. Following his capture, Brown was tried and sentenced to hang. While in a Charles Town jail awaiting execution, he was allowed to keep a copy of the King James Bible. As the clock ticked down to his hanging, Brown leafed through the sacred text, searching for divinely inspired words of justification, prophecy, and warning. He dog-eared the pages most dear to him and then highlighted key passages with pen and pencil marks, including this verse from Ezekiel 7:23: "Make a chain: for the land is full of bloody crimes, and the city is full of violence." On the morning he was hanged, on December 2, 1859, he handed to one of his jailers the last note he would ever write: "I, John Brown, am now quite *certain* that the crimes of this *guilty land* will never be purged away but with *blood*."

On March 4, 1865, Abraham Lincoln delivered his second inaugural address. Although remembered today for its message of peace—"with malice toward none, with charity for all"—the speech had a dark side. In a passage often overlooked, Lincoln warned that slavery was a bloody crime that might not be expunged without the shedding of more blood: "Fondly do we hope—fervently do we pray—that this mighty scourge of war may speedily pass away. Yet, if God wills that it continue, until all the wealth piled by the bond-man's two hundred and fifty years of unrequited toil shall be sunk, and until every drop of blood drawn with the lash, shall be paid by another drawn with the sword, as was said three thousand years ago, so still it must be said 'the judgments of the Lord, are true and righteous altogether.' "

Within days of Lincoln's assassination on April 14, 1865, a Boston photographer published a fantastical carte de visite image to honor the fallen president. That was not unusual; printers, photographers, and stationers across the country produced hundreds of thousands, if not millions, of ribbons, badges, broadsides, poems, and photographs to mourn Lincoln. But the image from Boston was different, for it expressed a sentiment not of mourning but of vengeance. In

"MAKE A CHAIN, FOR THE LAND IS FULL OF BLOODY CRIMES."

this carte de visite, a stern-faced woman, crowned and draped as Columbia, accompanied by her servant, a screaming eagle about to take flight in pursuit of its prey, keeps a vigil over a portrait of the martyred president and echoes John Brown's old warning: "Make a chain, for the land is full of bloody crimes." Soon, in the aftermath of the chase for Jefferson Davis and the Lincoln assassination and death pageant, manacles and chains became symbols of the spring of 1865.

Northerners believed that Jefferson Davis and the Confederacy had committed many bloody crimes, including the assassination of Abraham Lincoln, the torture, starvation, and murder of Union prisoners of war, and the battlefield slaughter of soldiers. In the South, Lincoln and his armies were seen as perpetrators, not victims, of great crimes. In the climate of these dueling accusations, the people of the Union and the Confederacy both shared a common belief and could agree upon one thing. In the spring of 1865, an era of bloody crimes had reached its climax.

The spring of 1865 was the most remarkable season in American history. It was a time to mourn the Civil War's 620,000 dead and to bind up the nation's wounds. It was a time to lay down arms, to tally plantations and cities that had been laid to waste, and to plant new crops. It was a time to ponder events that had come to pass and to look forward to those yet to be. It was the time of the hunt for Jefferson Davis and of the funeral pageant for Abraham Lincoln, each a martyr to his cause. And it was the time in America, wrote Walt Whitman, "when lilacs last in the door-yard bloom'd."

BLOODY CRIMES

PROLOGUE

WASHINGTON, D.C.

If you go there today, and walk to the most desolate corner of the cemetery, and then descend the half-hidden, decaying black slate steps, past all the other graves, down toward Rock Creek and the trees, you will find the tomb, now long empty. No sign remains that he was ever here. His name was never chiseled into the stone arch above the entry. But here, during the Civil War, in the winter of 1862, eleven-year-old Willie Lincoln, his father's best-beloved son, was laid to rest. Here his ever-mourning father returned to visit him, to remember, and to weep. And here, the boy waited patiently behind the iron gates, locked inside the marble vault that looked no bigger than a child's playhouse, for his father to claim him and carry him home.

That appointment, like his tiny coffin, was set in stone: March 4, 1869, the day Abraham Lincoln would complete his second term as president of the United States, leave Washington, and undertake the long railroad journey west, to Illinois. But in the spring of 1865, in the first week of April, that homecoming seemed a long way off. President Lincoln still had so much more to do.

RICHMOND, VIRGINIA

If you visit his home today, you will find no sign that he ever left. The exterior of the house looks almost exactly like it does in the Civil War–era photographs. In his private office, documents still lie on his desk, as if awaiting his signature. His presidential oil portrait hangs on a wall. Maps chart the once mighty territorial expanse of the antebellum South's proud agricultural empire. Books line the shelves. Children's toys lie scattered across the floor. The house is furnished as it was April 2, 1865, the day he last walked out the door, never to return.

In the spring of 1865, in the first week of April, he also had much to do. The future was uncertain. His capital city could no longer be defended and might fall to invading Union armies within days, even hours. To save his country, he had to abandon the president's mansion and flee Richmond. He could take little with him. Soon he would leave behind almost all he loved, including his five-year-old son, Joseph, who had died in his White House in 1864 and now rested in the sacred grounds of the city's Hollywood Cemetery, where many Confederate heroes, including General J. E. B. Stuart, were also buried. Perhaps one day Jefferson Davis would return to claim the boy, but for now, he had to go on ahead.

CHAPTER ONE

"Flitting Shadows"

On the morning of Sunday, April 2, 1865, President Jefferson Davis walked, as was his custom, from the White House of the Confederacy to St. Paul's Episcopal Church, where Robert E. Lee and his wife worshipped and where Davis was confirmed as a member of the parish in 1861. Everything that day appeared beautiful and serene. The air smelled of spring, and the fresh green growth promised a season of new life. One of the worshippers, a young woman named Constance Cary, recalled that on this "perfect Sunday of the Southern spring, a large congregation assembled as usual at St. Paul's."

Richmond did not look like a city at war, but it had become a symbol of the conflict. As the capital city of the Confederate States of America, it was the seat of slavery's and secession's empire, one of the loveliest cities in the South, the spiritual center of Virginia's aristocracy and of the rebellion, and, for the entire bloody Civil War that had cost the lives of more than 620,000 men, a strategic obsession in the popular imagination of the Union.

Jefferson Davis at the height of his power.

Despite Richmond's vulnerable proximity to Washington, D.C.—the White House of the Confederacy stood less than one hundred miles from Lincoln's Executive Mansion—the Confederate capital had defied capture. Unlike the unfortunate citizens of New Orleans, Vicksburg, Atlanta, Savannah, Mobile, and Charleston, whose homes had been besieged and prostrated, the people of Richmond had never suffered bombardment, capture, or surrender. In the spring of 1861, Yankee volunteers had naively and boastfully cried, "On to Richmond," for it seemed, at the beginning, that victory would be so easy. Many in the North believed that Richmond would fall quickly, ending the rebellion before it could even achieve much momentum.

But four years and oceans of blood later, the fighting continued and no Yankee invaders had breached Richmond's defenses. Not one enemy artillery shell had bombarded its stately residences, war factories, and government buildings. No blackened, burned-out ruins marred the handsome architectural streetscapes. And from the highest point in the city of the seven hills, no advancing federal armies were visible on the horizon. No, Richmond had been spared many of the horrors of war, the physical devastation and humiliating enemy occupation that had befallen many of the great cities of the South.

This morning as the Reverend Dr. Charles Minnigerode, a larger-than-life figure in Richmond society, was conducting services, a messenger entered the church. He carried a dispatch to the president that had arrived in Richmond at 10:40 A.M. It was a telegram from General Lee, bringing to the president's church pew news of a double calamity: The Union army was approaching the city gates, and the glorious Army of Northern Virginia was powerless to stop them.

Davis described the scene: "On Sunday, the 2d of April, while I was in St. Paul's church, General Lee's telegram, announcing his speedy withdrawal from Petersburg, and the consequent necessity for evacuating Richmond, was handed to me."

The telegram was not addressed to Davis, but to Confederate secretary of war John C. Breckinridge, vice president of the United

States from 1857 to 1861 during James Buchanan's administration. On March 4, 1861, Breckinridge's fellow Kentuckian Abraham Lincoln took the oath of office as president, and he heard the new commander in chief deliver his inaugural address. "We are not enemies, but friends. We must not be enemies," Lincoln had said to the South that day. Now Breckinridge had received a telegram warning him that the Union army was approaching and the government would likely have to abandon the capital that very night, in less than fourteen hours.

Headquarters,
April 2, 1865

General J. C. Breckinridge:

I see no prospect of doing more than holding our position here till night. I am not certain that I can do that. If I can I shall withdraw tonight north of the Appomattox, and if possible it will be better to withdraw the whole line tonight from James River. Brigades on Hatcher's Run are cut off from us. Enemy have broken through our lines and intercepted between us and them, and there is no bridge over which they can cross the Appomattox this side of Goode's or Beaver's, which are not very far from the Danville Railroad. Our only chance, then of concentrating our forces, is to do so near Danville Railroad, which I shall endeavor to do at once. I advise that all preparation be made for leaving Richmond tonight. I will advise you later, according to circumstances.

R. E. Lee

On reading the telegram, Davis did not panic, though the distressing news drained the color from his face. Constance Cary, who would later marry the Confederate president's private secretary, Colonel Burton Harrison, watched Davis while he read the telegram:

"I happened to sit in the rear of the President's pew, so near that I plainly saw the sort of gray pallor that came upon his face as he read a scrap of paper thrust into his hand by a messenger hurrying up the middle aisle. With stern set lips and his usual quick military tread, he left the church."

Davis knew his departure would attract attention, but he noted, "the people of Richmond had been too long beleaguered, had known me too often to receive notes of threatened attacks, and the congregation of St. Paul's was too refined, to make a scene at anticipated danger."

"Before dismissing the congregation," Cary remembered, "the rector announced to them that General Ewell had summoned the local forces to meet for the defence of the city at three in the afternoon . . . a sick apprehension filled all hearts."

Worshippers, including Miss Cary, gathered in front of St. Paul's: "On the sidewalk outside the church, we plunged at once into the great stir of evacuation, preluding the beginning of a new era. As if by a flash of electricity, Richmond knew that on the morrow her streets would be crowded by her captors, her rulers fled, her government dispersed into thin air, her high hopes crushed to earth. There was little discussion of events. People meeting each other would exchange silent hand grasps and pass on. I saw many pale faces, some trembling lips, but in all that day I heard no expression of a weakling fear."

Davis's calm notwithstanding, news of Lee's imminent retreat alarmed the people of Richmond. Many denied it credence. General Lee would not allow it to happen, they told themselves. He would save the city, just as he had repelled all previous Union efforts to take it. In the spring of 1865, Robert E. Lee was the greatest hero in the Confederacy, more popular than Jefferson Davis, whom many people blamed for their country's misfortunes. This news was not completely unexpected by Davis and others in his government, who had even begun making preparations for it. But there were no outward signs of danger and the people of Richmond had their judgment clouded by their faith in General Lee.

Now gloom seized the capital. A Confederate army officer, Captain Clement Sulivane, noted the change: "About 11:30 a.m. on Sunday, April 2d, a strange agitation was perceptible on the streets of Richmond, and within half an hour it was known on all sides that Lee's lines had been broken below Petersburg; that he was in full retreat . . . and that the city was forthwith to be abandoned. A singular security had been felt by the citizens of Richmond, so the news fell like a bomb-shell in a peaceful camp, and dismay reigned supreme."

Davis made his way from St. Paul's to his office at the old customs house. He summoned the heads of the principal government departments—war, treasury, navy, post office, and state—to meet with him there at once. "I went to my office and assembled the heads of departments and bureaus, as far as they could be found on a day when all our offices were closed, and gave the needful instructions for our removal that night, simultaneously with General Lee's withdrawal from Petersburg. The event was not unforeseen, and some preparation had been made for it, though, as it came sooner than was expected, there was yet much to be done."

Davis assured his cabinet that the fall of Richmond would not signal the death of the Confederate States of America. He would not surrender. No, if Richmond was doomed, then the president, his cabinet, and the government would evacuate the city, travel south, and establish a new capital in Danville, Virginia, one hundred and forty miles to the southwest, and, for the moment, beyond the reach of Yankee armies. The war would go on. Davis told them to pack their most vital records, only those necessary for the continuity of the government, and send them to the railroad station.

The train would leave that night, and he expected all of them—Secretary of State Judah Benjamin, Attorney General George Davis,

Secretary of the Treasury George Trenholm, Postmaster John Reagan, and Secretary of the Navy Stephen Mallory—to be on that train. Secretary of War John C. Breckinridge would stay behind in Richmond to oversee the evacuation and then follow the cabinet to Danville. What they could not take, they must burn. Davis ordered that the train take on other cargo too, more valuable than the dozens of document-crammed trunks: the Confederate treasury, several million dollars in gold and silver coins, plus Confederate currency.

Davis spent most of the afternoon working at his office with his personal staff. His circle of talented and devoted aides included Francis R. Lubbock, a former governor of Texas; William Preston Johnston, son of the president's old friend General Albert Sidney Johnston, who had been killed in 1862 at the battle of Shiloh; John Taylor Wood, U.S. Naval Academy graduate, who was Davis's nephew by marriage and a grandson of Mexican War general and later president of the United States Zachary Taylor; and Micajah H. Clark, Davis's chief clerk.

"My own papers," recalled Davis, "were disposed as usual for convenient reference in the transaction of current affairs, and as soon as the principal officers had left me, the executive papers were arranged for removal. This occupied myself and staff until late in the afternoon."

Davis then walked home to the presidential mansion at Twelfth and Clay streets to supervise the evacuation of the White House of the Confederacy. Worried citizens stopped him on his way: "By this time the report that Richmond was to be evacuated had spread through the town, and many who saw me walking toward my residence left their houses to inquire whether the report was true. Upon my admission . . . of the painful fact, qualified, however, by the expression of my hope that we would under better auspices again return, the ladies especially, with generous sympathy and patriotic impulse, responded, 'If the success of the cause requires you to give up Richmond, we are content.' The affection and confidence of this noble people in the hour of disaster were more distressing to me than complaint and unjust censure would have been."

When Davis arrived home, an eerie stillness possessed the mansion. His wife, Varina, and their four children were gone. He had foreseen this day. Hoping for the best but anticipating the worst, he had evacuated them from Richmond three days earlier, on Thursday, March 30. The president knew what could happen to civilians when cities fell to enemy armies. If Richmond fell, he wanted his family far removed from the scenes of that disaster.

Varina remembered their conversation before her departure: "He said for the future his headquarters must be in the field, and that our presence would only embarrass and grieve, instead of comforting him." The president decided to send his family to safety in Charlotte, North Carolina, which was farther south than Danville. They would not travel alone. He assured Varina that his trusted private secretary, Colonel Burton Harrison, would escort and protect her during the journey.

Until the end, the first lady begged to stay with her husband in Richmond, come what may: "Very averse to flight, and unwilling at all times to leave him, I argued the question . . . and pleaded to be permitted to remain." Davis said no—she and the children must go. "I have confidence in your capacity to take care of our babies," he told her, "and understand your desire to assist and comfort me, but you can do this in but one way, and that is by going yourself and taking our children to a place of safety."

Then the president spoke ominous words. "If I live," Davis promised his beloved companion and confidante of more than twenty years, "you can come to me when the struggle is ended."

If he lived? Varina could not admit that it was possible he might not. But Jefferson prepared her for the worst: "I do not expect to survive the destruction of constitutional liberty."

Varina did not want to leave behind all that she owned in Richmond, confessing a feminine attachment to her possessions. "All women like bric-a-brac, which sentimental people call 'household goods,' but Mr. Davis called it 'trumpery.' I was no superior to my sex in this regard. However, everything which could not be readily transported was sent to a dealer for sale."

Varina wanted to ask friends and neighbors to hide her large collection of silver from the Yankee looters, but her husband vetoed her scheme, explaining that enemy troops might punish anyone who helped them. "They may be exposed to inconvenience or outrage by their effort to serve us."

The president did insist that she carry with her on the journey something more practical than bric-a-brac. On March 29, the day before Varina and the children left Richmond, he armed his wife with a percussion-cap, black-powder .32- or .36-caliber revolver. "He showed me how to load, aim, and fire it," she said. The same day, Davis dispatched a written order for fresh pistol ammunition to his chief of ordnance, Josiah Gorgas: "Will you do me the favor to have some cartridges prepared for a small Colt pistol, of which I send the [bullet] moulds, and the form which contained a set of the cartridges furnished with the piece—The ammunition is desired as promptly as it can be supplied." Gorgas endorsed the note and passed it on to a subordinate: "Col. Brown will please order these cartridges at once and send them here. 50 will be enough I suppose."

The image was rich with irony. In the endangered war capital, home to the great Tredegar Iron Works, the principal cannon manufactory of the Confederacy, at a time when tens of thousands of battling soldiers were expending hundreds of thousands of rifle cartridges in a single battle, an anonymous worker in the Confederate ordnance department collected a handful of lead, dropped it into a fireproof ladle, melted the contents over a flame, poured the molten metal into a brass bullet mold, and cooled the silver-bright conical bullets in water. Then he took black powder and paper and formed finished, ready-to-fire cartridges for the first lady of the Confederacy.

She needed to be able to protect herself. The president feared that roving bands of undisciplined troops or lawless guerillas might seek to rob, attack, or capture his family.

He told Varina: "You can at least, if reduced to the last extremity, force your assailants to kill you, but I charge you solemnly to leave when you hear the enemy approaching; and if you cannot remain undisturbed in our own country, make for the Florida coast and take a ship there to a foreign country."

Davis gave Varina all the money he possessed in gold coins and Confederate paper money, saving just one five-dollar gold piece for himself. She would need money to pay—or bribe—her family's way south. Varina and the children left the White House on Thursday, March 30. "Leaving the house as it was, and taking only our clothing, I made ready with my young sister and my four little children, the eldest only nine years old, to go forth into the unknown."

Food was scarce in Richmond—there had been bread riots during the war—and it might prove rarer on the road, so Varina had ordered several barrels of flour loaded onto a wagon assigned to transport her trunks to the railroad station. When the president discovered the flour hoard, he forbade her to take it. "You cannot remove anything in the shape of food from here. The people want it, and you must leave it here." The sight of a wagon loaded with food ready to be shipped out of Richmond might have provoked a riot.

The children did not want to leave their father, and it was hard for Varina to part them from him. "Mr. Davis almost gave way, when our little Jeff begged to remain with him," she wrote. "And Maggie clung to him convulsively, for it was evident he thought he was looking his last upon us." Davis escorted his family to the depot and put them aboard the train. "With hearts bowed down by despair . . . ," Varina remembered, "we pulled out from the station and lost sight of Richmond, the worn-out engine broke down, and there we sat all night. There were no arrangements possible for sleeping, and at last, after twelve hours' delay, we reached Danville."

On the night of March 30, Davis returned home to his empty mansion and his imperiled city. There was much to do. He knew that over the next few days the fate of his capital was beyond his control. It was in the hands of the Army of Northern Virginia, which was engaged in a series of desperate battles to save Richmond.

On Saturday, April 1, Robert E. Lee sent word to Davis that the federal army was tightening the vise:

> The movement of Gen. Grant to Dinwiddie C[ourt] H[ouse] seriously threatens our position, and diminishes our ability to maintain our present lines in front of Richmond and Petersburg . . . it cuts us off from our depot at Stony Creek . . . It also renders it more difficult to withdraw from our position, cuts us off from the White Oak road, and gives the enemy an advantageous point on our right and rear. From this point, I fear he can readily cut both the south side & the Danville Railroads being far superior to us in cavalry. This in my opinion obliged us to prepare for the necessity of evacuating our position on the James River at once, and also to consider the best means of accomplishing it, and our future course. I should like very much to have the views of your Excellency upon this matter as well as counsel.

Lee's use of the phrase "future course" might seem vague or open-ended, suggesting that he felt they would be making a choice from many options. But he knew there was just one course of action—the abandonment of Richmond. At the end of the dispatch, Lee advised Davis that the situation was too dire for him to leave the front and come to Richmond to confer with the president. The Union forces, with their superior strength, could break through the Army of Northern Virginia's thin lines at any moment, without warning. If that happened, Lee must be in the field leading his men in battle, not idling and stranded in the capital, miles from the action.

Davis replied by telegraph, agreeing with his general that it was all in the hands of Lee and the army now: "The question is often asked of me 'will we hold Richmond,' to which my only answer is, if we can, it is purely a question of military power."

Lee invited the president and the secretary of war to visit his headquarters to discuss war planning, but Davis was too occupied with official business to leave Richmond, and so, during the next crucial days that might determine the fate of the Confederacy, the president and his general in chief never met in person. They communicated only through written dispatches and telegrams. Indeed, for the remainder of the Civil War, Jefferson Davis and Robert E. Lee would not meet again.

While Davis awaited news of further developments from Lee, he took stock of his armies in other parts of the country. In addition to Lee's army in the field in Virginia, there was General Joseph E. Johnston's army in North Carolina and General Kirby Smith's forces west of the Mississippi River in Texas. With these forces, the cause was not lost. Davis would not sit passively in Richmond and surrender the city, his capital, and his government to the Yankees. If the Army of Northern Virginia, in order to save itself from annihilation and live to fight another day, had to move off and uncover the city, then the government would move with it. Indeed, on April 1, Davis wrote to General Braxton Bragg, revealing his dreams of future Confederate attacks and a war of maneuver:

> My best hope was that Sherman while his army was worn and his supplies short would be successfully resisted and prevented from reaching a new base or from making a junction with Schofield. Now it remains to prevent a junction with Grant, if that cannot be done, the Enemy may decide our policy . . . Our condition is that in which great Generals have shown their value to a struggling state. Boldness of conception

> and rapidity of execution has often rendered the smaller force victorious. To fight the Enemy in detail it is necessary to outmarch him and surprise him. I can readily understand your feelings, we both entered into this war at the beginning of it, we both staked every thing on the issue and have lost all which either the public or private Enemies could take away, we both have the consciousness of faithful service and may I not add the sting of feeling that capacity for the public good is diminished by the covert workings of malice and the constant irritations of falsehood.

On April 1, Davis also received a message that, unlike the military dispatches that brought only news of military setbacks, offered some relief. It was from his wife, telegraphing from Greensboro, North Carolina, where she had gone after Danville. Varina's text was brief, written in haste, but precious to him: "Arrived here safely very kindly treated by friends. Will leave for Charlotte at Eight oclock tomorrow Rumors numerous & not defined have concluded that the Raiders are too far off to reach road before we shall have passed threatened points Hope hear from you at Charlotte all well." Lee's army was on the verge of destruction, Richmond in danger of occupation, and his own fate unknown, but Davis went to bed that night knowing that his family was safe from harm. What he did not know was that this was his last night in the White House.

Nor did Davis know that his nemesis, Abraham Lincoln, was on the move. Lincoln had left his White House several days earlier and was now traveling in Virginia, in the field with the Union army. The president of the United States wanted to witness the final act. Lincoln did not want to go home until he had won the war. He did not say it explicitly in conversation, nor did he reveal his desire by committing it to paper, but he wanted to be there for the end. And he dreamed of seeing Richmond fall.

In March 1865, Abraham Lincoln was restless in his White House. A number of times during the war, he had gone to the field to see his generals and his troops, and he had seen several battlefields, among them Antietam and Gettysburg. He had cherished these experiences and regretted that he could not visit his men more often. But the dual responsibilities of directing a major war and administering the civil government of the United States anchored him to the national capital. He always enjoyed getting away from the never-ending carnival parade of special pleaders, cranks, favor beggars, and office-seekers who were able to enter the White House almost at will. He had endured their impositions for four years, and now that he had won reelection, they tasted fresh spoils. Lincoln knew the war had now turned to its final chapter. It could be over within a few weeks. He had alluded to it in his inaugural address on March 4 when he said: "The progress of our arms, upon which all else chiefly depends, is as well known to the public as to myself; and it is, I trust, reasonably satisfactory and encouraging to all."

Relief came in the form of an invitation from General Grant that sent Lincoln on a remarkable journey.

On March 23 at 1:00 P.M., Lincoln left Washington from the Sixth Street wharf, bound on the steamer *River Queen* for City Point, Virginia, headquarters of the armies of the United States. His party included Mrs. Mary Lincoln and their son Tad, Mary's maid, White House employee W. H. Crook, and an army officer, Captain Charles B. Penrose. The warship *Bat* accompanied the presidential vessel. The next day the *River Queen* anchored off Fortress Monroe, Virginia, around 12:00 P.M. to take on water, and at 9:00 P.M. anchored off City Point, Virginia.

Lincoln rose early on the twenty-fifth, and after receiving a briefing from his son Robert, a captain on Grant's staff, the president went ashore and walked to Grant's headquarters. Lincoln wanted to see the battlefield. At 12:00 P.M. a military train took him to General Meade's headquarters. From there, Lincoln rode on horseback and watched reverently as the dead were buried. On the way back to City Point, he rode aboard a train bearing wounded soldiers from the field. He saw prisoners too. As Lincoln gazed upon their faces, he saw the costs of war. That night he was supposed to have dinner with General Grant but said he was too tired and returned to the *River Queen*. Later, he sent a message to Secretary of War Edwin Stanton: "I have seen the prisoners myself."

The next morning Lincoln went up the James River and then went ashore at Aiken's Landing. On March 27, he met with Generals Grant and Sherman and Admiral Porter on the *River Queen*. This conference carried over to the next day.

Their conversation was free-ranging and off the record, and General Sherman asked Lincoln about his plans for his rebel counterpart, Jefferson Davis. Many in the North had demanded vengeance if Davis was captured, and they wanted him to be hanged. Did Lincoln share that opinion, Sherman wondered, and did he approve of trials and executions not only of Davis, but of the entire Confederate military and political hierarchy?

"During this interview I inquired of the President if he was all ready for the end of the war," Sherman remembered.

> What was to be done with the rebel armies when defeated? And what should be done with the political leaders, such as Jefferson Davis . . . ? Should we allow them to escape . . . ? He said he was all ready; all he wanted of us was to defeat the opposing armies, and to get the men composing the Confederate armies back to their homes, at work at their farms and in their shops. As to Jeff. Davis, he was hardly at liberty to speak his mind fully, but intimated that he ought to clear out, "escape the country," only it would not do for him to say so openly. As usual, he illustrated his meaning by a story: "A man once had taken the total-abstinence pledge. When visiting a friend, he was invited to

> take a drink, but declined, on the score of his pledge; when his friend suggested lemonade, which was accepted. In preparing the lemonade, the friend pointed to the brandy-bottle, and said the lemonade would be more palatable if he were to pour in a little brandy; when his guest said, if he could do so 'unbeknown' to him, he would not object." From which illustration I inferred that Mr. Lincoln wanted Davis to escape, "unbeknown" to him.

This was a stunning revelation. Yes, Lincoln had promised "malice toward none" and "charity for all" in his inaugural address, but no one expected him to extend such mercy to the archtraitor and war criminal Jefferson Davis. But Lincoln was not a vengeful man. During the war, his private letters, public papers, and speeches had foreshadowed how he would treat his defeated enemies. "I shall do nothing in malice," he once said of his plans, "what I deal with is too vast for malicious dealing."

Lincoln was still in the field on March 31 when he received a telegram from Edwin Stanton. Some members of the cabinet wanted the president to return to Washington to take care of official business, but Stanton urged him to remain with the army: "I hope you will stay to see it out, or for a few days at least. I have strong faith that your presence will have great influence in inducing exertions that will bring Richmond; compared to that no other duty can weigh a feather. There is . . . nothing to be done here but petty private ends that you should not be annoyed with. A pause by the army now would do harm; if you are on the ground there will be no pause."

At City Point on April 1, Lincoln received reports and sent messages. He haunted the army telegraph office for news of the battles raging in Virginia. He was addicted to this technology. It was an impatient habit he had formed in Washington. He did not like to wait for important news. To his delight, the War Department telegraph office was a short walk from the Executive Mansion. He became a habitué of the office, befriending the men employed there, to whom he often made surprise visits at any time of the day or night. Now he was standing over the telegraph operators at City Point, and as soon as they transcribed the reports as they came off the wire, the president snatched the hurried scribblings from their hands.

Lee and his army were fighting a series of skirmishes and battles to save Richmond and themselves. Union forces pressed Lee's lines at multiple points, probing for weaknesses and forcing on Lee a major decision: Would he sacrifice the remnants of his once great and still proud army in a final battle of annihilation before Richmond, or would he abandon the capital in order to save his soldiers to fight again? Telegrams from the front kept Lincoln apprised of Lee's every move. Mary Lincoln and Secretary of State William H. Seward, who had joined the president in the field, returned to Washington, D.C., that day. The president kept Tad with him. He wanted his little companion to share in the historic days to come. That night he walked the deck of the *River Queen*, anxious about what the next day might bring.

When Davis and Lincoln awoke on the morning of Sunday, April 2, 1865, neither man knew this was the day. As Davis dressed for church, he did not know he would have to leave Richmond that night. Yes, he was aware of the danger facing the capital and that he might have to evacuate it soon. But he was not expecting to flee that night. Like the citizens of Richmond, like the entire Confederacy, he expected the impossible of Robert E. Lee.

Like Davis and Robert E. Lee, Lincoln spent part of April 2 reading and sending telegrams. Lincoln guessed that this was the Army of Northern Virginia's last act. Although he did not know that Richmond would be evacuated that night, he knew the citadel of the Confederacy must fall soon. The Union had too many men, too many cannons, too many guns, and limitless supplies. The Confederacy, starving and outnumbered, could not repel a Union advance. Today

Lincoln would send six important telegrams, two to Mary Todd Lincoln, three to Edwin Stanton, and one to Ulysses Grant.

At 11:00 A.M., around the time Jefferson Davis sat in St. Paul's Church reading the fateful telegram from General Lee, Lincoln telegraphed Stanton in Washington. A flurry of messages had come in from the front to City Point, and after Lincoln read them all, he summarized their contents.

City Point, Va.
April 2, 1865—11:00 a.m.

Hon. Edwin M. Stanton,
Secretary of War:

Dispatches frequently coming in. All going finely. Parke, Wright, and Ord, extending from the Appomattox to Hatcher's Run, have all broken through the enemy's intrenched lines, taking some forts, guns, and prisoners. Sheridan, with his own cavalry, Fifth Corps, and part of the Second, is coming in from the west on the enemy's flank, and Wright is already tearing up the South Side Railroad.
A. Lincoln

In Richmond, the doomsday clock ticked past the noon hour. Like a convict on death row awaiting his midnight execution, Confederate Richmond knew it had fewer than twelve hours to live. Between 2:00 P.M. and 3:00 P.M., a formal announcement was made to the public that the government would evacuate that evening. But the people already knew. Piles of burning documents in the street said it all. Captain Clement Sulivane remembered the scene: "All that Sabbath day the trains came and went, wagons, vehicles, and horsemen rumbled and dashed to and fro . . ."

In the midst of this frenzy, people had to decide whether to stay or to flee. The occupant of one house had no choice. She was an invalid and could not leave Richmond. President Davis sent over his most comfortable chair for Mrs. Robert E. Lee.

In midafternoon, Lee telegraphed another warning to Richmond.

> *Hd. Qrs Petersburg*
> *3. p.m. 2nd. April 1865*
>
> *MR. PRESIDENT*
>
> *. . . I do not see how I can possibly help withdrawing from the city to the north side of the Appomattox to night. There is no bridge over the Appomattox above this point nearer than Goode's & Bevill's over which the troops above mentioned could cross to the north side & be made available to us—Otherwise I might hold this position for a day or two longer, but would have to evacuate it eventually & I think it better for us to abandon the whole line on James river tonight if practicable—I have sent preparatory orders to all the officers & will be able to tell by night whether or not we can remain here another day; but I think every hour now adds to our difficulties—I regret to be obliged to write such a hurried letter to your Excellency, but I am in the presence of the enemy endeavoring to resist his advance—I am most respy & truly yours*
>
> *R.E. Lee*
> *Gnl.*

There was no denying it now. Lee's telegram could not have been clearer and he'd written it while in battle. If he failed to move his army by that night, it faced destruction. In either case, the Union army would take Richmond sometime the next day, Monday, April 3. Davis replied, seeming to underestimate the danger.

Richmond, Va.,
April 2, 1865

General R.E. Lee, Petersburg, Va.:

To move to night will involve the loss of many valuables, both for the want of time to pack and of transportation. Arrangements are progressing, and unless you otherwise advise the start will be made.

Jeff'n Davis

This was not the answer Lee expected. At this moment, his men were fighting and dying to save Richmond, while President Davis was fretting about the loss of valuables. Davis's telegram exasperated Lee. After he read it, he crumpled it into a ball, tossed it to the ground, and complained to his staff: "I am sure I gave him sufficient notice." Lee replied at 3:30 P.M. "Your telegram recd. I think it will be necessary to move tonight. I shall camp the troops here north of the Appomattox the Enemy is so strong that they will cross above us to close us in between the James & Appomattox Rivers—if we remain."

From City Point, Virginia, Lincoln telegraphed his wife.

City Point, Va.,
April 2, 1865

Mrs. Lincoln:

At 4:30 p.m. to-day General Grant telegraphs that he has Petersburg completely enveloped from river below to river above, and has captured, since he started last Wednesday, about 12,000 prisoners and 50 guns. He suggests that I shall go out and see him in the morning, which I think I will do. Tad and I are both well . . .

A. Lincoln

In Richmond, Davis received yet another urgent telegram from Lee, this one more insistent than his last. The general informed the president that he had ordered an officer to rush to the capital to escort him safely out of the city. This was the end.

Petersburg,
April 2, 1865

His Excellency President Davis, Richmond, Va.:
I think it absolutely necessary that we should abandon our position tonight. I have given all the necessary orders on the subject to the troops and the operation though difficult I hope may be successful. I have directed Genl Stevens to send an officer to your Excellency to explain the routes to you by which the troops will be moved to Amelia C[ourt] H[ouse] & furnish you with a guide and any assistance that you may require for yourself.
R. E. Lee

So there could be no doubt of the imminent peril, Lee dispatched a similar telegram to Secretary of War John C. Breckinridge. If Davis could not appreciate the danger, then perhaps Breckinridge, a major general in the Confederate army, would.

Petersburg,
April 2, 1865

General J. C. Breckinridge:
It is absolutely necessary that we should abandon our position tonight, or run the risk of being cut off in the morning. I have given all the orders to officers on both sides of the river, & have taken every precaution that I can to make the movement successful. It will be a difficult operation, but I hope not impracticable. Please give all orders that you find necessary in & about Richmond. The troops will all be directed to Amelia Court House.
R. E. Lee

At 7:00 P.M. Lee sent a final telegram to Davis and Breckinridge, letting them know he had given the order and was sending the president a rider to inform him of the safest routes west to link up with the Army of Northern Virginia.

Abraham Lincoln recognized the significance of the day's developments. If General Grant could crush Lee's army, or drive it off from Petersburg, then the road to Richmond would lie open. Lincoln relished every new piece of good news. Before he went to bed, he sent a telegram to his commanding general, congratulating him on the successes of this day.

Head Quarters Armies of the United States
City-Point,
April 2. 8/15 P.M. 1865.

Lieut. General Grant.

Allow me to tender to you, and all with you, the nations grateful thanks for this additional, and magnificent success. At your kind suggestion, I think I will visit you to-morrow.

A. Lincoln

That evening, Davis, unlike Lincoln, could not defer his travel plans until the morning. As Lincoln settled in for the night on the *River Queen,* Davis prepared to abandon his home. Davis packed some clothes, retrieved important papers and letters from his private office, and saved a few personal effects.

He sat down and wrote a letter to his housekeeper Mary O'Melia and to the mayor of Richmond, Joseph Mayo. His instructions to O'Melia were: "The furniture in the executive mansion it would be well to pack and store as your discretion may indicate and if any one should dispute your authority this will be your warrant—The Mayor will give you aid and protection." On another page Davis added a note to Mayo. "His honor the Mayor will find on the previous page that I have referred my house keeper to him, and will I hope allow me to commend her specially to his kind care."

With that taken care of, Davis had nothing left to do but wait, and he was joined by members of his inner circle: his old friend Clement Clay, and his aides Frank Lubbock, William Preston Johnston, and John Taylor Wood. It was dark now. Earlier in the day, the train had been scheduled to depart at 8:30 P.M., but the crowds and confusion at the station slowed the preparations. It had taken hours to pack the railroad cars.

Then a messenger brought word to Davis that the cabinet had assembled at the station and the train was ready to depart. With memories of all its joys and sorrows, Davis left his White House for the last time, mounted his favorite horse, Kentucky, and rode to the railroad station. He was always an elegant horseman, and during his final ride through the streets of Richmond, sitting in the saddle with ramrod-straight military bearing, he was a sight to behold.

Frank Lubbock never forgot that ride: "This was the saddest trip I had ever made, for I could feel but grieved—sorely distressed; a sorrow that was ominous of the future." Tumultuous crowds did not line the streets to cheer their president during his last ride through the capital or to shout best wishes for his journey to save the Confederacy. The citizens of Richmond were too swept up with their own concerns—locking up their homes, hiding their valuables, or fleeing the city before the Yankees arrived—to give their president a proper send-off.

As Davis readied for the train to leave the city, he knew he had made the right decision. "Richmond would be isolated, and it could not have been defended. Its depots, foundries, workshops, and mills could have contributed nothing to the armies outside, and its possession would no longer have been to us of military importance. Ours being a struggle for existence, the indulgence of sentiment would have been misplaced."

Not all of the residents dreaded the fall of the city. Bands of thieves, drunkards, and worse were waiting for the moment when the last Confederate troops would leave Richmond and take with them all vestiges of law and order. Mallory called them "the rabble who stood ready to plunder during the night." From darkness until Union troops arrived at dawn, the capital would be theirs.

Among the blacks of Richmond, the mood on the eve of their day of anticipated liberation was electric. At the African church it was a day of jubilee. Worshippers poured into the streets, congratulated each other, and prayed for the coming of the Union army. The next morning they would be slaves no more.

When Davis got to the station, he declined to board the train. He wanted to delay the departure until the last possible moment. Perhaps the fortunes of war had turned in the Confederacy's favor that night. Perhaps Lee had confounded the enemy as he had done so many times before and reestablished defensive lines protecting Richmond. At 10:00 P.M. Davis and Breckinridge walked into the office of the Richmond and Danville Railroad and waited for a miracle—a telegram from Robert E. Lee retracting his counsel to evacuate Richmond. For an hour Davis held the loaded and waiting train in the hope of receiving good news from Lee.

Nothing—no telegram ever came. The Army of Northern Virginia could not save the beleaguered city. It would be imprudent, even dangerous, to tarry any longer. The Yankees could arrive in just six or seven hours, and further delay might allow them to cut the railroad line below Richmond, blocking the only route for Davis's escape train.

Dejected, Davis and Breckinridge left the railroad office and the president boarded his car. Captain William Parker, a naval officer on special duty at the train depot that night, observed the scene: "While waiting at the depot I had an opportunity of seeing the President and his Cabinet as they went to the cars. Mr. Davis preserved his usual calm and dignified manner, and General Breckinridge . . . Who had determined to go out [of Richmond] on horseback, was as cool and gallant as ever—but the others . . . Had the air (as the French say) of wishing to be off. General Breckinridge stayed with me some time after the President's train had gone, and I had occasion to admire his bearing under the circumstances."

This was not a private, luxurious sleeping car constructed for a head of state. The Confederate railroad system had never equaled the scale, resources, and power of the United States Military Railroad. Jefferson Davis took his seat in a common coach packed with the heads of the cabinet departments, key staff members, and other selected officials. The departure was without ceremony. No honor guard, no well-wishers, and no martial band playing "Dixie" bade the president's train farewell. Jefferson Davis gave no speech from the station platform, or from the rear of the last car, as Abraham Lincoln had done on February 11, 1861, the morning he left Springfield, Illinois, for his journey to Washington to become president.

Captain Parker watched the train gather steam and creep out of the station at a slow speed, no more than ten miles per hour. The train groaned down the track. Parker noticed that it was loaded: "Not only inside, but on top, on the platforms, on the engine,—everywhere, in fact, where standing room could be found; and those who could not get that 'hung on by their eyelids.'" It was a humbling, even ignominious departure of the Confederate president from his capital city.

Postmaster John Reagan, riding on that train, pitied those left behind in Richmond. The fleeing government had abandoned not only a place but its people. A number of citizens had asked for his advice: Should they remain in their homes and "submit to the invading army," or should they flee? Reagan knew that most had no choice anyway and could not escape if they wanted to.

Throughout the day and into the night, countless people had fled the doomed city by any means possible—on foot or horseback; in carriages, carts, or wagons. Some rushed to the depot, but there was only a single rail line left open, and the small number of locomotives and cars had been commandeered by the government to transport the president, the cabinet, various officials, the Confederate archives, and the funds of the Confederate treasury to safety. The postmaster general knew the truth: Circumstances had "left but small opportunity for the inhabitants to escape."

As Davis's train rolled out of Richmond, most of the passengers were somber. There was nothing left to say. Mallory captured the mood around him. "Silence reigned over the fugitives. All knew how the route to Danville approached the enemy's lines, all knew the activity of his large mounted force, and the chances between a safe passage of the Dan [River] and a general 'gobble' by Sheridan's cavalry seemed somewhat in favor of the 'gobble.' "

Few spoke as the government in exile crossed the bridge leaving the city, and Richmond came into panoramic view. Mallory studied his fellow fugitives. "The terrible reverses of the last twenty-four hours were impressed upon the minds and hearts of all as fatal to their cause . . . Painful images of the gigantic efforts, the bloody sacrifices of the South, all fruitless now, and bitter reflections upon the trials yet to come, were passing through the minds of all, and were reflected . . . upon every face."

The city was still dark, but the fires would come soon. They would not be set by Union troops. The Confederates would, by accident, set their own city ablaze when they burned supplies to keep the goods out of Union hands. This improvident decision would reduce much of the capital to ruins. The flames would spread out of control and devour Richmond, enveloping it with a bright, unnatural, yellow-orange glow that would illuminate the heavens. Jefferson Davis was spared, at least, from witnessing the conflagration.

"It was near midnight," John Reagan remembered, "when the President and his cabinet left the heroic city. As our train, frightfully overcrowded, rolled along toward Danville we were oppressed with sorrow for those we left behind us and fears for the safety of General Lee and his army."

The presidential train was not the last to leave Richmond that night. A second one carried another precious cargo from the city—the financial assets of the Confederacy, in the form of paper currency, and gold and silver coins, plus deposits from the Richmond banks. Earlier on April 2, Captain Parker had received a written order from Secretary Mallory to have the corps of midshipmen report to the railroad depot at 6:00 P.M. When Parker learned that Richmond was to be evacuated, he went to the Navy Department office, where Mallory ordered him to take charge of the Confederate treasure and guard it during the trip to Danville. Men desperate to escape Richmond and who had failed to make it onto Davis's train climbed aboard their last hope, the treasure train. The wild mood at the depot alarmed Parker, and he ordered some of his men—some of them only boys—to guard the doors to the depot and not allow "another soul to enter."

Once Davis was gone, and the night wore on, Parker witnessed the breakdown of order. "The scenes at the depot were a harbinger of what was to come that night. The whiskey . . . was running in the gutters, and men were getting drunk upon it . . . Large numbers of ruffians suddenly sprung into existence—I suppose thieves, deserters . . . who had been hiding. To add to the horror of the moment . . . we now heard the explosions of the vessels and the magazines, and this, with the screams and yells of the drunken demons in the streets, and the fires which were now breaking out in every direction, made it seem as though hell itself had broken loose."

If the rampaging mob had learned what cargo Parker and his midshipmen guarded, these looters, driven mad by greed, would have descended upon the cars like insatiable locusts gorging themselves, not on grain but on gold. Parker was prepared to order his men to fire on the crowd. Before that became necessary, the treasure train got up steam and followed Davis, and the hopes of the Confederacy, into the night.

Back in Richmond, the darkness loosened the restraints of civilization, and the looters went wild. One witness recalled the mood: "By nightfall all the flitting shadows of a Lost Cause had passed away under a heaven studded by bright stars. The doomed city lay face to face with what it knew not." And, in the evening, "ominous groups of ruffians—more or less in liquor—began to make their appearance on the principal thoroughfares of the city . . . as night came on pillage and rioting and robbing took place . . . Richmond saw few sleeping eyes during the pandemonium of that night."

Union troops outside Richmond saw the flames and heard the explosions. An army officer, Captain Thomas Thatcher Graves, observed: "About 2 o'clock on the morning of April 3d bright fires were seen in the direction of Richmond. Shortly after, while we were looking at these fires, we heard explosions."

At about 3:00 A.M., while en route to Danville, Davis's train stopped at Clover Station, about sixty miles northeast of its destination. A young army lieutenant, eighteen-year-old John S. Wise, saw the train pull into the station. Through one of the train's windows, he spotted Davis, waving to the people gathered at the depot. Later, Wise witnessed the train carrying the Confederate treasury pass, and others too. "I saw a government on wheels," he said, "the marvelous and incongruous debris of the wreck of the Confederate capital . . . indiscriminate cargoes of men and things. In one car was a cage with an African parrot, and a box of tame squirrels, and a hunchback." From one train a man in the rear car cried out, to no one in particular, "Richmond's burning. Gone. All gone."

As Davis continued his journey, Richmond burned and Union troops approached the city. Only a few miles away, Lincoln and Admiral Porter heard incredible explosions, and Lincoln feared that some U.S. Navy guns had exploded. But Porter reassured him that the thundering booms were far off, evidence, no doubt, that the Confederates were blowing up their own ironclad warships to save them from capture. U.S. Army scouts closing in on Richmond noticed a bright glow painting the sky above the city like a luminous dome. The scouts knew it was too pronounced to be the result of a celestial phenomenon or gaslight streetlamps. The city must have been on fire.

Around dawn a black teamster who had escaped Richmond reached Union lines and reported what Lincoln, Porter, Grant, and others suspected. The Confederate government had abandoned the capital during the night, and the road to the city was open. There would be no battle for Richmond. The Union army could march in and seize the rebel capital without firing a shot.

The first Union troops entered the outskirts of Richmond shortly after sunrise on Monday, April 3. They marched through the streets, arrived downtown, and took hold of the government buildings. They also began the work of extinguishing the fires, which still burned in some sections of the city. When Lincoln's army raised the Stars and Stripes over the fallen capital, it signaled, for many Southerners, the end of the world. And then, scant hours after Davis had left it, the Union seized the White House of the Confederacy, pressing it into service as the new headquarters for the army of occupation.

The population of Richmond had endured a night of terror. The ruins and the smoke presented a terrible sight. "By daylight, on the 3d," witnessed Captain Sulivane, "a mob of men, women, and children, to the number of several thousands, had gathered at the corner of 14th and Cary streets, and other outlets, in front of the bridge, attracted by the vast commissary depot at that point; for it must be remembered that in 1865 Richmond was a half-starved city, and the

Confederate Government had that morning removed its guards and abandoned the removal of the provisions, which was impossible for want of transportation. The depot doors were forced open and a demoniacal struggle for the countless barrels of hams, bacon, whisky, flour, sugar, coffee . . . raged about the buildings among the hungry mob. The gutters ran whisky, and it was lapped up as it flowed down the streets, while all fought for a share of the plunder. The flames came nearer and nearer, and at last caught in the commissariat itself."

Union officer Thomas Thatcher Graves entered the city in the early morning, when it was still burning. "As we neared the city the fires seemed to increase in number and size, and at intervals loud explosions were heard. On entering the square we found Capitol Square covered with people who had fled there to escape the fire and were utterly worn out with fatigue and fright. Details were at once made to scour the city and press into service every able-bodied man, white or black, and make them assist in extinguishing the flames."

Constance Cary ventured outside to bear witness to the ruined and fallen city. Horrified, she discovered that Yankees had desecrated the Confederate White House merely by their presence. "I looked over at the President's house, and saw the porch crowded with Union soldiers and politicians, the street in front filled with curious gaping negroes who have appeared in swarms like seventeen year locusts." The sight of ex-slaves roving freely about disgusted her. "A young woman has just passed wearing a costume composed of United States flags. The streets fairly swarm with blue uniforms and negroes decked in the spoils of jewelry shops . . . It is no longer our Richmond . . . one of the girls tells me she finds great comfort in singing 'Dixie' with her head buried in a feather pillow."

The gloom suffocating President Davis's train vanished with the morning sun. Secretary of State Benjamin, Secretary of the Treasury Trenholm, and the president's aides all contributed to the change in mood. Benjamin, a larger-than-life epicurean and bon vivant, talked about food and told stories. "His hope and good humor [were] inexhaustible," recalled Secretary Mallory.

With a playful air, Benjamin discussed the fine points of a sandwich, analyzed his daily diet given the food shortages that plagued the South, and showed off, as an example of his "adroit economy," his coat and pants, both tailored from an old shawl that had kept him warm through three winters. Mallory, who appointed himself unofficial chronicler of the evacuation train, admired Benjamin for his "never give up the ship air" and took notice when Benjamin "referred to other great national causes which had been redeemed from far gloomier reverses than ours."

Colonels John Taylor Wood and William Preston Johnston, "gentlemen of high character, cultivated minds, and pleasing address," agreed. "They were quiet," recalled Mallory, "but professed to be confident, saw much to deplore, but no reason for despair or for an immediate abandonment of the fight." Colonel Frank Lubbock, the former Texas governor and Davis's favorite equestrian riding partner during the war, entertained his fellow travelers with wild Western tales. Mallory summed him up with a pithy character sketch: "An earnest, enthusiastic, big-hearted man was Colonel Lubbock, who had seen much of life, knew something of men, less of women, but a great deal about horses, with a large stock of Texas anecdotes, which he disposed of in a style earnest and demonstrative."

The most important contributor to the jolly mood might have been Secretary of the Treasury Trenholm. Although ill, semi-invalid, and unable to regale his colleagues, Trenholm contributed an essential ingredient to the cheery mood aboard the train: a seemingly inexhaustible supply of "old Peach" brandy. "As the morning advanced our fugitives recovered their spirit," testified Mallory, "and by the time the train reached Danville . . . all shadows seemed to have departed."

Lincoln was unaware of what was happening in Richmond

throughout the night. But in the morning he telegraphed Edwin M. Stanton in Washington, informing him that the Confederacy had evacuated Petersburg, and probably Richmond too.

> *Head Quarters Armies of the United States*
> *City-Point*
> *April 3. 8/00 A.M. 1865*
>
> *Hon. Sec. of War*
> *Washington, D.C.*
> *This morning Gen. Grant reports Petersburg evacuated; and he is confident Richmond also is. He is pushing forward to cut it off if possible, the retreating army. I start to him in a few minutes.*
> *Lincoln*

Soon Lincoln would go to Petersburg to meet with General Grant. But before he departed, he received a telegraph from his son Robert, informing him that he was awaiting his father at Hancock Station.

News traveled quickly. In New York City, Southern sympathizers, including the celebrated Shakespearean actor John Wilkes Booth, mourned the fall of the Confederate capital. New York had never been a stronghold for Lincoln or the Union. Indeed, in 1863, rioting New Yorkers who opposed the military draft had lynched blacks in the streets and become so depraved that they attacked and burned the Negro orphanage. Troops had to put down this mad rebellion. Walt Whitman scorned Gotham as a disloyal citadel of "thievery, druggies, foul play, and prostitution gangrened."

All day April 3, Washington, D.C., celebrated the fall of Richmond. The *Evening Star* captured the joyous mood: "As we write Washington city is in such a blaze of excitement and enthusiasm as we never before witnessed here . . . The thunder of cannon; the ringing of bells; the eruption of flags from every window and housetop,

Broadside from the first week of April 1865.

the shouts of enthusiastic gatherings in the streets; all echo the glorious report. RICHMOND IS OURS!!!"

The news spread through the capital, reported the *Star,* by word of mouth: "The first announcement of the fact made by Secretary of War Stanton in the War Department building caused a general stampede of the employees of that establishment into the street, where their pent-up enthusiasm had a chance for vent in cheers that would assuredly have lifted off the roof from that building had they been delivered with such vim inside . . . The news caught up and spread by a thousand mouths caused almost a general suspension of business, and the various newspaper offices especially were besieged with excited crowds."

Lincoln's favorite newsman, Noah Brooks, who was set to begin his new job soon as private secretary to the president, recorded the moment: "The news spread like wildfire through Washington . . . from one end of Pennsylvania Avenue to the other the air seemed to burn with the bright hues of the flag. The sky was shaken by a grand salute of eight hundred guns, fired by order of the Secretary of War—three hundred for Petersburg and five hundred for Richmond. Almost by magic the streets were crowded with hosts of people, talking, laughing, hurrahing, and shouting in the fullness of their joy . . ."

The Union capital celebrated without President Lincoln, who was still in the field. While Washington rejoiced, Edwin Stanton worried about Lincoln's security. He had always believed that Lincoln did not do enough to assure his own safety. No longer so eager to have Lincoln at the front, Stanton urged him to return to Washington. Lincoln's telegram advising Stanton that he would visit General Grant at City Point on April 3 alarmed the secretary of war, who urged the president not to go to the front lines. "I congratulate you and the nation on the glorious news in your telegram just recd. Allow me respectfully to ask you to consider whether you ought to expose the nation to the consequence of any disaster to yourself in the pursuit of a treacherous and dangerous enemy like the rebel army. If it was a question concerning yourself only I should not presume to say a word. Commanding Generals are in the line of duty in running such risks. But is the political head of a nation in the same condition."

Davis did not arrive in Danville until 4:00 P.M. on April 3. Much had happened in Richmond, and elsewhere, while he was languishing aboard the train. It had taken eighteen hours to travel just 140 miles. The train had averaged less than ten miles per hour, and at some points, especially on uphill grades, it could barely move at all. Sometimes the train stopped completely. This sorry performance demonstrated the lamentable state of the Confederate railroad system. The plodding journey from Richmond to Danville made clear another uncomfortable truth. If Jefferson Davis hoped to avoid capture, continue the war, and save the Confederacy, he would have to move a lot faster than this. Still, the trip had served its purpose. It had saved, for at least another day, the Confederate States of America. If Davis had allowed himself to be captured or killed on April 2 or 3, the Civil War might have ended the day Richmond fell. By fleeing, Davis ensured that the fall of the city was not fatal to the cause. As Mrs. Robert E. Lee, who endured the occupation, said, "Richmond is not the Confederacy."

The government on wheels unpacked and set up in Danville on the afternoon and evening of April 3. For two reasons, Davis hoped to remain there indefinitely, depending upon word of General Lee's prospects. First, as long as Davis stayed in Danville, he was maintaining a symbolic toehold in Virginia. He abhorred the idea of abandoning the Confederacy's principal state to the invaders. If circumstances forced a retreat from Danville, Davis would have no choice but to continue south and cross the North Carolina state line. Second, Danville put him in a better position to send and receive communications. Without military intelligence, it would be difficult to issue orders or coordinate the movements of his armies. It would be hard for his commanders to telegraph the president or send dispatch riders with the latest news if he stayed on the move, if they did not know where to find him, and if they had to chase him from town to town. In Danville he had everything he needed to continue the war.

Secretary Mallory rattled off a checklist of their resources: "Heads of departments, chief clerks, books and records; Adjutant General of the army Samuel Cooper with the material and personnel of his office, all the essential means for conducting the government, were here; and so able and expert were these agents, that a few hours only, and a half-dozen log cabins, tents, and even wagons, would have sufficed for putting it in fair working order."

The inhabitants of Danville had received advance word that their president was coming, and a large number of people waited at the station for his train. They cheered Jefferson Davis when he disembarked from his railroad car. With fine Virginia hospitality, leading citizens opened their homes to the president and the other dignitaries. Colonel William T. Sutherlin, a local grandee, considered it an honor for his town to host the Confederate government, and he offered the president, Trenholm, and Mallory accommodations in his own home.

The remaining cabinet members received similar offers from other citizens. Soon, refugees from Richmond and elsewhere flooded into Danville, overtaxing the town's hospitality and housing. Many of them, including women, slept in railroad cars parked on sidings near the tracks, obtained their food from Confederate commissaries, and cooked their meals in the open. The new, temporary capital could never match the grandeur of Thomas Jefferson's city of the seven hills, but Danville symbolized the Confederacy's resilience. As long as Davis was able to sustain his government from this humble outpost, the cause was not lost.

As Davis settled into Danville, Abraham Lincoln reassured his secretary of war that he was safe and had survived the day. Lincoln had ignored Stanton's warning. The end of the war was near. Nothing could stop him from traveling to the front lines of the Union army. But he thanked Stanton for his concern—after the trip:

> *Head Quarters Armies of the United States*
> *City-Point,*
> *April 3. 5 P.M. 1865*
>
> *Hon. Sec. of War*
> *Washington, D.C.*
> *Yours received. Thanks for your caution; but I have already been to Petersburg, stayed with Gen. Grant an hour & a half and returned here. It is certain now that Richmond is in our hands, and I think I will go there to-morrow. I will take care of myself.*
> *A. Lincoln*

If Lincoln's visit to Grant had worried Stanton, this proposed trip by the president to Richmond, a city still smoldering, inhabited by thousands of secessionists who hated Lincoln, must have given him fits.

As darkness approached Washington, D.C., the celebrations became more intense, even wild. Lincoln missed it all, but his friend Noah Brooks recorded his memories of the evening. "The day of jubilee did not end with the day, but rejoicing and cheering were prolonged far into the night. Many illuminated their houses, and bands were still playing, and leading men and public officials were serenaded all over the city. There are always hosts of people who drown their joys effectually in the flowing bowl, and Washington on April third was full of them. Thousands besieged the drinking-saloons, champagne popped everywhere, and a more liquorish crowd was never seen in Washington than on that night."

In Richmond, a different sun set on the first day of Union occupation. The ruins still smoked. As in Washington, people filled the streets. Soon the printmaker Currier & Ives, which had made a specialty of publishing images of American urban disasters—especially great conflagrations—would immortalize this night of fire and destruction in an oversized, full-color panoramic print suitable for framing. For customers on a budget, the Currier firm also published a less expensive, smaller version of Richmond in flames.

Constance Cary reflected on all she had seen that day. "The ending of the first day of occupation was truly horrible. Some negroes of the lowest grade, their heads turned by the prospect of wealth and equality, together with a mob of miserable poor whites, drank themselves mad with liquor scooped from the gutters. Reinforced, it was said by convicts escaped from the penitentiary, they tore through the streets, carrying loot from the burnt district. (For days after, even the kitchens and cabins of the better class of darkies displayed handsome oil paintings and mirrors, rolls of stuff, rare books, and barrels of sugar and whiskey.) One gang of drunken

THE FAMOUS CURRIER & IVES PRINT OF RICHMOND BURNING, APRIL 2, 1865.

rioters dragged coffins sacked from undertakers, filled with spoils . . . howling madly."

In Danville, Davis and his cabinet's first night of exile began a strange, anticlimactic interlude. These high officials might have regained their morale, but they had little to do. No Confederate army had encamped there, awaiting Davis's orders to go into battle. Yes, Davis and his department heads could busy themselves writing dispatches, issuing orders, and seeking news, and so they did. But the future course of the war in Virginia depended upon Robert E. Lee. Davis expected news from Lee on April 4 but none came. Speculation abounded, remembered Stephen R. Mallory, "more or less based upon a thousand rumors."

Davis hated being in this passive position. He craved action: He wanted to rally armies, choose points of strategic concentration, and continue resistance. Instead, he busied himself inspecting the defensive earthworks being thrown up around Danville and awaited word from the Army of Northern Virginia. Had Davis been interested in fleeing to save his own life, this was the time to push deeper south, into North Carolina and beyond. Tarrying in Danville increased his risk of capture or death. But he still hoped to win the war, not just save his own skin. Should Lee send news of victory, or just even word that he had escaped Yankee encirclement by swinging his army toward Danville, Davis could reunite with his most reliable commander and continue the fight. But if Davis left Virginia, he might be too far from the action to rejoin Lee.

"April 4 and the succeeding four days passed," noted Mallory, "without bringing word from Lee or Breckinridge, or of the operations of the army; and the anxiety of the President and his followers was intense." Judah Benjamin tried again to revive the spirits of his colleagues. "No news is good news," he chimed, but Mallory could see that the secretary of state's optimism had been unconvincing this time.

Refugees from Richmond carried to Danville, along with their most precious remaining possessions, wild rumors. "Some asserted, upon the faith of a very reliable gentleman 'just in,' that Lee had won a glorious victory, held his army well in hand, and was steadily pursuing Grant. Others declared that Lee was too busy fighting to send couriers, and that the wires were down." Davis ignored the rumors. Meanwhile, more refugees from Richmond clogged Danville. Many, including a bridal party, lived in railroad cars. Mallory recalled the surreal scenes: "Thus were passed five days. To a few, a very few, they were days of hope; to the many they were days of despondence, if not of despair; and to all, days of intense anxiety."

CHAPTER TWO

"In the Days of Our Youth"

On April 4, Abraham Lincoln experienced one of the most thrilling days of his life, one that was a culmination of his work and his presidency. But he did not gloat. "Thank God that I have lived to see this! It seems to me that I have been dreaming a horrid dream for four years, and now the nightmare is gone. I want to go to Richmond."

Admiral Porter agreed to take him there on the *River Queen*, "if there is any of [Richmond] left. There is black smoke over the city." Porter told the president that before they could go up the river, he must order all the "torpedoes" (mines) removed from the water so that they did not blow up the *River Queen*. Her sister ship had already struck a mine and been blown up. The admiral described the journey: "Here we were in a solitary boat, after having set out with a number of vessels flying flags at every mast-head, hoping to enter the conquered capital in a manner befitting the rank of the President of the United States, with a further intention of firing a national salute in honor of the happy result."

Oil portrait of Lincoln as he appeared on the eve of victory.

Porter was embarrassed that he could not deliver his commander in chief to the captive city in style. Lincoln said not to worry, and he told a funny story to make the admiral feel better. "Admiral, this brings to mind a fellow who once came to me to ask for an appointment as minister abroad. Finding he could not get that, he came down to some more modest position. Finally he asked to be made a tide-waiter. When he saw he could not get that, he asked me for an old pair of trousers. But it is well to be humble."

The river leading into Richmond had become too shallow for big boats, and as they got close to the city, Porter transferred the president, Tad, and Captain Penrose to his personal craft, the "admiral's barge." Despite the fancy name, it was no more than a spacious, glorified rowboat Porter used to travel between his flagship and other U.S. Navy warships, or between ship and shore. When they reached the riverfront, Porter and his crew had trouble spotting a landing and they had to continue along the edge. The city looked eerie. Lincoln and Porter peered at the rebel capital but saw no one—only smoke from the fires. The only sound was the creaking of the oars. "The street along the river-front was as deserted," Porter observed, "as if this had been a city of the dead." The Union army had occupied the city for a day, but "not a soldier was to be seen."

Then the current lodged the barge on a rock, and the oarsmen rowed for the first landing they saw. Lincoln stepped onto the wharf. They had landed at Rocketts, a shady waterfront district. Admiral Porter described what happened next:

> There was a small house on this landing, and behind it were some twelve negroes digging with spades. The leader of them was an old man sixty years of age. He raised himself to an upright position as we landed, and put his hands up to his eyes. Then he dropped his spade and sprang forward. "Bress de Lord," he said, "dere is de great Messiah! I knowed him as soon as I seed him. He's bin in my heart fo' long yeah, an' he's cum at

> las' to free his chillum from deir bondage? Glory, Hallelujah!" And he fell upon his knees before the President and kissed his feet. The others followed his example, and in a minute Mr. Lincoln was surrounded by these people, who had treasured up the recollection of him from a photograph, and had looked up to him for four years as the one who was to lead them out of captivity.

The adulation embarrassed Lincoln. He was a simple man with plain tastes who had, during his entire presidency, eschewed pomp and circumstance. He had no patience for politicians who behaved like royalty. He did not want to enter Richmond like a king. He spoke to the throng of slaves. "Don't kneel to me. That is not right. You must kneel to God only, and thank him for the liberty you will hereafter enjoy. I am but God's humble instrument; but you may rest assured that as long as I live no one shall put a shackle on your limbs, and you shall have all the rights which God has given to every other free citizen of this Republic."

Before allowing Lincoln to leave them and proceed on foot into Richmond, the freed slaves burst into joyous song:

> Oh, all ye people clap your hands,
> And with triumphant voices sing;
> No force the mighty power withstands
> Of God, the universal King.

The hymn drew hundreds of blacks to the landing. They surrounded Lincoln, making it impossible for him to move. Admiral Porter recognized how foolish he had been to bring the president ashore without a proper military escort. "The crowd immediately became very oppressive. We needed our marines to keep them off. I ordered twelve of the boat's crew to fix bayonets and surround the President . . . but the crowd poured in so fearfully that I thought we all stood a chance of being crushed to death. I now realized the imprudence of landing without a large body of marines; and yet this seemed to me . . . the fittest way for Mr. Lincoln to come upon the people he had redeemed from bondage."

The crowd became increasingly wild. Some rushed forward, laid their hands upon the president, and collapsed in ecstatic paroxysms. Some, too awed to approach Father Abraham, kept their distance and just stared at him. Others yelled for joy and performed acrobatic somersaults. Admiral Porter said the people were so excited that some of them appeared "demented." Lincoln spoke to them: "My poor friends, you are free—free as air. You can cast off the name of slave and trample upon it . . . Liberty is your birthright . . . But you must try to deserve this priceless boon. Let the world see that you merit it, and are able to maintain it by your good works. Don't let your joy carry you into excesses. Learn the laws and obey them . . . There, now, let me pass on; I have but little time to spare. I want to see the capital."

Porter ordered six men to march ahead of the president and Tad, and six behind them, and with that the landing party walked toward downtown Richmond. Lincoln stopped briefly to look at the notorious Libby Prison, a place of suffering for thousands of Union prisoners of war. "We will pull it down!" screamed voices in the crowd. But Lincoln said no, that they should "leave it as a monument." The streets were dusty and smoke from the fires still hung in the air. Lincoln could smell Richmond burning as he walked through it. By now thousands of people, blacks and whites, crowded the sidewalks.

A beautiful girl, about seventeen years old, carrying a bouquet of roses, stepped into the street and advanced toward the president. Porter watched her struggle through the crowd. "The mass of people endeavored to open to let her pass, but she had a hard time in reaching him. Her clothes were very much disarranged in making the journey across the street. I reached out and helped her within the circle of the sailors' bayonets, where, although nearly stifled with dust, she gracefully presented her bouquet to the President and made a neat little speech, while he held her hand . . . There was a card on the bouquet with these simple words: 'From Eva to the Liberator of the slaves.'"

Porter spotted a sole cavalryman and called out to him: "Go to the general, and tell him to send a military escort here to guard the president and get him through this crowd!"

"Is that old Abe?" the trooper asked, before galloping off.

Thomas Thatcher Graves, aide-de-camp on the staff of General Weitzel, approached the president and his group, and Lincoln asked him, "Is it far to President Davis's house?"

Graves accompanied the president to the Confederate White House. "At the Davis house, [Lincoln] was shown into a reception-room, with the remark that the housekeeper had said that the room was President Davis's office. [It was Davis's first-floor study, not his second-floor office.] As he seated himself he remarked, 'This must have been President Davis's chair,' and, crossing his legs, he looked far off with a serious, dreamy expression."

This was the closest Lincoln had ever come to Jefferson Davis during the war. Confederate vice president Alexander Stephens, and not Davis, had represented the Confederacy at the Hampton Roads peace conference in February 1865, where Stephens and Lincoln discussed how to end the war.

Lincoln knew the Confederate president had been in this room no more than thirty-six hours earlier. As one witness remembered, Lincoln "lay back in the chair like a tired man whose nerves had carried him beyond his strength." The journalist Charles C. Coffin observed on the president's face a "look of unutterable weariness, as if his spirit, energy and animating force were wholly exhausted." Sitting in the quiet study of the Confederate president, perhaps Lincoln weighed the cost—more than 620,000 American lives—paid to get there. He did not speak. Then he requested a glass of water.

It is not surprising that the paths of the two presidents had not crossed before the Civil War, even though they both had lived briefly in Washington, D.C., at the same time. Davis and Lincoln lived very different lives and moved in different circles. Lincoln became a giant, but in antebellum America he was considered inferior to Jefferson Davis in education, social status, military and political experience, national reputation, influence, fame, and prospects. Indeed, before Lincoln's run for Senate and the Lincoln-Douglas debates of 1858, few Americans north or south knew anything about him. Many had never heard the name Abraham Lincoln. Most knew the name of Jefferson Davis, a man who many people expected would be a future president of the United States. Today, nearly one hundred and fifty years after the Civil War, Lincoln's fame obscures Jefferson Davis. Davis's presidency of a slave empire that fought the deadliest war in American history has tainted, even swept away, the memory of anything that was good about him.

Davis is often remembered as a grotesque caricature: a humorless, arrogant, inflexible, racist, slave-owning traitor who tried to overthrow the Constitution but failed to win Southern independence and who then vanished from history. In reputation, Lincoln and Davis stand as polar opposites, as emancipator and slave master, as two men who could not have been more different from each other. The truth is more complex. In some ways, Lincoln and Davis had nothing in common. In others, some profound, they shared striking similarities and experiences.

Born in 1808, Jefferson Davis attended private academies and universities, and then, with the sponsorship of his prosperous older brother, Joseph, attended the U.S. Military Academy at West Point. He was a fine equestrian, and he cut a splendid, elegant figure in the saddle. Serving as an army officer on the western frontier, he undertook long and arduous cross-country journeys that gave him great knowledge of the American continent. In Mississippi, his brother set him up as a planter. He was elected to the U.S. Congress; fought gallantly in the Mexican War as colonel of an infantry regiment, the Mississippi Rifles; was wounded in battle but refused to leave the field; and then came home a hero. He was a fervent nationalist who believed that North and South, by working together, could conquer the continent.

Serving in President Franklin Pierce's administration, he became one of the greatest secretaries of war in American history. Highly innovative, he pursued advanced weapons systems, tried to modernize the command structure of the army, and, in a little-remembered program, introduced military camels into the vast, parched Western territories. He was instrumental in founding the Smithsonian Institution, and he supervised the expansion of the U.S. Capitol building. He revered the Revolutionary generation, the founders, and the Constitution, and, as a planter and slave owner, he believed that the founders, as part of the great compromise to create the new nation, had embedded and protected the "peculiar institution" in American law forever.

Like the framers of the Constitution, like most eighteenth- and nineteenth-century Americans in the South and the North, like almost every previous president of the United States, and like many antislavery leaders and abolitionists, Davis believed in white racial superiority. He admired the North and praised its industry, traveled there often, sought to know its mind, and developed many friends in New England and the Northeast. He even vacationed there. He gave well-received speeches in several of the great cities of the North. And, as the political crisis over slavery and the admission of new states into the Union grew more divisive through the 1850s, he refused to join the shrill ranks of the Southern fire-eaters and rabid secessionists. As a U.S. senator, he admonished radical hotheads in both parties and abhorred the idea of disunion. He favored comity, not confrontation.

In 1858, Abraham Lincoln, in his seventh and final debate with Stephen A. Douglas at Alton, Illinois, referred to Davis as "that able and eloquent statesman." Davis had earned a reputation as one of the first men of the South and as one of the chief spokesmen not only for his home state of Mississippi but for half the nation. He was a friend to many of the great statesmen of his age, the widower son-in-law of President Zachary Taylor, and a confidant and counselor to other presidents. At his Washington, D.C., town house, he and his sparkling second wife, Varina Howell Davis, presided over a brilliant salon that welcomed leaders from North and South. *Harper's Weekly* published a front-page woodcut portrait of him. He looked like a statesman. Some said he resembled an eagle, and many viewed him as a plausible Democratic candidate for the presidential election of 1860. He favored logic and reason over undisciplined passion, and his beautiful speaking voice was one of his most powerful political attributes.

By any measure, Jefferson Davis was one of the most well-known, respected, admired, and influential political leaders of pre–Civil War America. His achievements were all the more remarkable because, beneath this shining surface of privilege, talent, and success, he had suffered through all of his adult life from a collection of serious, chronic, and sometimes disabling illnesses, which had brought him near death more than once. He was plagued by, among other things, malaria, neuralgia, and progressive blindness in one eye. But his resilient body and will to live had kept him alive. Indeed, his unconquerable life force suggested that perhaps fate and destiny had preserved him for some great task.

After secession began, he neither campaigned for, nor even desired, the presidency of the new Confederate States of America. Only after his adopted state of Mississippi seceded on January 9, 1861, did Davis resign his Senate seat. His farewell speech from the Senate floor on January 21 moved observers to tears, caused a sensation, and won him praise from both Southerners and Northerners. Davis was chosen by acclamation as the provisional president of the Confederate States of America on February 9, 1861. Later, he was elected to a six-year term as president and was inaugurated on February 22, 1862, George Washington's birthday.

Lincoln and Davis were both born in rustic Kentucky cabins, one year and one hundred miles apart, but their early years could not have been more different. Born February 12, 1809, Lincoln lacked family sponsors. His father, Thomas, an uneducated, illiterate, restless manual laborer who seemed proud of his limitations, made no effort to educate his son. Thomas Lincoln moved his family from Kentucky to Indiana and then to Illinois, but wherever he lived, success and prosperity eluded him. After young Abe's mother died when he was nine years old, he lived in the squalor of his father's cabin like a wild, feral child. When Thomas brought home a new wife, Sarah Bush Johnson, the rough, downtrodden state of Abe and his siblings horrified her. But she grew to love them and, though uneducated herself, took a special interest in her tall, awkward stepson. Abe had less than a year of formal schooling, but he learned to read and write and perform elementary mathematics. "Abraham Lincoln, his hand and pen, he will be good, but God knows when," he inscribed in his boyhood sum book. While his indifferent father exploited him as a manual laborer—Abe had a rail-splitting axe thrust into his hands at age nine—his stepmother encouraged his learning. Years later, after being elected president, Lincoln would not leave Illinois without paying her an emotional—and perhaps final, he thought—visit.

When young Abe Lincoln reached maturity he had no connections, no money, no proper education, and no prospects beyond following his father's footsteps into a lifetime of physical toil. But he was driven by ambition for a better life. He widened his world through a variety of occupations: flatboat river pilot, surveyor, storekeeper, and postmaster. He studied law on his own, became a member of the Illinois bar, and joined a two-man firm. He prospered in that trade, earning a reputation for honesty and ability while he rode the circuit from courthouse to courthouse. Unlike Jefferson Davis, who possessed a large private library and who studied all manner of subjects,

Lincoln owned few books, but he read narrowly and deeply in politics, Shakespeare, the Bible, and history.

Elected to Congress for a single term in 1846, Lincoln made little impression on official life in the nation's capital. When war broke out between the United States and Mexico, President James K. Polk and Senator Thomas Hart Benton viewed the conflict as an opportunity to pursue America's Manifest Destiny and create an empire that stretched from sea to sea. Congressman Abraham Lincoln opposed the war, said so on the floor of the House, and quibbled with President Polk about whether hostilities had begun on American or Mexican soil. Lincoln implied that the president had provoked the war to justify an unlawful land grab. In stark contrast, Jefferson Davis resigned his seat in Congress, led Mississippi troops in combat against superior numbers of enemy infantry and deadly cavalry lancers, and distinguished himself in the Battles of Monterrey and Buena Vista. Had Davis not left Washington on July 4, 1846, he would likely have met Lincoln there in 1847. Lincoln and Davis were in Washington at the same time in December 1848, and also in early 1849, after Davis had been elected to the Senate, but they did not meet then.

At the end of Lincoln's undistinguished single term, he went home to Illinois and rose to prominence in the Illinois bar. Never a lawyer of national renown, like Daniel Webster, William Wirt, or the other great Supreme Court and constitutional advocates of his day, Lincoln practiced in the local, state, and federal courts and handled a diverse mix of criminal and civil cases, with collections work occupying a significant portion of his practice. He did not travel widely beyond Illinois, possessed little firsthand knowledge of the South, and did not cultivate friendships with influential Southerners.

He was headed for a life of local celebrity, prosperity, and respectability—and national obscurity—until he was aroused in 1854 by the repeal of the Missouri Compromise and the possibility of admitting new slave states to the Union. Between his famous Peoria speech in

1854 and the Lincoln-Douglas debates during the Illinois Senate campaign of 1858, he emerged as a major antislavery voice. Lincoln lost that election but the debates, published in book form for the presidential campaign of 1860, made him a national political figure and helped him capture the nomination and then the White House.

When Lincoln took the presidential oath of office on March 4, 1861, he was, on paper, one of the least-qualified chief executives in the nation's history. Of his fifteen predecessors, any comparison to the first five—Washington, Adams, Jefferson, Madison, and Monroe—would have been considered absurd. Of the following ten, not all enjoyed successful presidencies, but every one surpassed Lincoln in raw qualifications for the office. Perpetually disorganized, Lincoln had never administered anything bigger than a two-man law office, and he had done a poor job of that, often unable to keep track of essential paperwork. And the myth is true—he often stuffed important documents into his stovepipe hat.

Davis, in addition to his other military and political merits, had held an important cabinet post and had overseen the administration of the U.S. Army. In November 1860, a majority of Americans would have said that Jefferson Davis was far more qualified than Abraham Lincoln to occupy the White House. If the Democratic Party had not split and produced two rival candidates, Stephen A. Douglas and John C. Breckinridge, and if Jefferson Davis had been nominated in 1860 as the sole Democrat to run against Lincoln, Davis might have been elected the sixteenth president of the United States. Indeed, Lincoln won the race with less than 40 percent of the popular vote. He may have secured an electoral majority, but 60 percent of the country voted for someone other than him.

Lincoln, who was not an abolitionist, agreed with Jefferson Davis that the Constitution protected slavery. Thus, the federal government had no power to interfere with it wherever it existed. And like Davis, Lincoln—at least the Lincoln of the 1840s and 1850s—accepted white racial superiority. But Lincoln parted ways with Davis and the South over the morality of slavery and the right to introduce it into new states and territories.

Lincoln believed that slavery was a moral crime—"If slavery is not wrong, nothing is wrong." He argued that even if blacks were not "equal" to whites, they should enjoy the equal right to liberty and the fruit of their labor. Lincoln insisted that the founders had allowed slavery with the uncomfortable understanding that it was an unholy compromise necessary to create the new nation, and that the founders had envisioned, at some future time, slavery's natural and ultimate extinction. Lincoln also opposed the expansion of slavery into new territories and states, fearing that its spread would give it a second wind, thus perverting the intentions of the founders and the true meaning of the Declaration of Independence.

Davis and his fellow Southerners rejected that ideology, insisting that slavery was not a necessary evil but something good that benefited both masters and slaves. The "peculiar institution," they argued, civilized, westernized, and Christianized a primitive, heathen African people. Southern leaders resented the accusation that slavery was a moral evil and not a positive good, and they interpreted the rising antislavery movement in the North as part of a conspiracy to outnumber the slave states with new free states to strip the South of its political power in Congress, especially in the Senate.

Despite their differences, Davis and Lincoln had many things in common. They possessed striking physical similarities. Both were tall, thin—even cadaverous-looking—men. At six feet, four inches, Lincoln was the bigger man, but Davis's erect military bearing, a disciplined posture drilled into him at West Point and that he maintained into old age, made him appear taller than his five feet, eleven inches. Both had angular, craggy faces, and both men looked underfed. Lincoln and Davis were sparse eaters and indifferent to the pleasures of the table.

Both men had lean builds, but they had been strong as young

men. When Davis was a boy, he learned to wrestle with slaves, and as a young man he possessed quickness and strength, gaining the better of several men in fights. Lincoln was a spectacular wrestler, contending in legendary matches with the Clary's Grove Boys in New Salem, Illinois. Years of manual labor with the axe and the maul had given him prodigious strength. In their later years, both men displayed astonishing moments of physical power.

Neither man was distracted by luxuries. Yes, Davis cherished his fine and extensive library—it was one of his proudest possessions—but, like Lincoln, he had no taste for fine antiques, furniture, or artworks. Neither man was a Beau Brummell, but Lincoln's indifference to his personal appearance—wrinkled, ill-fitting shirts and suits and wild, uncombed hair—outdid Davis's, who had at least learned during military service how to dress. Each man possessed an inner confidence, a belief that he was, somehow, different from other men. Both men shared memorable appearances. When Davis or Lincoln appeared in public, people noticed them and talked about it. Frozen nineteenth-century photographic daguerreotypes, ambrotypes, albumens, or tintypes failed to capture it, but in person Jefferson Davis and Abraham Lincoln were charismatic, captivating, and unforgettable.

In youth, Davis and Lincoln could be fun-loving, high-spirited, and undisciplined. More than once, Davis was almost expelled from West Point for carousing, drinking, and other forbidden behavior. Lincoln was not a drinker, but he possessed a natural talent for jokes and storytelling, many of them dirty. They were also young men of mirth and love, and sorrow and longing.

Davis met her in August 1832, when she was eighteen years old and he twenty-four. Sarah Knox Taylor was the daughter of General Zachary Taylor. Jefferson and Knox, the name she went by, fell in love, but her parents, seeking to protect her from the hard life of an army officer's wife, and due to a possible misunderstanding between Davis and the general, denied permission to marry. Undeterred, the young lovers persisted for more than two years until Jeff won the Taylors over. They married on June 17, 1835, and journeyed by steamboat to Davis Bend, the site of brother Joseph Davis's plantation on the Mississippi River. In August, Jefferson and Knox traveled south to visit his sister Anna Smith at Locust Grove, in Louisiana. There, just three months later, Jefferson endured an unspeakable loss that nearly killed him and changed his life forever.

It was the hot season, when the mosquitoes reigned over the plantation fields of the Deep South. They spread a dangerous form of malaria that the first generations of slaves had brought over from Africa almost two centuries earlier. Jeff and Knox contracted the disease. He almost died, but then, after suffering days of fever, chills, delirium, and nausea, he rallied to live. But Knox, her new husband beside her, succumbed. On September 15, 1835, Jefferson Davis surrendered to the grave the body of his twenty-one-year-old bride of twelve weeks. He was a lost man. When she died, something in him died too. He retreated into a private inner world, a "great seclusion," he once called it, with slaves and crops and books and his protective mentor Joseph. When Jefferson Davis emerged from that self-imposed isolation several years later, he was a different, reserved, harder, more mysterious man. Knox survives in a single letter to her family, in one to her from Jefferson, and in a lone image, a portrait painted in oils. She was a gorgeous, spirited girl with dark hair and generous eyes. Jefferson Davis cherished the memory of her for the rest of his life.

Abraham Lincoln met Ann Rutledge, four years his junior, in 1831 in the small village of New Salem, Illinois, where he worked as a surveyor, storekeeper, and postmaster. Ann, engaged to a ne'er-do-well sharp operator who left town and never returned to claim her, grew close to the awkward but interesting Lincoln. He exhibited little of the confidence and extraordinary powers that would reveal themselves later, but the core of his character had already formed. What happened next has been mostly suppressed by historians and belittled by Mary Lincoln apologists for the past one hundred and seventy-five years, but its truth can no longer be denied. In that tiny, isolated village on the Illinois frontier, Abraham Lincoln and Ann Rutledge formed a deep emotional bond. "Ann M. Rutledge is now learning grammar," reads Lincoln's affectionate handwritten inscription in an ancient, tattered book on that subject. Family and neighbors noticed the connection, heard the talk, and observed, by 1834 and early 1835, the familiar, age-old rituals of courtship. No documents survive to prove their love. No letters between them exist. But decades later, after Lincoln's assassination, his last law partner, William Herndon, collected evidence of Lincoln's early life from people who knew him, including the old villagers from the New Salem ghost town. They remembered everything.

They told Herndon that it was common knowledge that Abraham and Ann were in love, and that friends and family had expected them to marry until Ann became ill, probably from typhoid fever, and died on August 25, 1835. They could not forget how her death shattered Lincoln, how he visited her grave during thunderstorms and collapsed upon it, embracing her in death, how they feared for his mind and suspected that he might take his own life. In time, he walked in the world again. But after he left New Salem, he never, as far as anyone can tell, spoke or wrote of their bond or her death. Three decades later, when he was president of the United States, no one heard him mention her name. He possessed no portrait of her but for the one locked in his memory. Her death predated the introduction of photography, and anyway who would have thought to make a photograph or paint a portrait of a simple, poor young girl who lived in a little river town on the Illinois frontier? A physical description of her survives. Years later, one of her brothers described to William Herndon the girl Lincoln once knew: "She had light hair, and blue eyes."

Lincoln served as a volunteer captain in the Black Hawk Indian war in Illinois and Wisconsin in the early 1830s but never faced the enemy. Still, his election by the men of his company as their leader gave him immense pleasure. Later, while serving in Congress, Lincoln made light of his brief military career, joking that he had fought many bloody battles with the mosquitoes. This was to be Lincoln's only military experience—until nineteen years later, when he commanded great armies and navies.

Davis also served in the Black Hawk War, holding a superior and more prestigious rank as an officer in the regular army. He earned a singular honor. On a journey from Fort Crawford, Wisconsin, to St. Louis, he was placed in charge of the captured Native American warrior Black Hawk. When a group of white visitors taunted the shackled captive in his cell, Davis rebuked their lack of respect toward his prisoner. Impressed, Black Hawk praised Jefferson Davis as a great warrior and man of honor.

Davis and Lincoln each revered the federal union, but Davis, who had traveled the country and territories more extensively than had Lincoln, possessed far more personal knowledge of its varied geography and vast territorial expanse. Both honored the law, Lincoln pursued it as a calling and Davis thought of doing the same. He wrote a letter saying that he might purchase the books, read law, and become an attorney. Law was a well-established route to political advancement. As young men, each feared that he might amount to nothing. Lincoln wrote that if he failed to make something of himself, it would be as though he had never lived. Davis, on his twenty-second birthday, expressed similar fears. Both yearned to be remembered.

In time, both men rediscovered love. Ten years after the death of Sarah Knox Taylor, Jefferson Davis, thirty-seven years old, married Varina Howell, an eighteen-year-old, educated, savvy, and independent daughter of a fine Mississippi family. When Joseph Davis decided that his long-mourning brother should end his brooding and rejoin society, he introduced Jefferson to Varina. Their first meeting did not go well. She wrote a letter to her mother about it the next day. Mr. Davis, she confided, had an "uncertain temper, and . . . a way of taking for granted that everybody agrees with him when he expresses an opinion, which offends me." And then there was the age difference: "He is old, from what I hear he is only two years younger than you are." But Varina also recognized his qualities and was intrigued. Jefferson had an "agreeable manner" and possessed "a peculiarly sweet voice and winning manner of asserting himself." Her parents did not favor the match, knowing how deeply Knox's death had hurt—and changed—him. They feared that he would never recover from the loss and that no woman, even their daughter, could win his love. But Jefferson made peace with Knox Taylor's ghost and married Varina Howell on February 26, 1845. Theirs was an adoring marriage, a passionate and intellectual union that produced six children. For the rest of his life Jefferson depended upon Varina's love, advice, support, and loyalty. Indeed, two decades later, during his greatest trial and most profound despair, Varina rallied to save her husband, and she rose to the historic occasion.

Abraham Lincoln was not as lucky in marriage. After a few bungled attempts at courting other women, he married Mary Todd of Lexington, Kentucky, on November 4, 1842. They had met in Springfield, Illinois, when Lincoln was a young, hungry lawyer on the rise and Mary was a well-educated, politically savvy woman searching for a husband. It was an ill-starred union that plagued Abraham's peace of mind and domestic happiness for much of his life, and it climaxed finally in epic conflicts during his presidency.

Mary possessed few of the qualities that defined Varina Davis, who was everything—honest, dignified, courteous, and brave—that Mary Lincoln was not. Perhaps the only thing they had in common was a dressmaker, a free black woman, Elizabeth Keckly, who made frocks for Varina in antebellum Washington and later for Mary during her White House years, and who went on to write a controversial book about her strange wartime role as Mrs. Lincoln's confidante.

During the war Varina sold many of her fine garments for the sake of the cause, while Mary's extravagant purchases, the trademark of a compulsive shopper, failed to satisfy her unquenchable taste for luxury. The real Mary Lincoln was mercurial, jealous, insulting, rude, selfish, deceitful, paranoid, financially dishonest, and, without doubt, mentally unbalanced. During the Civil War, Jefferson Davis's White House was a sanctuary for the beleaguered president. For Abraham Lincoln, his White House was often a place of unrest and unpredictable marital discord.

During the Civil War both presidents had trouble with several of their general officers who sought to embarrass, defy, and undermine them. Both men mourned their losses in battle, and each experienced death in his White House. Jefferson Davis had lost one son in infancy, and in 1864, another son fell to his death at the Richmond White House. Lincoln too lost a young son long before the war, and his favorite boy, Willie, had succumbed to illness in the Washington White House in 1862.

Neither president enjoyed the universal love and support of his people. Both Davis and Lincoln experienced savage attacks by opponents who criticized their every decision. Mocked, lampooned, caricatured, second-guessed, and despised by segments of their own electorate throughout the war, Davis and Lincoln persevered for four years, each seeking to win his war.

Both men loved books. Davis enjoyed the privilege of a man of his class and built an extensive library. Better educated than Lincoln, he read politics, history, literature, and science, and after Knox's death it was the companionship of his brother—and a rigorous reading program—that kept him sane. Books were rare in the world of Lincoln's youth. He did not come from a family of readers—he once wrote that his father could only "bunglingly" sign his own name. Lincoln treasured the few he could obtain. As an adult, he never read as widely as Davis, but he read his favorite texts—Shakespeare, the Bible, and others—deeply and many times to enjoy their language and decode their meaning.

On the nature of man, Lincoln and Davis would never agree. Davis believed that one race of people was fitted by nature for slavery and was destined to remain the inferior race for all future time. Blacks would never enjoy the same legal rights as white men. Lincoln rejected that cruel fatalism and came to believe that the institution of bondage itself, and not nature, had temporarily "clouded" the minds of its victims. Once freed, the former slaves, Lincoln believed, would rise through work, ambition, and talent as they enjoyed equal rights under the law.

Lincoln accepted mankind with all its faults. His years as a lawyer had schooled him in the book of human behavior. He had sued—and defended—liars, cheats, thieves, deadbeats, adulterers, slanderers, and murderers. For nearly a quarter of a century, he had immersed himself in a world of vexatious disputes. Lincoln's experiences had made him resigned and forgiving, not callous and bitter. He rarely held a grudge. He was a skeptic who believed that, with the possible exceptions of his heroes George Washington and Henry Clay, there were no perfect men. Of the first president, Lincoln had once said: "Let us believe, as in the days of our youth, that Washington was spotless. It makes human nature better to believe that one human being was perfect—that human perfection is possible." When Lincoln was president, he was willing to do business with imperfect men if they could serve his purpose. Lincoln employed and trusted men despite their high opinions of themselves.

Davis lived by a different code and judged men more harshly than Lincoln did. He defined himself as a man of integrity who had conducted himself in politics and on the battlefield in principled ways. Davis tried to be a courteous, loyal, honest man who never stole, accepted graft, or sold his office for personal gain. He was a gentleman proud of the fact that he had never raised the whip to or been cruel to a slave. He could not abide men who failed to live up to the standards he set for himself. As president of the Confederacy, his commitment to the cause was total. He sacrificed all he had—his mind, body, health, and wealth, and the life of one of his sons—to the South, and it was by those benchmarks that he judged others. Because he acted without self-interest, so should others, he believed, including his generals, his cabinet officers, and the state governors. To disagree with him was to question his integrity and devotion and to risk his wrath. Often, to his disadvantage, Davis interpreted criticism or even helpful advice as an attack on his personal integrity, or as evidence of disloyalty. In turn, critics accused him of being remote, stubborn, and proud. He could be all of those things, but he also brought to his office many superb talents, and a total and relentless commitment to Southern independence. Many men criticized Jefferson Davis, Robert E. Lee once noted, but no man, he said, could have done better.

These, then, are the stories that made the two men who would, in the spring of 1865, preside over the destiny of two nations. Both were fifty-six years old—born ten months apart—and for most of the Civil War, they had laid their heads on their pillows each night in mansions less than one hundred miles apart, each dreaming of saving his cause and country.

Abraham Lincoln told Thomas Graves that he wanted to see the rest of Jefferson Davis's mansion. "At length he asked me if the housekeeper was in the house. Upon learning that she had left he jumped up and said, with a boyish manner, 'Come, let's look at the house!' We went pretty much over it; I retailed all that the housekeeper had told me, and he seemed interested in everything. As we came down the staircase General Weitzel came, in breathless haste, and at once President Lincoln's face lost its boyish expression as he realized that *duty* must be resumed."

Lincoln quickly put Davis's home to official use and conducted a meeting there with John Archibald Campbell, a former justice of the U.S. Supreme Court, and Joseph Reid Anderson, two leading Confederate citizens who arrived to see him.

Admiral Porter admired the mansion but judged it less grand than its Washington counterpart. It was "quite a small affair compared with the White House, and modest in all its appointments, showing that while President Davis was engaged heart and soul in endeavoring to effect the division of the States, he was not, at least, surrounding himself with regal style, but was living in a modest, comfortable way, like any other citizen."

With the fall of the city, Richmond's photographers had lost their prime business—portrait photographs of Confederate political leaders, government officials, generals, officers, and soldiers. Those customers had fled. But George O. Ennis, an enterprising photographer, figured out a way to make money under the Union occupation. Ennis set up his camera and photographed the White House of the Confederacy, and his publisher, Selden & Co., "news and book agents, dealers in photographic and stereoscopic views, fancy articles, & etc.," located at no. 836 Main Street, packaged it as a carte de visite with a caption sure to attract Union buyers. Selden promoted the former home of Jefferson Davis as the new headquarters of the Yankee occupiers: "JEFF. DAVIS MANSION. This building is beautifully situated, on the corner of Clay and 12th streets, and is noted as being the residence of the late Chief Magistrate of the Confederate States. It is now, and has been since the evacuation, the residence and headquarters of the General commanding this Department."

On this day, General Weitzel, who was now in command of the former Confederate capital, asked Lincoln what policies he should adopt in dealing with the conquered rebels. Thomas Graves overheard the conversation, and Lincoln's answer became an American legend. "President Lincoln replied that he did not want to give any orders on that subject, but, as he expressed it, 'If I were in your place I'd let 'em up easy, let 'em up easy.'"

This was one of the most remarkable statements ever spoken by a commander in chief. During his time in Richmond, Lincoln did not order the arrests of any rebel leaders who remained there, nor did he order their property seized. And he uttered no words of vengeance or punishment. Even while he sat in Jefferson Davis's own home, he did not disparage or defame the Confederate president. Nor did he order an urgent manhunt for Davis and the cabinet officers who had evacuated the city less than two days before. It was a moment of singular greatness. It was Abraham Lincoln at his best.

After Lincoln left the Confederate White House, he toured Richmond in a buggy. Blacks flocked to him and rejoiced, just as they had at the river landing and during his walk to Davis's mansion. His triumphant tour complete, he returned to the wharf for the journey back to City Point. As he left a black woman warned him to be careful. "Don't drown, Massa Abe, for God's sake!" If he had heard her, any man possessing Abraham Lincoln's sense of humor would have enjoyed laughing at that heartfelt, urgent, yet comical plea.

Death by drowning was not the greatest threat Lincoln faced that day. Not all of Richmond welcomed him to the ruined capital. Most whites stayed in their homes behind locked doors and closed shutters, with some glaring at the unwelcome conqueror through their windows. It was a miracle that not one embittered Confederate—not a single one—poked a rifle or a pistol through an open window and opened fire on the despised Yankee president. No one even shouted epithets. Lincoln knew the risk: "I walked alone on the street, and anyone could have shot me from a second-story window." The Richmond tour was one of Lincoln's most triumphant days—certainly the most important day of his presidency. But it was also one of the most dangerous days of his life. No American president before or since has ever placed himself in such a volatile and dangerous environment.

Lincoln left no written account of his journey to Richmond. He was a splendid writer with a fine analytical mind and keen powers of observation, but he did not possess a diarist's temperament nor had he ever kept a journal. It was unlikely that he would have written his memoirs after he left office in March 1869. With less than a year of formal schooling, he came to writing as a utilitarian, employing it to plead a legal case, convey information, make an argument, reply to an inquiry, propose a policy, justify an action, or persuade the reader. Only a few times in his life did he write to reminisce, to entertain, to regale, or to amuse with a story or a joke. His storytelling art was oral and ephemeral. Lincoln was a superb and—when the occasion demanded—eloquent writer, and an equally talented narrative speaker.

CHAPTER THREE

"Unconquerable Hearts"

While Lincoln toured her home on April 4, Varina Davis had just reached Charlotte, North Carolina. She had declined an invitation to remain in Danville, electing instead to press on. She remembered the journey as being incredibly miserable: "The baggage cars were all needing repairs and leaked badly. Our bedding was wet through by the constant rains that poured down in the week of uninterrupted travel which was consumed in reaching our destination. Universal consternation prevailed throughout the country, and we avoided seeing people for fear of compromising them with the enemy, should they overrun North Carolina."

Varina, her children, and their small group of traveling companions settled into a rented house in Charlotte, where they awaited word from Jefferson Davis. Colonel Burton Harrison, his escort mission accomplished, headed north to Danville to rejoin his chief in the new, temporary Confederate capital.

For Davis to maintain command over the forces of the Confederacy, and to order them into action, he needed military intelligence, especially from General Lee. The sudden evacuation of Richmond had disrupted Davis's regular channels of communication and had left him blind. He spent much of April 4 sending and receiving messages. He wrote to General P. G. T. Beauregard: "Please give me any reliable information you have as to *movements of enemy* and dispositions to *protect the Piedmont R.R.* I have no communication from *Gen'l Lee since Sunday.*"

Beauregard replied at 3:30 P.M. from Greensboro, North Carolina. He knew nothing of Lee. "I consider R.R. from Chester to Danville safe at present. *Will send today 600 more men to latter point. Twenty-five hundred* more could be sent if absolutely needed but they are returned men from various commands in Army of Tennessee temporarily stopped here & organized here. General *Johnston* has ordered here *some cavalry* which I have diverted *from Hillsborough to Danville.* No news from Lee or Johnston."

Davis replied promptly. "The reports in regard *to the raiders* very contradictory. But evidence indicates that they have not been at *Madison. The cavalry* you have ordered here, will be of especial value at this time, and with the Infantry en route will probably serve the immediate necessity. Have sent courier *to Gen'l Lee* from whom I have no communication." The present status of Lee's Army of Northern Virginia two days after the evacuation of Richmond was central to Davis's plans, and the lack of intelligence frustrated the president.

Later that day, bad news arrived from other regions of the Confederacy. Howell Cobb, former governor of Georgia and a major Confederate leader, sent word to Davis of multiple disasters: "Selma has fallen—The Enemy threatens Montgomery and it is believed will march upon Columbus Georgia. I submit for your consideration that Woffords command should be kept in Georgia & ordered to report to me. Please answer as Wofford is preparing to move towards Chattanooga and Knoxville Road East Tennessee & Georgia Railroad."

Davis heard about more setbacks from his nephew Joseph R. Davis: "My Brigade was lost except about twenty men all captured; I went to Richmond to join you—arrived too late. I came to this place [Powhatan Courthouse] on foot. On the capture of my command lost everything. I will join the army and remain with it in some capacity. I deeply regret having missed you as I hoped in an humble way to have served you. Remember me in love to aunt and the children."

Davis knew he had to inspire the people of the Confederacy and make them realize that his move to Danville was not a shameful flight to save himself but instead was a strategic retreat. He sought, by personal example, to make them believe that he had not abandoned them, that the cause was not lost, that he would never surrender, and that he would lead them to victory and independence. He drafted a presidential proclamation for the whole South to read. Issued the same day that Abraham Lincoln toured Richmond, the text was published as a one-page broadside on the printing press of the local Danville newspaper. Remembered only by students of the Civil War, and rarely quoted in full, the remarkable Danville Proclamation provides unfiltered insights into the mind of the retreating but unbowed president.

To the People of the Confederacy
Danville, Va.,
April 4, 1865

The General-in-Chief of our Army has found it necessary to make such movements of the troops as to uncover the Capital, and thus involve the withdrawal of the Government from the city of Richmond.

It would be unwise, even if it were possible, to conceal the great moral, as well as material injury to our cause that must result from the occupation of Richmond by the enemy. It is equally unwise and unworthy of us, as patriots engaged in a most sacred cause, to allow our energies to falter, our spirits to grow faint, or our efforts to become relaxed, under reverses however calamitous. While it has been to us a source of national pride, that for four years of unequalled warfare, we have been able, in close proximity to the centre of the enemy's power to maintain the seat of our chosen Government free from the pollution of his presence; while the memories of the heroic dead, who have freely given their lives to its defence, must ever remain enshrined in our hearts; while the preservation of the capital, which is usually regarded as the evidence to mankind of separate existence, was an object very dear to us, it is also true, and should not be forgotten, that the loss which we have suffered is not without compensation.

For many months the largest and finest army of the Confederacy, under the command of a leader whose presence inspires equal confidence in the troops and the people, has been greatly trammeled by the necessity of keeping constant watch over the approaches to the capital, and has thus been forced to forego more than one opportunity for promising enterprises.

The hopes and confidences of the enemy have been constantly excited by the belief, that their possession of Richmond would be the signal for our submission to their rule, and relieve them from the burthen of a war which, as their failing resources admonish them, must be abandoned if not brought to a successful close.

It is for us, my countrymen, to show by our bearing under reverses, how wretched has been the self-deception of those who have believed us less able to endure misfortune with fortitude, than to encounter danger with courage.

We have now entered upon a new phase of the struggle, the memory of which is to endure for all ages, and to shed ever increasing lustre upon our country. Relieved from the necessity of guarding cities and particular points, important but not vital to our defence with our army free to move from point to point, and strike in detail the detachments and garrison of the enemy; operating in the interior of our own country, where supplies are more accessible, and where the foe will be far removed from his own base, and cut off from all succor in case of reverse, nothing is now needed to render our triumph certain, but the exhibition of your own unquenchable resolve. Let us but will it, and we are free; and who in the light of the past, dare doubt your purpose in the future?

Animated by that confidence in your spirit and fortitude, which never yet has failed me, I announce to you, fellow countrymen, that it is my purpose to maintain your cause with my whole heart and soul; that I will never consent to abandon to the enemy one foot of the soil of any one of the States of the Confederacy; that Virginia, noble State, whose ancient renown has eclipsed by her still more glorious recent history; whose bosom has been bared to receive the main shock of this war; whose sons and daughters have exhibited heroism so sublime as to render her illustrious in all time to come; that Virginia, with the help of the people, and by the blessing of Providence, shall be held and defended, and no peace ever be made with the infamous invaders of her homes by the sacrifice of any of her rights or territory.

If by stress of numbers, we should ever be compelled to a temporary withdrawal from her limits, or those of any other border State, again and again will we return, until the baffled and exhausted enemy shall abandon in despair his endless and impossible task of making slaves of a people resolved to be free.

Let us not then respond, my countrymen, but, relying on the never failing mercies and protecting care of our God, let us meet the foe with fresh defiance, with unconquered and unconquerable hearts.

Jefferson Davis

April 4 was a day of two different messages from two different men. One man, Lincoln, wanted to end the war and appealed to his people to "let 'em up easy." The other man, Davis, anticipating a "new phase of the struggle," beseeched his people to "let us meet the foe with fresh defiance."

On April 5 Davis wrote a letter to Varina, revealing the details of his last few hours in the city:

Danville Va
5 April 65

My Dear Wife

. . . I made the necessary arrangements at my office and went to our house to have the proper dispositions made there—Nothing had been done after you left and but little could be done in the few hours which remained before the train was to leave—I packed the bust [of his deceased son, Samuel] and gave it to Jno. Davis who offered to take it & put it where it should never be found by a Yankee—I also gave him charge of the painting of the heroes of the valley—Both were removed after dark—The furniture of the house was left and very little of the things I directed to be put up—beddings and groceries were saved. Mrs. Omelia behaved just as you described her, but seemed anxious to serve and promised to take care of every thing which may mean some things. The Auctioneer returned acct of sale 28,400 dollars—could not dispose of the carriages—Mr. Grant was afraid to take the carriage to his house—&c. &c. I sent it to the Depot to be put on a flat, at the moment of starting it was said they could not take it in that train but would bring it on the next one—It has not been heard from since—I sent a message to Mr. Grant that I had neglected to return the cow and wished him to send for her immediately—

Called off on horseback to the Depot, I left the servants to go down with the boxes and they left Tippy—Watson came willingly, Spencer came against my will, Robert Alf. V.B. & Ives got drunk—David Bradford went back from the Depot to bring out the spoons and forks which I was told had been left—and to come out with Genl. Breckinridge since then I have not heard from either of them—I had short notice, was interrupted so often and so little aided that the results are very unsatisfactory.

The people here have been very kind, and the Mayor & Council have offered assistance in the matter of quarters and have very handsomely declared their unabated confidence—I do not wish to leave Va, but cannot decide on my movements until those of the Army are better developed—I hope you are comfortable and trust soon to hear from you.

Kiss my dear children—I weary of this sad recital and have nothing pleasant to tell—May God have you in his holy keeping is the fervent prayer of your ever affectionate

Husband

It was an odd letter. Davis had just lost his capital, the military situation was dire, and yet he wrote of personal things—carriages, paintings, a sculpture, and silver spoons. At first, this checklist makes Davis seem out of touch, oblivious to the danger, even foolish. Davis had never indulged such petty concerns during the war. Even when Union forces closed in on his beloved Mississippi plantation, Brierfield, and when Confederate soldiers were placed at his disposal to rescue his possessions, he declined the offer, proclaiming that the army does not act for the president's personal convenience. The purpose of Davis's letter was to calm Varina, to reassure her that the world had not yet turned upside down, that he had left their home with a sense of order—and that he had rescued her favorite painting of Stonewall Jackson and the precious marble bust of her dead son, Samuel. All was not chaos, at least not yet.

After Lincoln's tour of Richmond, he returned to City Point, not to Washington. He still did not want to leave the field. He wanted to be there, with his army, for the end. His visit to the rebel capital had given him a taste of victory. On April 5, Secretary of State William Seward sent a telegraph to Lincoln: "We need your personal sanction to several matters here which are important and urgent in conducting the Government but not at all critical or serious. Are you coming up or shall I go down to you with the papers. The public interest will not suffer by you remaining where you are."

But Lincoln did not want to go home: "Yours of to-day received. I think there is no probability of my remaining here more than two days longer. If that is too long come down. I passed last night at Richmond and have just returned."

In Danville, Jefferson Davis did not know that Abraham Lincoln was using the White House of the Confederacy as an office to conduct peace negotiations with officials in Richmond. On April 6, Davis wrote another letter to Varina. "In my letter of yesterday I gave you all of my prospects which can now be told, not having heard from Genl. Lee and having to conform my movements to the military necessities of the case. We are now fixing an Executive office where the current business may be transacted here and do not propose at this time to definitely fix upon a point for seat of Govt. in the future. I am unwilling to leave Va. and do not know where within her borders the requisite houses for the Depts. and the Congress can be found . . . Farewell my love, may God bless preserve and guide you."

Many Southerners agreed that the loss of Richmond did not signify the total defeat of the Confederacy. On April 6, Eliza Frances Andrews, the twenty-four-year-old daughter of Judge Garnett Andrews, a lawyer in Washington, Georgia, and the owner of Maywood plantation and its two hundred slaves, wrote in her diary: "I took a long walk through the village with Capt. Greenlaw after dinner, and was charmed with the lovely gardens and beautiful shade trees. On coming home, I heard of the fall of Richmond. Everybody feels very blue, but not disposed to give up as long as we have Lee."

On April 6, Robert E. Lee telegraphed Davis from his headquarters at Rice's Station, Virginia, South Side Railroad: "I shall be tonight at Farmville. You can communicate by telegraph to Meherrin and by courier to Lynchburg." The Army of Northern Virginia was, President Davis believed, still in the game.

From Charlotte, Varina Davis wrote to her husband again on April 7. Their exchange of letters after the fall of Richmond was the beginning of a correspondence that evolved into one of the great collections of American love letters. "The news of Richmond came upon me like the 'abomination of desolation,'" she wrote. "... I who know that your strength when stirred up is great, and that you can do with a few what others have failed to do with many am awaiting prayerfully the advent of time when it is God's will to deliver us through his own appointed agent ... Numberless surmises are hazarded here as to your future destination and occupation—but I know that wherever you are, and in whatever engaged, it is an efficient manner for the country." She ended her letter intimately: "Our little ones are all well, but very unruly ... Li Pie [their infant daughter Varina Anne] is sweet and pink, and loving her hands and gums are hot, and swollen, and I think she is teething ... Write to me my own precious only love, and believe me as ever your devoted Wife."

On April 7, Abraham Lincoln, still at City Point, continued to follow the telegraph and dispatch traffic. Reading between the lines, he sensed that victory was imminent. He had become an expert at reading the dry words of a military communication and then interpreting the unsaid meaning behind the text. He had read several thousand of them during the war and knew how to take their pulse. Now, on April 7, when he held them with his fingers, Lincoln could feel victory resonating from the sheets of paper. Then General Phil Sheridan gave the president a military assessment that inflamed his taste for victory so much that it provoked him to send a telegraph to General Grant. He ordered his commanding general of the armies of the United States to close in for the kill and win the war.

> *Head Quarters Armies of the United States*
> *City-Point,*
> *April 7. 11 A.M. 1865*
>
> *Lieut. Gen. Grant.*
> *Gen. Sheridan says "If the thing is pressed I think that Lee will surrender." Let the thing be pressed.*
>
> *A. Lincoln*

That day in Washington, Secretary of the Navy Gideon Welles wrote in his diary: "It is desirable that Lee should be captured. He, more than any one else, has the confidence of the Rebels, and can, if he escapes, and is weak enough to try and continue hostilities, rally for a time a brigand force in the interior. I can hardly suppose he would do this, but he has shown weakness, and his infidelity to the country which educated, and employed, and paid him shows great ingratitude. His true course would be to desert the country he has betrayed, and never return." Perhaps, Welles suspected, so would Jefferson Davis, and he expressed the same wish for the rebel chief.

In City Point, as Lincoln prepared to board the *River Queen* and return to Washington, a U.S. Army band serenaded him with a farewell evening concert. The president had spent eighteen days and seventeen nights with his men: The long visit had invigorated him and increased the bonds of affection between them. During the war the common soldiers had always been happy to see their president and cheered him on sight. In the election of 1864, it was the soldiers' vote that kept Lincoln in office when their former commander, General George McClellan, tried to unseat him. Lincoln enjoyed military music, and during summers in Washington, the U.S. Marine Corps band had played concerts on the White House grounds. At 11:00 P.M. the *River Queen* steamed away from City Point and headed for Washington. Lincoln did not know it, but he was leaving a day too early. If only he could have read Robert E. Lee's mind, he would never have returned to Washington that night.

While Lincoln was en route to Washington on April 8, Davis had been in Danville for five days. He still refused to believe that the Army of Northern Virginia was in danger of immediate collapse, even though Secretary of War Breckinridge had given him a report that day saying the war was lost. But Davis was far from the front lines and could not receive telegrams or couriers in anything close to real time. At the front, events were in flux, with the situation changing hourly. Far away in the new capital, Davis did not learn of battlefield events before or while they were occurring, but only after they had already happened. And Lee was fighting for his life. He did not have time to dispatch a series of detailed telegraphic or courier messages. And so the president of the Confederacy did not know what his most important general was thinking.

Lee considered the possibility of continuing the fighting, but he had hardly any men left and fit for battle—no more than several thousand. His thoughts, and loyalty, turned to his surviving soldiers. The postwar South would need them—the country had lost so many boys already. In many ways *they* were the South, not cities like Richmond, Atlanta, Vicksburg, New Orleans, Savannah, and the rest. If the Confederacy was doomed to lose these final battles, suffering great loss of life with no hope of victory, was it right to sacrifice any more lives? More fighting might have been suicidal, even criminal. Lee sent a courier to Danville bearing a message for the president:

Surrender was inevitable. Lee knew what he must do. He composed a letter to General Ulysses S. Grant, asking that they meet the next day at a little place called Appomattox Court House.

In the morning a great controversy erupted in Richmond, on the first Sunday since the burning of the city and the beginning of Union occupation. In church services during the war, it was the custom of the ministers to ask God's blessing for President Davis and the Confederate cause. Now Yankee officials demanded that ministers bless not Davis but Lincoln. This was too much for the downtrodden citizens to bear. The dispute made it all the way to the ears of Lincoln, who found the whole episode embarrassing.

In Danville, Davis, ignorant of Lee's appointment with Grant later in the day, continued to make war plans. He sent a telegram to his top general to plan the next phase of the struggle: "Your dispatch of the 6th . . . received. Hope the line of couriers established will enable you to communicate safely and frequently . . . You will realize the reluctance I feel to leave the soil of Virginia . . . the fall of Selma and the reported advance of the enemy on Montgomery, and the fears expressed for the safety of Columbus, Georgia, caused me to direct Gen'l Cobb to aid in resisting the enemy in Alabama . . . I hope to hear from you soon at this point, where offices have been opened to keep up the current business, until more definite knowledge would enable us to form more permanent plans. May God preserve, sustain and guide you."

It happened on April 9 around 1:00 P.M., without the participation of Abraham Lincoln or Jefferson Davis. While Lincoln sailed back to Washington, and while Davis waited in Danville for news, Ulysses S. Grant and Robert E. Lee met at the McLean house at Appomattox Court House, Virginia.

Grant treated Lee with the highest military courtesy and, after reminiscing with his foe about their common service in the Mexican War, offered to accept the surrender of the Army of Northern Virginia on generous terms. Once the men laid down their arms and signed their paroles, they could return to their homes. They could wear their Confederate uniforms, take their horses, and just go home. They would not be made prisoners of war nor be punished as traitors. And before the men of the Army of Northern Virginia left the field for the final time, the boys in blue paid honor to them. It was as Lincoln would have wished.

Lincoln arrived in Washington at 6:00 P.M. and went from the wharf straight to William Seward's home in Lafayette Square, across from the White House. Seward, bedridden from terrible injuries he had suffered in a recent carriage accident, lay still while Lincoln stretched his long frame across the foot of Seward's bed and brought him encouraging news from the front and tales of his wondrous visit to Richmond. The president was ecstatic: The war would be over soon; he could feel it. Lincoln and Seward did not know that Lee had surrendered several hours earlier. After an hour of quiet, intimate talk, Lincoln went home.

Crowds at the White House demanded that the president show himself—the people had missed him and were disappointed that he had not been in Washington on April 3 to celebrate the fall of Richmond with them. He stepped to a window beneath the north portico and spoke an inconsequential greeting. News from Appomattox did not arrive at the War Department until later on the night of the ninth, too late for Washington to celebrate en masse, but Lincoln was told. No one knows what he did after he heard the news: Was he too overjoyed to sleep that night? Did he walk the halls or go to his office and stare through the window into the night? Did he haunt the telegraph office? Did he know that tomorrow morning would begin the greatest day in the history of Washington?

Washington awoke the next morning to the sound of an artillery barrage. If this was 1861, not 1865, Lincoln might have concluded that the national capital was under rebel bombardment. But, as one of the few people who had learned the previous night about Lee's surrender, Lincoln knew better.

The president ate breakfast with Noah Brooks, who described how the inhabitants of the national capital learned of the surrender at Appomattox:

> Most people were sleeping soundly in their beds when, at daylight on the rainy morning of April 10, 1865, a great boom startled the misty air of Washington, shaking the very earth, and breaking windows of houses about Lafayette Square . . . Boom! Boom! went the guns, until five hundred were fired. A few people got up in the chill twilight of the morning, and raced about in the mud to learn what the good news might be . . . but many lay placidly abed, well knowing that only one military event could cause all this mighty pother in the air of Washington; and if their nap in the gray dawn was disturbed with dreams of guns and terms of armies surrendered to Grant by Lee, they awoke later to read of these in the daily papers; for this was Secretary of War Stanton's way of telling the people that the Army of Northern Virginia had at last laid down its arms, and that peace had come again.

Welles delighted in the moment: "Guns are firing, bells ringing, flags flying, men laughing, children cheering; all, all are jubilant." Welles, like many others, believed that Lee's surrender meant the war was over now, and he made no mention in his diary that day about the retreating Jefferson Davis: "This surrender of the great Rebel captain [Lee] and the most formidable and reliable army of the Secessionists virtually terminates the Rebellion. There may be some marauding, and robbing and murder by desperadoes, but no great battle, no conflict of armies, after the news of yesterday reaches the different sections. Possibly there may be some stand in Texas or at remote points beyond the Mississippi."

On this day of victory no one in Washington was dwelling upon Jefferson Davis, his government in exile, or his last-ditch plans. It was seven days after the fall of Richmond, and Lincoln had still not issued any orders to capture Davis or the top Confederate political and military leaders. He had his reasons. The *New York Times* speculated that the rebel chief had already escaped but called for his death anyway. "It is doubtful whether Jeff Davis will ever be captured. He is, probably, already in direct flight for Mexico . . . but if he is caught he should be hung." Indeed, on this day of jubilee, the predominant popular image of the fleeing president was one of dismissive bemusement rather than one of avenging pursuit. Soon that would change.

Robert E. Lee was preparing to leave his army and travel to Richmond, where he would reunite with his wife, Mary Custis Lee. His house had survived the fire and was now under guard to protect her property from looters. But first he wanted to thank his men and say good-bye. He did so by drafting a document that was meant as a personal, heartfelt tribute to be read aloud to the soldiers under his personal command who had surrendered with him. But soon it became known to a wider audience and spread throughout the South, where the people of the Confederacy embraced it as the thanks of the nation to all the men, living and dead, who had fought in the war.

General Order, No. 9

Headquarters, Army of Northern Virginia

April 10, 1865

After four years of arduous service, marked by unsurpassed courage and fortitude, the Army of Northern Virginia has been compelled to yield to overwhelming numbers and resources.

I need not tell the brave survivors of so many hard fought battles, who have remained steadfast to the last, that I have consented to the result from no distrust of them.

But feeling that valor and devotion could accomplish nothing that would compensate for the loss that must have attended the continuance of the contest, I determined to avoid the useless sacrifice of those whose past services have endeared them to their countrymen.

By the terms of the agreement officers and men can return to their homes and remain until exchanged. You will take with you the satisfaction that proceeds from the consciousness of duty faithfully performed, and I earnestly pray that a Merciful God will extend to you His blessing and protection.

With an increasing admiration of your constancy and devotion to your country, and a grateful remembrance of your kind and generous considerations for myself, I bid you an affectionate farewell.

R. E. Lee

Genl

While Washington began a week of rejoicing, word traveled to Danville that there had been a great disaster, the worst possible news. A courier from Lee's army reached Jefferson Davis. The intelligence he carried, remembered Navy Secretary Mallory, "fell upon the ears of all like a fire-bell in the night." The rider delivered the message to the president's office, where Davis and several cabinet and staff members had gathered. Davis read the dispatch, did not speak, and passed it on. "They carefully scanned the message as it passed from hand to hand," Mallory recalled, "looked at each other gravely and mutely, and for some moments of silence."

Robert E. Lee had surrendered on April 9. The Army of Northern Virginia, one of the greatest military forces in history, was no more. The war in Virginia was over.

Lee's surrender made Davis's position in Danville dangerous. The news from Appomattox devastated the president. He questioned whether Lee should have surrendered. Couldn't his best general have somehow disengaged from the Union army, charted a route south, and escaped to fight another day? Or could he have dispersed his men to reassemble at a designated point of concentration? Davis also feared that Lee's capitulation would set a surrender precedent that other Confederate armies would follow. Such a chain reaction would be a catastrophe and would surely cause the total collapse and defeat of the Confederacy. Davis could not fight on alone, without troops to sustain the cause. With the Army of Northern Virginia now lost, it was urgent that the Confederate government increase the distance between it and the Union armies by retreating at once, deeper into the southern interior.

If Davis did not order that everyone evacuate the city at once, enemy cavalry could swoop in and capture what remained of the Confederate government. That would end the war. Leaving Danville meant not only fleeing one town but abandoning the state. To Davis, fleeing the principal state of the Confederacy was a terrible psychological blow. First he had lost his capital, Richmond; he had just lost his greatest general and his best army; and now he was about to lose all of old Virginia. This series of three staggering blows, all within one week, jeopardized Davis's ability to rally the people and save the nation.

He ordered the immediate evacuation of Danville by a night train to Greensboro, North Carolina. Burton Harrison, back at the president's side after escorting Varina Davis to safety in Charlotte, took control of the train: "We set to work at once to arrange for a railway train to convey the more important officers of the Government and such others as could be got aboard, with our luggage and as much material as it was desired to carry along, including the boxes and papers that had belonged to the executive office in Richmond." The boxes were an important symbol because Davis felt that as long as he kept his cabinet intact and did not abandon the archives and working documents necessary for the continued operation of the government, the Confederate States of America lived.

Davis could not leave this place without thanking the people of Danville. He drafted a letter to the mayor.

> *To Mayor J. M. Walker*
> *Danville, Va.,*
> *April 10, 1865*
>
> *Sir:*
>
> *Permit me to return to yourself and council my sincere thanks for your kindness shown to me when I came among you, under that pressure of adversity which is more apt to cause the loss of friends than to be the occasion for forming new ones.*
>
> *I had hoped to have been able to maintain the Confederate Government on the soil of Virginia, though compelled to retire from the Capital. I had hoped to have contributed somewhat to the safety of your city, the desire to the last was rendered more than a mere sense of public duty, by your generous reception of myself and the Executive officers who accompanied me. The shadows of misfortune which were on us when I came have become darker, and I trust you accord to me now as then your good wishes and confidence in the zeal and singleness of heart with which I have sought to discharge the high trust which the people of the Confederate States conferred upon me.*

May God bless and preserve you, and grant to our country independence and prosperity.

Very truly yours,

JEFFN. DAVIS

Jefferson Davis's "Danville Farewell" communicated a message very different from Robert E. Lee's "General Order #9." Lee told his men that continuing the war would have resulted in the "useless sacrifice" of their lives. He advised them that it was time to "return to [their] homes" and fight no more. In Danville, Davis expressed contrary sentiments. He regretted only that he could not "maintain the Confederate Government on the soil of Virginia" and called upon God to grant the Confederacy its independence. Davis and his government headed for their next destination, Greensboro, North Carolina. When he would cross the state line the next day on April 11, he would have to concede an awful fact. Virginia, queen of the Confederacy, was lost.

While Jefferson Davis and the cabinet packed up in Danville, in Washington Lincoln was treated to an evening of White House serenades that featured a boisterous performance of "Dixie." Lincoln had loved the tune from the moment he heard it performed before the war at a theater in Chicago. The Confederacy's adoption of the song as its anthem failed to diminish Lincoln's enjoyment of it. When he spotted a band among a crowd of torch-bearing well-wishers who had gathered on his lawn, he made the people laugh by telling them that "Dixie" was one of the captured spoils of war and that he wanted to hear it right then. The band obliged, and the music of the Lost Cause echoed through Lincoln's White House, drifted across the grounds and into the streets of the Union capital.

Meanwhile, Harrison posted guards to prevent unauthorized persons or baggage from coming aboard the train. The sentinels had their hands full. "Of course," recalled Harrison, "a multitude was anxious to embark, and the guards were kept busy in repelling them." Dozens of people beseeched him for passes. One general from the "torpedo bureau" claimed that he possessed valuable fuses and explosives vital for the war effort. Dubious, Harrison told the general there was no room aboard the train for him and his collection. Undeterred, the general got access to President Davis, with whom he had served in the army years ago. Davis told Harrison to find a place for the man and his daughters, and the ever-courteous president invited one of the women to share his seat.

Mallory painted a railroad station scene more chaotic than the night Richmond was evacuated:

> Much rain had fallen, and the depot could be reached only through mud knee deep. With the utter darkness, the crowding of quartermasters' wagons, the yells of their contending drivers, the curses, loud and deep, of soldiers, organized and disorganized, determined to get upon the train in defiance of the guard, the mutual shouts of inquiry and response as to missing individuals or luggage, the want of baggage arrangements, and the insufficient and dangerous provision made for getting horses into their cars, the crushing of the crowd, and the determination to get transportation at any hazard, together with the absence of any recognized authority, all seasoned by *sub rosa* rumors that the enemy had already cut the Greensboro road, created a confusion such as it was never before the fortune of old Danville to witness.

Burton Harrison marveled at the mad scene. He watched as the guards "excluded all persons and material not specially authorized by me to go aboard." Harrison was not above taking advantage of the situation, if it was for the good of the cause. As he stood in front of the government's headquarters supervising the removal of baggage and boxes of documents, two mounted officers—one a colonel—rode into town from Richmond. Harrison told them that Lee had surrendered and that the government was about to abandon Danville. Then Harrison eyed the colonel's mount.

"I remarked on the freshness and spirit of his horse, and asked where he had got so good a steed," he recalled. Harrison knew Davis could not count on obtaining uninterrupted railroad transportation for their entire journey. At some point, circumstances would dictate that they continue the retreat on horseback. The president and his aides could use all the good horses they could get for the next stages of the trip, and Harrison proposed a trade. He said he "should be glad to have the horse" in exchange for passage on the train.

Reluctant to surrender the animal, the colonel rode off and tried to board one of the cars, but the guards told him he could not without a written order from Harrison. He returned to Harrison, "whereby he remarked," said Harrison, "that, if I would furnish such an order, he would accept my proposition about the horse. The arrangement was made immediately, and the colonel became a passenger on the train, which also conveyed my horse, with others belonging to the President and his staff."

Mallory watched his colleagues gather near the train: "At ten o'clock, Cabinet officers and other chiefs of the government, each seated upon or jealously guarding his baggage, formed near the cars a little silent group by themselves in the darkness, lighted only by Mr. Benjamin's inextinguishable cigar. It was nearly eleven o'clock when the president took his seat and the train moved off. The night was intensely dark, and with a slight rain, the road in wretched condition, and the progress was consequently very slow."

It didn't take long before Davis began to regret the invitation he had extended for the torpedo general's daughter to sit beside him. "That young lady," complained Colonel Harrison, "was of a loquacity irrepressible; she plied her neighbor diligently—about the weather, and upon every other topic of common interest—asking him, too, a thousand trivial questions." Until the train could get up steam, the passengers crowded together in the cars, according to Harrison, "waiting to be off, full of gloom at the situation, wondering what would happen next, and all as silent as mourners at a funeral." The exception was the general's daughter, "who prattled on in a voice everybody heard."

Then an explosion close to the president rocked the car. No one knew what had just happened. Had Union troops intercepted the slow-moving train and tossed a grenade into Davis's car? Or had a traitor sitting in the car tried to assassinate the president with a suicide bomb?

Burton Harrison saw it all: "A sharp explosion occurred very near the President, and a young man was seen to bounce into the air, clapping both hands to the seat of his trowsers. We all sprang to our feet in alarm." The car smelled of black gunpowder, but no one had seen the telltale flash of the explosion. Harrison quickly discovered that this was not an attack but an absurd accident. One of the torpedo general's officers, carrying explosive detonation fuses in the coattail pocket of his long frock coat, had sat down atop a flat-bottomed stove. His weight crushed one of the fuses, setting off the explosion, and nearly blowing off his backside. Davis and the other occupants of the car were unharmed.

Davis's train arrived in Greensboro, North Carolina, at around 2:00 P.M. on April 11. He conferred with General Beauregard that day, and on the following day General Joseph E. Johnston, commander of the Confederate army in North Carolina, joined them to discuss Davis's desire to continue the war. Davis also learned that a unit of federal cavalry had cut the road at a point where his train had passed only five minutes before. This was the closest he had come to capture since he had left Richmond, and from this point on, the government in exile was in danger of encountering Union troops at any moment.

Davis was not greeted with open arms by the citizens of Greensboro, as he had been in Danville. This time the local dignitaries did not come forward to offer food and lodging to their president and his cabinet. The unfriendly reception outraged Stephen Mallory. "No provision had been made for the accommodation of the President and staff, or for his Cabinet . . . Greensboro had been a flourishing town, and there were many commodious and well-furnished residences in and about it, but their doors were closed and their 'latch-strings pulled in' against the members of the retreating government." Colonel John Taylor Wood from Davis's staff invited the president to share his family's modest quarters, which Wood had rented for them after moving them away from Richmond to safety. Colonel Harrison commented on Davis's reception there: "[The owners] of the house continuously and vigorously insist[ed] to the colonel and his wife . . . that Mr. Davis must go away, saying they were unwilling to have the vengeance of Stoneman's [Union] cavalry brought upon them by his presence in their house."

Mallory denounced the people of Greensboro as "pitiable" and ill-mannered. "Generous hospitality has ever been regarded as characteristic of the South, and had such a scene as this been predicted of any of its people, it would have encountered universal unbelief." But Greensboro had denied the president the "uniform kindness, courtesy, and hospitality" which he had received elsewhere. Harrison echoed Mallory's opinion of Greensboro. "The people in that part of North Carolina had not been zealous supporters of the Confederate Government; and, so long as we remained in the State, we observed their indifference to what should become of us. It was rarely that anybody asked one of us to his house; and but few of them even had the grace even to explain their fear that, if they entertained us, their houses would be burned by the enemy, when his cavalry should get there." While in Greensboro, the horses belonging to Davis, his personal aides, and the cabinet were kept under twenty-four-hour guard to prevent their theft by townspeople or refugees.

The members of the Confederate cabinet, just as they had made the best of their two train rides from Richmond and from Danville, endured their Greensboro humiliation with good humor. Upon their humble quarters, they bestowed the exalted nickname the "Cabinet Car" and made the best of the situation. It was, said Mallory, "a very agreeable resort" during the "dreary days" in the unfriendly town. "Its distinguished hosts did the honors to their visitors with a cheerfulness and good humor, seasoned by a flow of good spirits, which threw a charm around the wretched shelter and made their situation seem rather a matter of choice than of necessity. The navy store supplied bread and bacon, and by the active foraging of Paymaster Semple and others of the party, biscuits, eggs, and coffee were added; and with a few tin cups, spoons, and pocket knives, and a liberal use of fingers and capital appetites, they managed to get enough to eat, and they slept as best they could." Unashamed, the highest officials of the Confederacy ate like common soldiers.

"The curious life of the fleeing Confederate Government in the 'Cabinet Car' at Greensboro continued for nearly a week, and was not all discomfort," Mallory insisted.

> Indeed, the difficulties of their position were minimized by the spirit with which these men encountered every trial. Here was the astute "Minister of Justice," a grave and most exemplary gentleman, with a piece of half-broiled "middling" in one hand and a hoe-cake in the other, his face bearing unmistakable evidence of the condition of the bacon. There was the clever Secretary of State busily dividing his attention between a bucket of stewed dried apples and a haversack of hard-boiled eggs. Here was the Postmaster-General sternly and energetically running his bowie knife through a ham as if it were the chief business of life, and there was the Secretary of the Navy courteously swallowing his coffee scalding hot that he might not keep the venerable Adjutant-General waiting too long for the coveted tin cup! All personal discomforts were not only borne with

cheerful philosophy, but were made the constant texts for merry comment, quaint anecdotes, or curious story.

As soon as Davis arrived in Greensboro on April 11, he wrote to Joe Johnston.

> The Secty. Of War did not join me at Danville, is expected here [Greensboro] this afternoon. As your situation may render best, I will go to your Hd. Qrs. immediately after your arrival of the Secty of War, or you can come here . . . I have no official report from Genl. Lee, the Secty. Of War may be able to add information heretofore communicated. The important question first to be solved is at what point concentration shall be made.

The president had visions of concentrating all available forces at a single strategic place from which he could smash the Union army.

As Davis dreamed of new victories, Richmond, the city from which he had been driven by force of arms, had become *the* tourist destination for the Washington elites, who pestered high government or military officers for written passes to enter the ruined city. Indeed, Mary Lincoln and a party of her guests had already toured Richmond, and on April 11 the president wrote out a pass authorizing his friend and marshal of the District of Columbia, Ward Hill Lamon, to enter that city. In the spring of 1865, it was the place to be.

On the afternoon of the eleventh Abraham Lincoln sat in his office and wrote out in his vigorous, clear hand the draft of an important speech he planned to deliver from the White House window that night. The president wanted to pay tribute to the armed forces that won the war, prepare the people for his postwar plans, and propose that blacks be given the right to vote. On April 12, General Lee wrote his penultimate letter to Jefferson Davis, telling his commander in chief what he already knew. This was Lee's official announcement to the president that he had surrendered.

Near Appomattox Court House, Virginia
April 12, 1865

Mr. President:

It is with pain that I announce to Your Excellency the surrender of the Army of Northern Virginia. The operations which preceded this result will be reported in full . . . The enemy was more than five times our numbers. If we could have forced our way one day longer it would have been at a great sacrifice of life; at its end, I did not see how a surrender could have been avoided. We had no subsistence for man or horse . . . the supplies could not reach us, and the men deprived of food and sleep for many days, were worn out and exhausted.

With great respect, yr obdt svt
R. E. Lee
Genl

Before receiving this communication, Davis gave a brief speech—no more than twelve or fifteen minutes long—in Greensboro. He boasted to his audience "how vast our resources still were, and that we would in a few weeks have a larger army than we ever had." Davis explained how such an army was to be raised. "There is Gen. Lee's army ought to be 140,000 strong—it is not 40,000—Gen. Johnston's army is only 15,000—it ought to be 100,000—Three fourths of the men are at home, absent without leave. Now we will collect them, and . . . then there are a great many conscripts on the rolls who have never been caught—we will get them—and with the 100,000 men from Gen. Lee's army and the 85,000 men from Gen. Johnston's, we will have such an army as we have never had before."

These remarks, more optimistic even than Davis's "Danville Proclamation" of April 4, rested on wishful thinking, not the situation on the ground. Lee did not have forty thousand fighting men; his effectives numbered fewer than twenty thousand, and Johnston's forces grew weaker every day. Furthermore, Davis had no real force to round up deserters by the bayonet and compel them to fight. And even if, by some miracle, the Confederacy massed nearly two hundred thousand men, Union forces would still have outnumbered them. And even if Davis could raise such numbers, they could not be fed or supplied.

General Lee's letter jolted Davis into reality. Robert E. Lee Jr. was present in Greensboro when Davis received it: "After reading it, he handed it without comment to us [Lee and John Taylor Wood]; then, turning away, he silently wept bitter tears. He seemed quite broken at the moment by this tangible evidence of the loss of his army and the misfortune of its general."

At least Davis knew his family was safe. Varina wrote on April 13, telling him she was now in Chester, South Carolina. She was staying ahead of Union cavalry raiding parties: "The rumors of a raid on Charlotte induced me to come to this side of Charlotte—A threatened raid here induces me to leave here without making an hours stay which is unnecessary—I go with the Specie train because they have a strong guard, and are attended by two responsible men—I am going somewhere, perhaps to Washington Ga . . . Would to God I could know the truth of the horrible rumors I hear of you—One is that you have started to Genl Lee, but have not been heard of . . . May God have mercy upon me, and preserve you safe for your devoted wife."

In Washington, Lincoln conducted a full day of business. The city was still celebrating Lee's surrender, but the president had plenty of work to do. The war was not over. And soon, when it was, he would have to implement his plan for the reconstruction of the South. He had visited the telegraph office early in the morning, then had meetings with General Grant and Edwin Stanton, and another with Gideon Welles.

The staff saddled Lincoln's horse at the White House stables, and he rode to his summer cottage at the Soldiers' Home. Maunsell Field, an assistant secretary of the Treasury, rode in a carriage beside Lincoln's horse and they talked along the way. Later, when Lincoln returned to his White House office he wrote out several passes allowing the bearers to visit various points south, including Richmond. Then the president, like other Washingtonians, enjoyed the grand illumination of the city.

Benjamin Brown French, commissioner of public buildings and grounds, enjoyed supervising the decoration and illumination of the public buildings and described the night: "The Capitol made a magnificent display—as did the whole city. After lighting up my own house and seeing the Capitol lighted, I rode up to the upper end of the City and saw the whole display. It was indeed glorious . . . *all of Washington* was in the streets. I never saw such a crowd out-of-doors in my life." French even designed one sign himself. "I had the 23rd verse of the 118th Psalm printed on cloth, in enormous letters, as a transparency, and stretched on a frame the entire length of the top of the western portico [of the Capitol building] . . . 'This is the Lord's doing; it is marvelous in our eyes.' It was lighted with gas and made a very brilliant display . . . as it could be read very far up the Avenue."

Not everyone in Washington relished the illumination. That night in his room at the National Hotel on Pennsylvania Avenue, John Wilkes Booth, the young stage star and heartthrob, wrote a letter to his mother. "Everything was bright and splendid," he said. But, he lamented, "more so in my eyes if it had been a display in a nobler cause."

The next day, on April 14, Jefferson Davis sent a hurried note to Varina.

Greensboro N.C.
14 April 65

Dear Winnie

I will come to you if I can. Every thing is dark.—you should prepare for the worst by dividing your baggage so as to move in wagons. If you can go to Abbeville it seems best as I am now advised—If you can send every thing there do so—I have lingered on the road and labored to little purpose—My love to the children and Maggie—God bless, guide and preserve you ever prays your most affectionate

Banny—

I sent you a telegram but fear it was stopped on the road. Genl. Bonham bears this and will [tell] you more than I can write as his horse is at the door and he waits for me to write this again and ever your's—

Lincoln began another busy day that included breakfast with his son Robert, just back from Appomattox; a cabinet meeting attended by General Grant; meetings with several congressmen; and letter writing, including one to a Union general about the future: "I thank you for the assurance you give me that I shall be supported by conservative men like yourself, in the efforts I may make to restore the Union, so as to make it, to use your language, a Union of hearts and hands as well as of States." He agreed to escort Mary to the theater that night—Laura Keene was playing in the comedy *Our American Cousin* at Ford's.

In the afternoon Abraham and Mary Lincoln went on a carriage ride to the Navy Yard. He told her that this day, he considered the war to be over. It was Good Friday, and in two days Washington would celebrate Easter. Lincoln wanted to laugh this night. That evening, just before he left the White House for the theater, a former congressman arrived and asked to see him on business. The president wrote a pass giving him an appointment at 9:00 A.M. the next day. As he was stepping into his carriage another former congressman, this one a friend from Illinois, approached him in the driveway. Lincoln said he couldn't talk then or he would be late for the play. Come back later, the president told him. We will have time to talk then. Lincoln closed the carriage door.

In Greensboro, Davis spent a quiet night wondering what events the coming days might bring. His journey, although difficult, had not been a complete disaster. Yes, he had fled Richmond, lost Lee and the Army of Northern Virginia, and abandoned the state of Virginia to the enemy. He did not deny that these disasters had inflicted catastrophic blows upon the cause. Indeed, in his letter to Varina he despaired, saying everything was "dark." But the situation was not all bad. During his twelve days on the run, he had escaped capture; relocated the Confederate capital twice, first to Danville, then Greensboro; kept the cabinet intact; retained the loyalty of a hand-picked inner circle of aides who vowed to never abandon him; protected his family; and prevented his strategic retreat from unraveling into a disorderly free-for-all. And he had maintained his dignity. He had fled Richmond not like a thief in the night, but as a head of state.

No one living in William Petersen's house across the street from Ford's Theatre ever claimed to have seen President Lincoln's carriage pull over and park across the street. No one in the handsome, three-story brick house watched the coachman, Francis Burke, tighten the slack in the reins, nor did anyone see the president's valet, Charles Forbes, jump down from the black, closed-top carriage to the dirt street, reach for the handle, and swing open the door for the passengers. Some of the Petersen boarders were out for the evening. The rest were occupied with other things.

They did not watch the president and Mrs. Lincoln or their companions, Major Henry Rathbone, an army officer, and Clara Harris, daughter of a U.S. senator, as they disembarked, walked several yards to the front door of Ford's Theatre, and disappeared inside. It was Good Friday, at approximately 8:30 P.M., April 14, and they were late. And no one from the Petersen house hurried across Tenth Street, or followed the Lincolns into the theater, and purchased a ticket to the play, as more than 1,500 other Washingtonians had done, to attend the tired old comic chestnut *Our American Cousin* in the company of the president of the United States.

Abraham Lincoln loved the theater, and during the Civil War he had attended many plays at Ford's and Grover's, Washington's two leading, and rival, playhouses. Tonight, twelve-year-old Tad Lincoln enjoyed *Aladdin* at Grover's, a few blocks away on Pennsylvania Avenue. The Lincolns' other surviving son, twenty-two-year-old Robert, home from his duties on Grant's staff, chose to stay at the White House to read.

The next few hours passed without incident. Passing by Ford's that night was the customary Friday-night foot and horse traffic, as well as revelers in the ongoing war's-end celebration. At Ferguson's restaurant, adjacent to the theater's north wall, patrons ate their meals without the owner, James Ferguson, who had gone to Ford's hoping to see General Grant. Earlier that day, newspaper ads had mistakenly touted the general as Lincoln's theater guest. They were wrong: He and his wife, Julia, had declined the invitation and left town.

At Taltavul's Star Saloon, the narrow brick building just south of Ford's, customers gulped their whiskeys and brandies and tossed their coins on the bar as payment. One patron—a handsome, pale-skinned, black-eyed, raven-haired, mustached young man—placed his order, drank it, and left the bar without speaking a word.

If anyone from the Petersen house had been watching the front door of the Star Saloon between 9:30 and 10:00 P.M., he might have recognized John Wilkes Booth, one of the most famous stage stars in America, as he emerged wearing a black frock coat, black pants, thigh-high black leather riding boots, and a black hat. Booth turned north up Tenth Street, observed the president's carriage parked several yards in front of him, and then turned right, toward the theater, passing under the white painted arch and through Ford's main door, the same one the president had passed through about an hour earlier. If his intention was to see the play, Booth was impossibly late.

CHAPTER FOUR

"Borne by Loving Hands"

The Petersen house was no different from hundreds of other boardinghouses that had enjoyed a thriving business during the last four years in overpopulated wartime Washington. Indeed, this style of urban living had been commonplace ever since the District of Columbia was established as the national capital. Military officers, cabinet members, senators, and congressmen—including a one-term representative named Abraham Lincoln elected from Illinois in 1846—were veterans of Washington's traditional boardinghouse culture.

William Petersen, like many homeowners in Washington, rented out extra rooms to boarders. Born in Hanover, Germany, William and his wife, Anna, had emigrated to the United States in 1841, when they were twenty-five and twenty-two years old. Landing in the port of Baltimore, they moved to Washington and on February 9, 1849, purchased the lot at 453 Tenth Street for $850. Petersen, a tailor, hired contractors to build him a large, attractive, four-level brown brick row house with a tall basement and three main stories. By 1860, the year Lincoln was elected president and South Carolina seceded from the Union, nine boarders resided there, along with the Petersens' seven children, bringing the household total to eighteen occupants living in eleven rooms.

At one moment the street between the Petersen house and Ford's Theatre was quiet. At the next, sometime between 10:15 and 10:30 P.M., dozens of playgoers rushed out the doors onto Tenth Street. This was not an audience's ordinary, leisurely exit at the end of a performance. And the play was not over yet—the last scene had not yet been performed. People began pushing one another aside and knocked one another down to squeeze through the exits, like a great volume of water bursting through a tiny hole in a dike. Some of the first men who escaped the theater fled in both directions on Tenth Street toward E and F streets, shouting crazy, unintelligible words as they ran. Within seconds they turned the corners and vanished from sight. Then hundreds of men, women, and children escaped Ford's and gathered in the street. Many screamed. Others wept. Soon more than one thousand panicked playgoers were crowded in front of the theater. Screaming, cursing, shouting, weeping, their voices combined into a loud and fearful roar. Something had gone terribly wrong inside Ford's Theatre.

At first it appeared that the theater might have caught fire. Fires were a constant and almost unpreventable danger in nineteenth-century urban America. Wood buildings, fabric drapes, errant candles, whale-oil lamps, primitive gas lighting, and the lack of effective firefighting equipment led to disastrous conflagrations that had nearly destroyed several major American cities. New York City, Philadelphia, Boston, and other urban centers had each suffered fantastic firestorms that spread from building to building and burned wide swaths through the hearts of their residential, commercial, and industrial districts.

Fires were so commonplace that Currier & Ives published numerous prints depicting American cities ablaze, meticulously hand-coloring each calamitous scene with menacing orange and yellow flames. Fire was such a source of dread that long before the Civil War, Thomas Jefferson compared the antebellum conflict over slavery to a "fire bell in the night" that might burn down the American house.

Theater fires were especially dangerous. Wood stages, huge fabric curtains, footlights of open gas flames, and large audiences seated in close quarters with few exits could prove a deadly combination. Fifty-four years earlier, in 1811, a horrible fire in Richmond killed more than seventy-five playgoers. Those not consumed by flames or smoke leaped to their deaths, according to a rare surviving print of the disaster. In Washington, in 1862, just three years before, Ford's Theatre had burned to the ground and a new one, guaranteed fireproof by the Ford brothers, arose in its place.

But no one fleeing Ford's shouted the terrifying word "fire!" Instead they screamed out other strange words such as "murder," "assassin," "president," and "dead" that pierced the din and could be heard above the general roar. Then random words formed into sentences: "Don't let him escape." "Catch him." "It was John Wilkes Booth!" "Burn the theater!" "The president has been shot." "President Lincoln is dead." "No, he's alive."

On Pennsylvania Avenue and Tenth Street, two blocks south of Ford's, Seaton Munroe, a treasury department employee, was walking with a friend when "a man running down 10th Street approached . . . wildly exclaiming: 'My God, the President is killed at Ford's Theatre!' " Monroe ran to Ford's, where he found "evidences of the wildest excitement."

In the Petersen house, Henry Safford, one of the renters, who shared a second-floor room facing Tenth Street, heard the disturbance outside. He was still awake, reading a book. From his window he had an unobstructed view of Ford's Theatre and the street below. He saw the crowd and heard its anger and fear. Something was wrong. He raced downstairs, unlocked the front door, and descended the curving staircase that led from the door to the street. He walked past the tall gaslight lamp in front of his house, stepped into the dirt street and tried to push through the crowd. Halfway across, the mob blocked his progress to Ford's. He could not take another step. He dared not fight his way through them. This crowd was angry, volatile, and potentially dangerous. But why?

Safford decided to return to the safety of the Petersen house. "Finding it impossible to go further, as everyone acted crazy or mad, I retreated to the steps of my house." Before he disentangled himself from the mob, he heard their news: Abraham Lincoln had just been assassinated in Ford's Theatre. He had been shot, the murderer had escaped, and the president was still inside.

An eyewitness from Ford's reached nearby Grover's Theatre by 10:40 P.M. In the audience was an employee from the War Department hardware shop, Mose Sandford:

> I was at Grover's . . . They were playing Aladdin or the Wonderful Lamp and had just commenced the fourth act . . . Miss German had just finished a song called "Sherman's March Down to the Sea" and was about to repeat it when the door of the theatre was pushed violently open and a man rushed in exclaiming "turn out for Gods sake, the President has been shot in his private box at Ford's Theatre." He then rushed out. Everybody seemed glued to the spot I for one and I think I was one of the first who attempted to move . . . Everybody followed. I made straight for Ford's and such another excited crowd I never before witnessed. I asked who did it and was informed Wilkes Booth. They were just bringing the President out when I arrived on the spot. The city was in one continued whirl of excitement. Crowds on every corner and 10th Street was one solid mass of excited men flourishing knives and revolvers and yelling "down with the traitors" instead of hunting for them.

Soon other boarders at Petersen's were aroused by the disturbance. George Francis and his wife lived on the first floor, and their two big front parlor windows faced the theater. "We were about getting into bed," Francis recalled. "Huldah had got into bed. I had changed my clothes and shut off the gas, when we heard such a terrible scream that we ran to the front window to see what it could mean."

Perhaps it was nothing more than an intoxicated reveler celebrating the end of the war, they thought. George had seen a lot of that: "For a week before the whole city had been crazy over the fall of Richmond and the surrender of Lee's army. Only the night before, the city was illuminated, and though it had been illuminated several times just before this time, it was more general, and was the grandest affair of the kind that ever took place in Washington." But tonight was different. They looked out their windows: "We saw a great commotion—in the Theater—some running in, others hurrying out, and we could hear hundreds of voices mingled in the greatest confusion. Presently we heard some one say 'the President is shot,' when I hurried on my clothes and ran out, across the street, as they brought him out of the Theatre—Poor man! I could see as the gas light fell upon his face, that it was deathly pale, and that his eyes were closed."

While George Francis, Mose Sandford, and more than a thousand other people loitered in the street, Henry Safford had returned to the Petersen house. He climbed the stairs and, at this moment, elevated above the heads of the people going mad in the streets, he observed from the first-floor porch the confusing scene. He noticed a knot of people at one of the theater doors and then watched as they pushed their way into the street. An army officer waved his unsheathed sword in the air, bellowing at people to step back and clear the way.

Someone suggested bringing Lincoln next door to Taltavul's. No, the owner pleaded, don't bring him in here. It must not be said later that the president of the United States died in a saloon. Someone else ran across the street and pounded on the door of a house to the south of Petersen's. No one answered. In command of that little group was Dr. Charles A. Leale, a U.S. Army surgeon attending the play who was the first doctor to enter the president's theater box.

Leale described the scene: "When we arrived to the street, I was asked to place him in a carriage and remove him to the White House. This I refused to do fearing that he would die as soon as he would be placed in an upright position. I said that I wished to take him to the nearest house, and place him comfortably in bed. We slowly crossed the street, there being a barrier of men on each side of an open passage towards the house. Those who went ahead of us reported that the house directly opposite was closed."

Safford watched the little group that was carrying the body of Abraham Lincoln. They were not going to the president's carriage. It looked like they wanted to bring him somewhere else, into a house on Tenth Street. "Where can we take him?" Safford heard one of the men shout.

Henry Safford seized a candle and held it up so the men could see it. "Bring him in here!" he yelled. He waved the light. "Bring him in here!" He caught their attention.

"I saw a man," said Dr. Leale, "standing at the door of Mr. Petersen's house holding a candle in his hand and beckoning us to enter."

George Francis, still outside, watched in amazement: "They carried him on out into the street and towards our steps . . . The door was open and a young man belonging to the house standing on the steps told them to bring him in there."

Lincoln's bearers changed direction and, turning slightly to their right, walked northwest from Ford's Theatre to the Petersen House. Huldah Francis watched them get closer and closer until they were right below her window. Transfixed by what she saw, "Huldah," George Francis explained, "remained looking out of the window" to the last possible moment, "until she saw them bringing him up our steps when she ran to get on her clothes." As she hurried to pull off her nightclothes and get dressed for the surprise visitors, the men, struggling to support Lincoln's limp body in a prone position, carried him up the curving staircase. George Francis raced back to the house to rejoin his wife. When

THE PETERSEN HOUSE, WHERE LINCOLN DIED.

he got there, he expected to find Abraham Lincoln lying in his bed.

The gas streetlamp in front of the house, just a few feet from the stairs, allowed the whole crowd to see what was happening. One man, an artist named Carl Bersch who lived one house north of Petersen's, watched from his room. "My balcony being twelve or fourteen feet above the sidewalk and street, I had a clear view of the scene, above the heads of the crowd. I recognized the lengthy form of the President by the flickering of the torches, and one large gas lamp post on the sidewalk. The tarrying at the curb and the slow, careful manner in which he was carried across the street, gave me ample time to make an accurate sketch of that particular scene."

From his all-seeing perch, Bersch watched and drew while Henry Safford invited Leale's party across the threshold.

"Take us to your best room," Dr. Leale commanded. All eyes in the street looked up to that doorway as the president's wounded, apparently unconscious body disappeared from sight, into William Petersen's boardinghouse. The time was 11:00 P.M.

Safford led Dr. Leale and the men carrying Lincoln into the front hall. The confined space could barely accommodate the horizontal president and his bearers. On the right, a narrow staircase led up to the second floor. On the left was a closed door. Leale had asked for the "best room." Obeying this criterion, Safford should have opened that door and burst into the two-room suite occupied by George and Huldah Francis. Their front parlor faced Tenth Street, and behind that room, separated from the parlor by folding wood doors, was a spacious bedroom. Safford clasped the handle, tried to turn it, but the door was locked. He then headed deeper into the dim hallway and stopped at a second door on the left, the one to the Francis's bedroom. Also locked! Behind that door, Huldah Francis was dressing.

Just one room was left, the smallest one on the first floor. If it was locked, they would have to carry the president up the cramped staircase to the second floor. When Stafford reached the door, he rotated the knob. It was unlocked and the room was unoccupied. The boarder, Private William Clarke, had gone out for the evening to celebrate the end of the war. Leale ordered the bearers to carry Lincoln into the room and lay him on the bed.

Lincoln's eyes betrayed the severity of his wound. Dr. Leale noticed it before they had undressed him: "When the President was first laid in bed a slight ecchymosis of blood was noticed on his left eye lid and the pupil of that eye was dilated, while the pupil of the right eye was contracted."

A few minutes after the president was laid in Willie Clarke's vacant bed, Mary Lincoln appeared in the doorway of the Petersen house. Her companions, Major Rathbone and Miss Harris, had pried her out of the theater box, helped her descend the same winding staircase that her husband had just been carried down, and escorted her through the wailing mob in the street and into the house. George Francis, now back home, witnessed her arrival: "She was perfectly frantic. 'Where is my husband! Where is my husband!' she cried, wringing her hands in the greatest anguish."

As Mary Lincoln scurried down the hall, the billowing skirt of her silk dress swished against the banister post and the narrowly spaced walls. Moments later, she reached the back room where her husband was lying prone on a bed. Dr. Leale and two other physicians, Dr. Charles Sabin Taft and Dr. Albert F. Africanus King, who were also in the audience at Ford's and who had rushed to the president's box, were bent over Lincoln, preparing to strip him of his clothes and conduct the kind of thorough examination impossible on the floor of a theater.

George Francis recalled the moment when Mary entered the room: "As she approached his bedside she bent over him, kissing him again and again, exclaiming 'How can it be so? Do speak to me!' "

Leale was reluctant to examine Lincoln in his wife's presence: "I went to Mrs. Lincoln and asked her if she would have the kindness to step into the next room for a few minutes while we examined him, removed his clothes, and placed him more comfortably on the bed. Mrs. Lincoln readily assented."

Henry Rathbone and Clara Harris escorted Mary to the front parlor and seated her on a large, wood-framed Victorian sofa upholstered with slick, shiny black horsehair. Rathbone felt light-headed. Moments after Booth shot the president, he stabbed the major in the arm. The wound was deep, and the cut would not stop bleeding. Rathbone sat down in the hall, and then he fainted. The alarming sight of his unconscious form lying faceup in the front hall of the Petersen house greeted the first visitors to the scene. He was in the way, and when he regained consciousness, he was picked up from the floor and delivered to his house. He would live.

The bed Lincoln was in was positioned in an awkward way, behind the door, shoved into the room's northeast corner. The doctors dragged the bed away from the walls to create space for them to surround the president. Then they pushed all the chairs close to the bed. The lone gas jet protruding from the south wall cast weird, moody shadows and exaggerated the pained countenances of the men in the room. It was like a theatrical light raking across the stage to emphasize the drama. But this tableau, like Booth's flamboyant scene at Ford's less than half an hour earlier, was real. Leale ordered everyone except his two medical colleagues to leave the room. Then they stripped their patient and searched his body for additional wounds.

In the front parlor, Mary Lincoln was coming apart. When Clara Harris sat beside her on the sofa and tried to comfort her, Mary could not take her eyes off Clara's bloodstained dress: "My husband's blood!" she cried. "My husband's blood." The first lady did not know it was Henry Rathbone's blood, not the president's. The major's wound had stained his fiancée's frock. If Mary had examined her own dress, she would have been more horrified, because it did bear the stains of her husband's blood.

As the crowd outside thickened and some in it approached the unguarded front door, Leale and company were in the back room, preoccupied with Lincoln. At that vulnerable moment, quick-thinking army junior officers and enlisted men, recognizing the danger, took the initiative and blocked the front doorway, commandeered the staircase, took positions in front of the house, and ordered the people back. Within fifteen minutes of his being carried into the Petersen house, the commander in chief was under the personal protection of the U.S. Army. All curiosity-seeking intermeddlers discovered in the house were ejected.

As soon as Maunsell Field, assistant to the secretary of the Treasury, arrived, he came face-to-face with Clara Harris: "The first person I met in the hall was Miss Harris. She informed me the President was dying but desired me not to communicate the fact to Mrs. Lincoln. I then entered the front parlor, where I found Mrs. Lincoln in a state of indecipherable agitation. She repeated over and over again, 'Why didn't he kill me? Why didn't he kill me?' "

Mary Lincoln needed help fast. Clara Harris was not suited for that delicate psychological role—Mary hardly knew her. The first lady had few friends in Washington and now she asked for them all: Mary Jane Welles, wife of Navy Secretary Gideon Welles; Elizabeth Keckly, her black dressmaker and confidante; and Elizabeth Dixon, wife of Senator James Dixon. Messengers ran off in search of these women. While she waited for them to arrive at her side, Mary, in torment, sat on the sofa. The crowd was just outside the windows. She could hear their voices.

What happened at the Petersen house over the next eight and a half hours was no less than the transfiguration of Abraham Lincoln from mortal man to martyred saint.

Leale and the other physicians examined the president's naked corpse: "After undressing him I found that his lower extremities were quite cold to a distance of several inches above his knees. I sent the Hospital Steward who had been of great assistance to us while removing him from the theatre, for bottles filled with hot water, hot blankets, etc. which we applied to his lower extremities."

Leale knew this case was too big for him so he sent messengers to locate his military superiors: "I asked again to have the Surgeon

General and also sent a special messenger for Surgeon D.W. Bliss then in command of Armory Square Hospital."

Dr. Taft recalled that "about twenty-five minutes after the President was laid on the bed, Surgeon-General Joseph K. Barnes and Dr. Robert King Stone, the family physician, arrived and took charge of the case." At once, Leale deferred to Stone: "I was introduced to Dr. Stone as having charge of him. I asked . . . if he would take charge of him [and] he said 'I will.' I then told Dr. Stone the nature of the wound and what had been done. The Surgeon General and Surgeon Crane arrived in a few minutes and made an examination of the wound." Dr. Stone and Surgeon General Barnes approved of everything Leale had done. They agreed that Leale's decisive actions had saved Lincoln from immediate death at Ford's Theatre. As Charles S. Taft testified, "It was owing to Dr. Leale's quick judgement in instantly placing the almost moribund President in a recumbent position the moment he saw him in the box, that Mr. Lincoln did not expire in the theater within ten minutes from fatal syncope."

Leale recalled: "About 11p.m. the right eye began to protrude, which was rapidly followed by an increase of the ecchymosis until it encircled the orbit extending above the supra orbital ridge and below the infra orbital foramen . . . The wound was kept open by the Surgeon General by means of a silver probe and as the President was placed diagonally on the bed his head was held supported in its position by Surgeon Crane and Dr. Taft."

All the doctors agreed with Leale's on-the-spot diagnosis at Ford's Theatre. This was no longer a medical emergency. There were no remedies or treatments, and nothing could save Lincoln. By midnight, it had become a death watch. More doctors arrived. They were superfluous, but out of professional courtesy, they were given the privilege of playacting in a charade of treatment: examining the wound, taking the pulse, and making somber, redundant, and useless pronouncements that in future would permit them to boast, "Yes, I was there.

I was one of the doctors at the Petersen house." And for years after, many of them did.

The death pageant for Lincoln had begun. It started while the president still lived, as soon as the doctors, in their collective wisdom, gave up all hope. There was nothing more they could do. An operation was impossible. Cranial surgery was in its infancy during the Civil War, and no doctor would risk removing a bullet embedded so deeply in the brain. Booth's Deringer pistol had performed superbly. The wound was fatal, the damage irreversible. Dr. Leale had known this while his patient still lay on the floor of the president's box at Ford's Theatre. Now the diagnosis was unanimous. The president would die. Indeed, some were surprised he was not already dead.

The doctors agreed that from then on, they would not tinker with Lincoln's body—no more brandy poured down his throat to see whether he would swallow it or almost choke to death; no more fruitless Nélaton-probe thrusts through the bullet puncture in his skull into his brain to trace, for curiosity's sake, the wound tunnel and locate the missile. No, all they would do now was watch and wait.

No one at the Petersen house was aware of this yet, but a second assassin had struck in Washington that night. At 10:15 P.M., a crazed man with superhuman strength had invaded the home of Secretary of State William H. Seward. The assailant stabbed and slashed Seward—who was bedridden from injuries he had suffered during a recent carriage accident—almost to death, wounded a veteran army sergeant serving as Seward's nurse, and knifed a State Department messenger. The unknown killer had also, while beating Seward's son Frederick with a pistol, crushed his victim's skull and rendered him senseless. Seward's home was off Lafayette Park near the White House, just a few blocks from Ford's Theatre.

Runners carried the news of the attack on Seward to Secretary of War Edwin M. Stanton and Secretary of the Navy Gideon Welles, who were at their homes preparing for bed and had not heard about the assassination of the president. Each man raced by carriage to Seward's mansion. There, they first heard rumors of another attack, this one upon the president at Ford's Theatre. Together, Stanton and Welles drove a carriage to Tenth Street and arrived at the Petersen house before midnight. Stanton barreled his way through the crowded hallway. Reeling at the sight of the wounded president, the secretary of war concluded that Lincoln was a dead man. There was nothing he could do for him. Except work. There was much to do. Stanton steeled himself for the long night ahead. He would not spend the night mourning at Lincoln's bedside.

Welles volunteered to play that role. As news of the assassination spread through Washington, many important public officials made pilgrimages to the Petersen house. Welles decided that at least one man should remain, never to leave Lincoln's side until the end, to bear witness to his suffering. Stanton could lead the investigation of the crime, interview witnesses, send telegrams, launch the manhunt for Booth and his accomplices, and take precautions to prevent more assassinations later that night.

Welles, on the other hand, would lead the death vigil. And he would record in his diary what he saw: "The giant sufferer lay extended diagonally across the bed, which was not long enough for him. He had been stripped of his clothes. His large arms were of a size which one would scarce have expected from his spare form. His features were calm and striking. I have never seen them appear to better advantage, than for the first hour I was there. The room was small and overcrowded. The surgeons and members of the Cabinet were as many as should have been in the room, but there were many more, and the hall and other rooms in front were full."

Could it be, Welles wondered, that something the president had said at the White House earlier that day prefigured the assassination? Had Lincoln's strange dream foretold this tragedy? At the 11:00 A.M. cabinet meeting, the president said that he expected important news

THE PETERSEN HOUSE DEATHBED VIGIL, SKETCHED BY AN ARTIST FROM THE ARMY MEDICAL MUSEUM.

soon. He had experienced, the previous night, a recurring dream that he believed always foretold the coming of great events. Welles preserved the remarkable story in his diary. Lincoln told his cabinet that "he had last night the usual dream which he had preceding nearly every great and important event of the War . . . the dream . . . was always the same." Welles asked Lincoln to describe it. "[I] seemed," the president recounted, "to be in some singular, indescribable vessel . . . moving with great rapidity towards an indefinite shore."

The president said he had this dream preceding the bombardment of Fort Sumter and the battles of Bull Run, Antietam, Gettysburg, Vicksburg, and more. Had a premonition of his own assassination come to Lincoln in a dream? As Gideon Welles sat beside his dying chief, he did not know that, several days earlier, Lincoln had dreamed a far more vivid nightmare of death.

By midnight the Petersen house had become the cynosure of official Washington. Like a major planet exerting an invisible but irresistible gravitational force, the brick home attracted the luminaries who orbited the national capital. Throughout the night, as word spread that the president had suffered a fatal wound, dozens of people—generals, army officers, cabinet secretaries, members of Congress, government officials, and personal friends—made pilgrimages to Tenth Street to augment the bedside vigil and to behold for the last time the still-living form of Abraham Lincoln. Throughout the night and into the early morning, a steady procession of mourners went to and from the Petersen house. Some, content to gaze upon the president's face for a few minutes, left shortly after they had done so. Others remained and would not leave until the end. They wanted to watch Abraham Lincoln die.

Elizabeth Dixon was the first of Mary's friends to arrive, and the gruesome scene horrified her: "On a common bedstead covered with an army blanket and a colored woolen coverlid lay stretched the murdered President his life blood slowly ebbing away. The officers of the government were there & no lady except Miss Harris whose dress was spattered with blood as was Mrs. Lincoln's who was frantic with grief calling him to take her with him, to speak one word to her . . . I held and supported her as well as I could & twice we persuaded her to go into another room."

Mary never saw many of the visitors who came to the Petersen house that night. The door on the left side of the front hall that opened to the Francises' front parlor remained half-closed through much of the vigil. Most visitors sped past it on their way to the room at the end of the hall. Some, out of respect, did not wish to intrude upon Mary's privacy and grief. Others, aware of her unpredictable volatility and proneness to anger, avoided her. After Edwin M. Stanton arrived and decided to occupy the Francis bedroom as his headquarters, and he shut the folding wood doors dividing that space from the front parlor, Mary was sealed off from the activity in the rest of the house. For most of the night, and through the early morning hours of the next day, Mary remained in semiseclusion, converting the front parlor into her private chamber of solitude, mourning, and, at times, derangement.

For Tad Lincoln, brought home from Grover's Theatre by the White House doorkeeper, the night of April 14 was filled with terrors. By the time Tad got there, Robert Lincoln had already left to join his parents. Without his mother or older brother to comfort Tad, or even explain to him what had happened to his father, the frightened little boy spent the night alone with servants in the near-empty mansion. All he knew was that a crazy man had burst into Grover's Theatre, screaming that President Lincoln had been shot, and that the theater manager had also announced the disaster from the stage.

Until the next morning, when Mary and Robert returned to the White House and informed him that his beloved "Pa" was dead, Tad relived the fear and pain he had suffered three years before, when his best companion, his brother Willie, had died. During the long Petersen house death vigil, not once did Robert or Mary go to Tad—even though the White House was just a five-minute carriage ride away and the coachman Francis Burke was ready to whip the president's carriage through the Tenth Street mob and gallop there. Nor did Robert or Mary order a messenger to retrieve Tad and carry him to the Petersen house and his dying father. It was the first troubling sign of how, in the days to come, Mary's crippling descent into a mad, gothic, self-absorbed grief caused her to neglect the needs of her inconsolable and lonely little boy.

When Senator Charles Sumner, never close to the president but a confidant of the first lady, heard the news he dismissed it as a wild rumor. When a messenger burst in on him and blurted out, "Mr. Lincoln is assassinated in the theater. Mr. Seward is murdered in his bed.

There's murder in the streets," Sumner said he did not believe the news about the president or the attack on Secretary of State Seward.

"Young man," he said, chastising the messenger, "be moderate in your statements. Tell us what happened."

"I have told you what has happened," the man said insistently, and then repeated his story.

When the senator arrived at the Petersen house, he sat down at the head of the bed, held Lincoln's right hand, and spoke to him. One of the doctors said, "It's of no use, Mr. Sumner—he can't hear you. He is dead."

The senator retorted: "No, he isn't dead. Look at his face; he is breathing."

That may be, the doctor admitted, but "it will never be anything more than this."

Sumner remembered the night's other victim and asked Major General Halleck, army chief of staff, to drive him to the secretary of state's mansion. There he found Mrs. Seward sitting on the stairs between the second and third floors. She seized him with both hands and spoke: "Charles Sumner, they have murdered my husband, they have murdered my boy." Sumner hoped it was not true. He knew firsthand that it was possible to survive a vicious assault. Before the war, a pro-slavery Southern congressman, Preston Brooks, had almost caned Sumner to death in a brutal surprise attack on the Senate floor. Trapped by his desk, Sumner could not rise to fight back or escape and he suffered grievous head wounds. He lived, but recuperation was long and painful, and he did not return to the Senate for three years.

Many people tried to get inside the Petersen house that night. They pressed against the front wall and stood on tiptoe to peek through the front windows, but the panes were set too high above street level to allow a clear view into the front parlor. Other people strained toward the stairs, tempted to ascend them and try to get inside. At any moment the crowd might have gone wild, with hundreds of people forcing their way through the doorway. The curved shape of the public staircase and its protective iron railing served as a barricade against frontal assault, impeding any mob attempt to rush the house head-on.

News of the assassination stunned Washington. The local *New York Times* correspondent said it best: "A stroke from Heaven laying the whole of the city in instant ruins could not have startled us as did the word that broke from Ford's Theatre a half hour ago that the President had been shot." The Petersen house had become an irresistible magnet, drawing people from all over the city. Soon Major General C. C. Augur, commander of the military district of Washington, and Colonel Thomas Vincent arrived and became impromptu doorkeepers. They admitted only a privileged few into 453 Tenth Street. They denied entry to many: citizen strangers, minor government officials, low-ranking military officers who had no business there, and newspaper reporters.

The motives of the callers varied. Some wanted to express their love for the president. Others wanted to help. Many sought a small place in history—to see the wounded president, to claim they had stood beside his deathbed, to boast to their children and grandchildren that they had been there, or, in the case of the journalists, to be the ones who first reported what happened there. The gatekeepers turned almost all of them away. Indeed, not one journalist made it past them.

Sometime around 2:00 A.M., the doctors decided to probe Lincoln's brain for the bullet. Dr. Leale described what happened:

> The Hospital Steward who had been sent for a Nelaton's probe arrived and an examination of the wound was made by the Surgeon General who introduced it to a distance of about two and a half inches when it came in contact with a foreign substance which laid across the tract of the ball, this being easily passed the probe was introduced further when it again touched a hard substance which was at first supposed to be the ball but the porcelain bulb of the probe did not show the stain of lead upon it after its withdrawal it was generally supposed to be another piece of loose bone. The probe was introduced a second time, and the ball was supposed to be distinctly felt by the Surgeon General, Dr. Stone and Dr. Crane. After this second exploration nothing further was done except to keep the opening free from coagula, which if allowed would soon produce signs of increased compression. The breathing became profoundly stertorous, and the pulse more feeble and irregular.

Throughout the night, Dr. Leale watched Mary stagger from the front parlor into the bedroom: "Mrs. Lincoln accompanied by Mrs. Senator Dixon came into the room several times during the course of the night. Mrs. Lincoln at one time exclaiming, 'Oh, that my Taddy might see his Father before he died' and then she fainted and was carried from the room."

Mary's pleadings moved Secretary of the Interior John P. Usher: "She implored him to speak to her [and] said she did not want to go to the theatre that night but that he thought he must go . . . She called for little Tad [and] said she knew he would speak to him because he loved him so well, and after indulging in dreadful incoherences for some time she was finally persuaded to leave the room."

One of the last Petersen house visitors was Benjamin Brown French, commissioner of public buildings and grounds. Operating from an office in the U.S. Capitol, French was in charge of all the major public federal buildings in Washington, including the White House and the Capitol. A larger-than-life personality and longtime veteran of the Washington political and social scene, he had, for decades, known everyone of importance in the national capital. Unbeknownst to them, he had recorded his impressions in a secret diary.

President Franklin Pierce, a friend of Jefferson Davis, had appointed French commissioner in 1853 but forced him out in 1855. Once Lincoln arrived in Washington in 1861, the savvy veteran lobbied for reappointment. He secured several inconclusive meetings with the president and Mary Lincoln, but the president nominated someone else. When the Senate failed to confirm that appointment, Lincoln, after dangling French in suspense, finally signed his commission on January 29, 1862. Although French owed his position to the president, he disliked the first lady and clashed frequently with Mary over her misuse of White House expense accounts. On April 14, French went to bed around 10:00 P.M., and during the night no one had thought to send a messenger to summon him to the Petersen house. French slept well until daylight: "I awoke and saw that the streetlamps had not been extinguished. I lay awake, perhaps ½ an hour, & seeing that they were still burning, I arose and saw a sentry passing before my house. I thought something wrong had happened, so I dressed & went down & opened the front door."

Downtown at the Petersen house, Dr. Leale knew Lincoln would not live much longer: "As morning dawned it became quite evident that he was gradually sinking and at several times his pulse could not be counted two or three feeble pulsations being felt and followed by an intermission when not the slightest movement of the artery could be felt. The inspiration now became very prolonged accompanied by a guttural sound. At 6:50am the respirations ceased for some time and all eagerly looked at their watches until the profound silence was disturbed by a prolonged inspiration, which was soon followed by a sonorous expiration."

As Benjamin French stood in front of his house on Capitol Hill, a soldier came along and said, "Are not the doings of last night dreadful?" French asked what he meant by that. The soldier replied, "Have you not heard?" and told French that the president had been shot in Ford's Theatre "and Secretary Seward's throat cut in his residence." French hurried to the East Front of the Capitol, ordered the building closed, and sped to the Petersen house. There he found Lincoln, who was still alive, in the back bedroom. "[He] was surrounded by the members of his cabinet, physicians, Generals, Members of Congress, etc. I stood at his bedside a short time. He was breathing very heavily, & I was told, what I could myself see, that there was no hope for him."

French gazed at his wounded president and patron with a mixture of personal and professional concern. If Lincoln died, he would have much to do in the next few days. It would be his responsibility to decorate all the public buildings in the city with the appropriate symbols of mourning. As he hovered over the deathbed in the crowded little bedroom, perhaps he already wondered where, in all of Washington, could he hope to find enough black mourning crepe and bunting.

French left Lincoln's bedside and entered the front parlor, where he found Mary and Robert Lincoln. "I took Mrs. Lincoln by the hand, and she made some exclamation indicating the deepest agony of mind. I also shook hands with Robert, who was crying audibly." French noticed three women who sat near Mary Lincoln: her friend Elizabeth Dixon, wife of Senator James Dixon of Connecticut; Elizabeth's sister, Mrs. Mary Kinney; and Kinney's daughter Constance.

After a few minutes somebody asked French to take the president's carriage and fetch Mary Jane Welles, wife of the secretary of the navy and another of Mary Lincoln's friends.

When he arrived at the Welles home, he could not persuade Mary Jane to come out. "Mrs. Welles was not up, & a lady at the house said she was too unwell to go, so I returned to the carriage, but, before we could get away, someone said from the upper window that Mrs. Welles would go. I returned to the house and waited for her to dress and take a cup of tea & some toast, & then the carriage took us round to the President's House—I, supposing she was to go there and be ready to see Mrs. Lincoln when she should get home. She thought I was mistaken, and that she was to go 10th Street." French got out of the carriage, ordered the coachman to drive Mary Jane Welles to the Petersen house, and entered the White House. He instructed the staff to close the house, and then he went home to Capitol Hill for breakfast.

Lincoln was close to death now, and Mary returned to the bedroom. Dr. Taft recalled the scene: "Her last visit was most painful. As she entered the chamber and saw how the beloved features were distorted, she fell fainting to the floor. Restoratives were applied, and she was supported to the bedside, where she frantically addressed the dying man. 'Love,' she exclaimed, 'live but one moment to speak to me once—to speak to our children.'" Secretary of the Treasury Hugh McCulloch said that Mary Lincoln's presence "pierced every heart and brought tears to every eye."

Elizabeth Dixon witnessed Mary's collapse: "Just as the day was struggling with the dim candles in the room we went in again. Mrs. Lincoln must have noticed a change for the moment she looked at him she fainted and fell upon the floor. I caught her in my arms & held her to the window which was open . . . She again seated herself by the President, kissing him and calling him every endearing name—The surgeons counting every pulsation & noting every breath gradually growing less & less—They then asked her to go into the adjoining room."

Dr. Leale noted how Mary's cries unnerved Edwin M. Stanton: "As Mrs. Lincoln sat on a chair by the side of the bed with her face to her husband's, his breathing became very stertorous and the loud, unnatural noise frightened her in her exhausted, agonized condition. She sprang up suddenly with a piercing cry and fell fainting to the floor. Secretary Stanton, hearing her cry, came in from the adjoining room and with raised arms called out loudly, 'Take that woman out and do not let her in again.' Mrs. Lincoln was helped up kindly and assisted in a fainting condition from the room. Secretary Stanton's order was obeyed and Mrs. Lincoln did not see her husband again before he died."

Jefferson Davis awoke on the morning of April 15 ignorant of last night's bloody crimes in Washington. There was no direct telegraph line between the capital and Greensboro. Davis did not know John Wilkes Booth and had not sent him to kill Lincoln. Davis did not know that Lincoln had been marked for death, that Booth had met with Confederate secret agents in Montreal, Canada, that the actor had assembled a list of Confederate operatives in Maryland and Virginia to help him, and that one of his soldiers, Lewis Powell, a brave combat veteran captured at Gettysburg, had joined Booth's plot and nearly killed the secretary of state. Nor did Davis know that Booth was on the run, fleeing for the heart of the Confederacy, the prey of what would soon become a nationwide manhunt.

That morning Davis had no idea that, last night in the Union capital, events beyond his knowledge or control would now reach out to affect his fate. Within hours his longtime archenemy, Vice President Andrew Johnson, an implacable foe of the planter class, would ascend to the presidency. The South could expect no mercy from him. Worse, this morning's newspapers accused Davis of being the mastermind behind the great crime. Many editorials demanded his death by hanging or horrible torture. A patriotic envelope, published as a souvenir, carried a blood-red vignette of Davis bound on a scaffold facing the guillotine. The stakes were higher now.

All of this had happened without Davis knowing about any of it. And for several more days, he would not know that Lincoln was dead or that the government of the United States would soon scheme to charge him with murder and put him on trial for his life. Lincoln's murder was like a violent storm on a distant horizon, its mighty thunderclap taking time to travel a great distance before it caught up with Davis.

Davis did evacuate Greensboro on April 15 but the move wasn't prompted by news of Lincoln's assassination. It was coincidence and the overall military situation. Secretary of the Navy Mallory tried to convince Davis that he should do more than relocate the temporary capital—he should flee the country: "It was evident to every dispassionate mind that no further military stand could be made . . . But it was no less evident that Mr. Davis was extremely reluctant to quit the country at all, and that he would make no effort to leave it so long as he could find an organized body of troops, however small, in the field. He shrank from the idea of abandoning any body of men who might still be found willing to strike for the cause, and gave little attention to the question of his personal safety."

If Davis's staff had known that Lincoln had just been murdered, they might have been even more forceful in demanding that Davis flee to Mexico, the Bahamas, or Europe to escape the North's vengeance. But they did not know and went about their packing up for the next stage of their journey south.

They would no longer enjoy the luxury of railroad transportation. There were no trains at Greensboro, so that afternoon Davis; Colonels Harrison, Lubbock, and Wood; and some of the cabinet members rode horses, while other dignitaries climbed aboard wagons and ambulances. "Heavy rains had recently fallen," Burton Harrison wrote, "the earth was saturated with water, the soil was a sticky red clay, the mud was awful, and the road, in places, almost impracticable." The presidential party plotted their route and planned to spend successive nights at Jamestown, Lexington, Salisbury, and Concord, where they would be guests of Victor C. Barringer.

Rough travel conditions would not intimidate Davis. He was not a creature softened by effete, cocooned salons. He was ready for the physical challenge that lay ahead. He had endured journeys far more arduous than this journey away from Richmond promised to be. As a seven-year-old child, he rode a pony 500 miles up the Natchez Trace from Natchez, Mississippi, to Nashville, Tennessee, where he met General Andrew Jackson; in 1833, while an army officer, he and his unit of dragoons (a heavy, mounted cavalry) traveled 450 miles through difficult territory to a remote post on the Arkansas frontier; in 1834, Davis and the dragoons made a 500-mile round trip from their fort into Comanche territory, enduring 100-degree heat, exhaustion, and dehydration; in 1845, Jefferson and Varina traveled from Vicksburg to Washington, D.C., through the northern route into Ohio, where severe winter weather and ice on the Ohio River required them to continue by sled; he traveled to Mexico for the war, experienced hard travels there, made a 1,000-mile trip home to Mississippi, and then returned to Mexico; in December 1862, as president of the Confederacy, he embarked on a twenty-seven day, 3,000-mile inspection tour of the South; later, he made other long, wartime journeys through his embattled country; and in Richmond he often went on dangerous, 20-mile night rides on horseback to visit Lee's headquarters and other military posts. A lifetime of difficult journeys had accustomed Davis to the hardships of the road.

At the Petersen house, the Reverend Dr. Gurley called everyone around the deathbed. "Let us pray," he said as all present in the room kneeled. "He offered a most solemn and impressive prayer," recalled Dr. Leale. "We arose to witness the struggle between life and death." Abraham Lincoln drew his last breath at 7:21 and 55 seconds. At 7:22 and 10 seconds his heart stopped beating. He was dead.

No one knew it yet, but the mourning that began in the back room of a boardinghouse in downtown Washington would continue well beyond Lincoln's death. What began there could not be contained. Soon, the assassination would set in motion strange forces, a national phenomenon, the likes of which America had never seen. In the days to come, the footsteps of millions of Americans would join the small procession that began, in the words of Walt Whitman, that "moody, tearful night" when a handful of their fellow citizens made a pilgrimage to look upon their dying president. By morning almost sixty people had come and gone from the Petersen house.

Dr. Taft recalled that "immediately after death, the Rev. Dr. Gurley made a fervent prayer, inaudible, at times, from the sobs of those present. As the surgeons left the house, the clergyman was again praying in the front parlor. Poor Mrs. Lincoln's moans, which came through the half-open door, were distressing to hear. She was supported by her son Robert, and was soon after taken to her carriage. As she reached the front door she glanced at the theater opposite, and exclaimed several times, 'Oh, that dreadful house! That dreadful house!'"

Lincoln's death was not the last sadness to haunt the Petersen house. By 1870, William and Anna Petersen's two youngest children, Anne and Julia, had died, and on June 18, 1871, the Metropolitan Police found William Petersen lying unconscious on the grounds of the Smithsonian. He had poisoned himself. He was taken to the hospital, where he died the same day from an overdose of the drug laudanum. Before succumbing, Petersen told the police he had been taking the substance "once or twice a week" for several years. The coroner ruled his death accidental. He was fifty-four years old. Given the notoriety of his house six years earlier, the *Washington Evening Star* noted his sad end.

Exactly four months later, Anna Petersen died. Her body was laid out in the house, and the funeral was held two days later. Just ten days after her funeral, the firm of Green & Williams sold at public auction the entire contents of the house. An ad in that day's *Evening Star* stated that the furnishings would be sold on the premises. Crowds assembled outside the Petersen house, just as they had on that terrible night six and one-half years before. Once again, strangers crowded the halls and first-floor rooms. The auctioneer led the customers and the curiosity seekers from room to room. In the front parlor, he sold "1 horsehair covered sofa" for $15.25. The price was high, up to three times the value for a like item. But this was the sofa where a shattered, sobbing Mary Lincoln spent most of the night of April 14 and the morning of April 15, 1865. In the back bedroom, the auctioneer put up a lot listed in the inventory as "1 bedstead & 2 Mattresses," appraised at $7. The bidding soared to $80, the highest price paid for any item in the house. This was the bed where Lincoln died.

The bed, along with most of the other contents of the death room, including the chairs, the washstand, and even the gaslight jet that was mounted to the wall, were purchased by Colonel William H. Boyd of Syracuse, New York, for his son Andrew, a young Lincoln enthusiast and early collector who had published, in 1870, a pioneering bibliography of early writings about the president.

In 1889 and 1890, Andrew Boyd corresponded with the Chicago candy millionaire Charles F. Gunther, an obsessive collector who would stop at almost nothing to acquire unique historical treasures like the Confederate Libby Prison in Richmond, which he purchased, dismantled stone by stone, and reassembled for the 1893 Columbian Exposition in Chicago. Gunther decided he had to possess Lincoln's deathbed and the accompanying furniture and paid Boyd one hundred thousand dollars for the bed alone.

In the 1920s, the Chicago Historical Society acquired Gunther's hoard and constructed an exact replica of the room in which Abraham Lincoln died, right down to the reproduction wallpaper and the prints hanging on the walls. It was a sensational attraction, and for decades awestruck Chicago schoolchildren pushed a button that triggered a dramatic sound recording which, from a hidden loudspeaker, narrated the events of April 14 and 15, 1865. Alas, several years ago, the museum broke up the riveting display, dismissing it as no longer in fashion.

Stanton ordered the army to remove Lincoln's corpse from the Petersen house and transport it to the White House. Soldiers brought a wood box and placed the president's body inside it. They carried the makeshift coffin into the street and placed it in a horse-drawn hearse. Major General C. C. Augur, head of the military district of Washington, D.C., and commander of the presidential escort, ordered all officers in the procession, including General D. H. Rucker, Colonel Louis H. Pelouze, and Captains Finley Anderson, C. Baker, J. H. Crowell, and D. C. Thomas, to march on foot and not ride horses. It was as if they were preparing to enter a battle. During the Civil War, officers, even generals, often led their troops forward into combat on foot, with swords drawn. They walked as a sign of respect for their fallen commander. They removed their hats and marched bareheaded. In the field officers always wore their hats into combat. Now they doffed them as an additional sign of deference.

Augur gave the command and the escort got under way. There was no band or drum corps to beat the slow tempo of the age-old military funeral march. The officers set the pace with the thud of their own steps on the dirt street. Corporal James Tanner, who had transcribed in shorthand the testimony Stanton had extracted from witnesses through the night, had gone home after the deathbed climax. About two hours after Lincoln died, Tanner was back in his room in the house one door south of the Petersen house. He looked outside. "I stepped to the window and saw the coffin of the dead President being placed in the hearse which passed up Tenth street to F and thus to the White House. As they passed with measured tread and arms reversed, my hand involuntarily went to my head in salute as they started on their long, long journey back to the prairies and the hearts he knew and loved so well, the mortal remains of the greatest American of all time."

On the street the scene was less solemn. Dr. Charles Sabin Taft had lingered at the Petersen house for two hours because he had not wanted to leave while the body still lay there. When the army officers and soldiers carried Lincoln's coffin outside, into view of the immense crowd, Taft followed them out the front door into the street, where he witnessed a violent, horrifying scene: "A dismal rain was falling on a dense mass of horror-stricken people stretching from F Street to Pennsylvania Avenue. As they made a passage for the hearse bearing the beloved dead, terrible execrations and mutterings were heard."

But not everyone in that crowd loved Abraham Lincoln. A few rebel sympathizers yelled insults at the president as the coffin passed them by, and enraged mourners turned on them and even killed some. According to Dr. Taft, "one man who ventured a shout for Jeff Davis was set upon and nearly torn to pieces by the infuriated crowd."

Noah Brooks did not learn of the assassination until the morning. He could not believe it—yesterday morning he had been at the White House having breakfast with the president. He began walking the streets of the gloomy capital, taking in the mood of the people and the sights of a city draping itself in mourning clothes. He felt himself drawn to the place of the great crime: "Wandering aimlessly up F Street toward Ford's Theatre, we met a tragical procession. It was headed by a group of army officers walking bareheaded, and behind them, carried tenderly by a company of soldiers, was the bier of the dead President, covered with the flag of the Union, and accompanied by an escort of soldiers who had been on duty at the house where Lincoln died. As the little cortege passed down the street to the White House, every head was uncovered, and profound silence which prevailed was broken only by sobs and by the sound of the measured tread of those who bore the martyred President back to the home which he had so lately quitted full of life, hope, and cheer."

Now that Lincoln's body had been taken away, the drama at the Petersen house was done. The house was empty now, but for the Petersen family and its tenants, and the evidence of what had happened there: bloody handkerchiefs, pillowcases, sheets, and towels, plus water pitchers, mustard plasters, and liquor bottles. And muddy

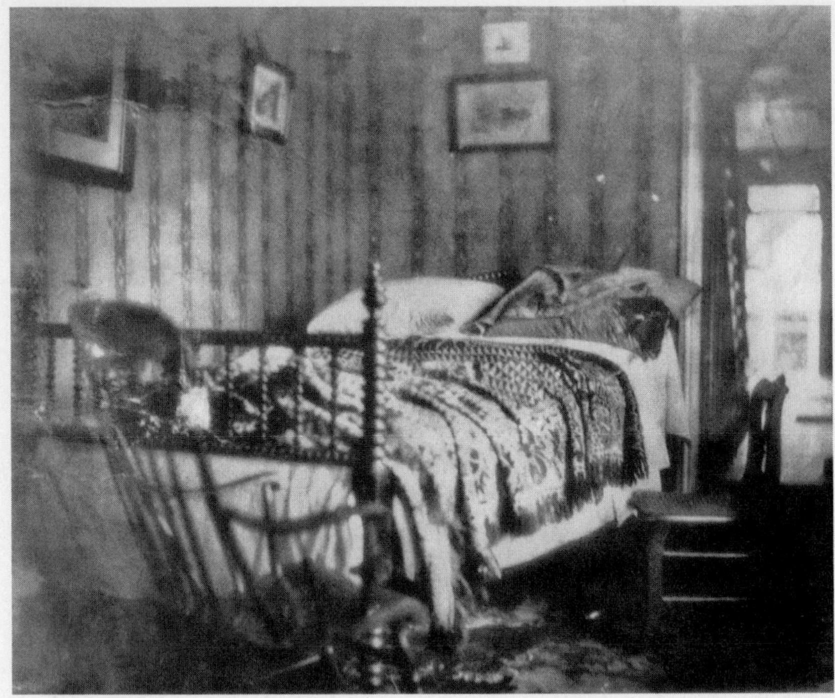

THE BLOODY DEATHBED SHORTLY AFTER LINCOLN'S BODY WAS REMOVED.

footprints. Disgusted by the mess made of his house, William Petersen collected some of the stained linens and heaved them out a rear window. The front door faced east, and the morning light flooded the hallway all the way to the back bedroom. Two of Petersen's tenants, Henry and Julius Ulke, brothers and artists, entered the empty death chamber. Bloodstained pillows, sheets, and a coverlet—later someone stole it and it was never seen again—lay on the bed. They were still wet. The Ulkes recognized a historic opportunity. They retrieved Henry's camera, set up its tripod at the southwest end of the room, and aimed the lens at the bed. To compose the best possible photo, they pushed the bed back to its original position in the northeast corner of the room. Henry Ulke uncovered the big lens and exposed his glass-plate negative for up to one minute, saturating it with the scene. Then he made one or two more plates.

Why did the Ulkes photograph the death room? Being commercial photographers, they must have intended to print multiple albumen-paper copies from their negatives and market them to the public. Soon, the Washington papers would be filled with advertisements offering photos of Lincoln, and John Wilkes Booth, for sale. An exclusive photograph of Abraham Lincoln's deathbed made shortly after his body had been removed, before the bloody sheets and pillows had been taken away, would be a commercial coup. Such an image would transport viewers into the Petersen house and allow them to imagine what it must have been like to be at the dying president's side.

Strangely, no evidence survives to suggest that the Ulkes ever attempted to market the photograph. No contemporary newspapers copied it as a woodcut, no carte de visite examples with letterpress-printed captions—a telltale sign of commercial exploitation—have ever been found, and only two or three original prints from the negatives have been located, the first one not until almost a century later.

Several artists sketched the death room, several others made oil paintings, and printmakers published more than fifteen different artworks depicting Lincoln in his deathbed, surrounded by mourners. Perhaps the Ulkes decided that their photograph of the empty bed was too stark and graphic, unlike the more romanticized prints that sanitized Lincoln's death. In the days to come, Stanton would suppress other photographs connected to the assassination, and it is possible he learned of this one and judged it too shocking a memento. Seaton Munroe might not have approved of the graphic image. In the days after the assassination, he complained about the lust for blood rel-

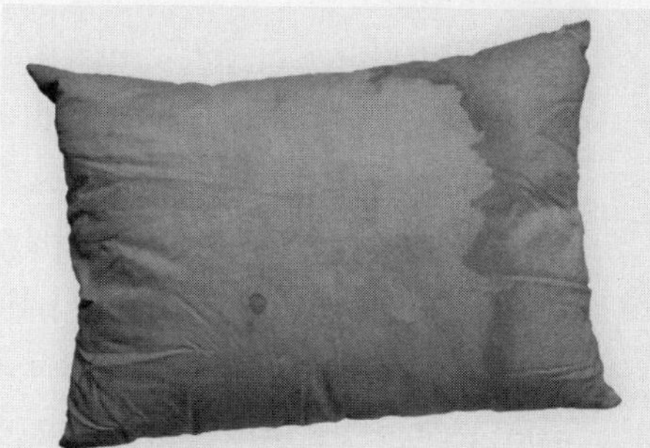

Blood relic: a pillow from Lincoln's deathbed.

ics: "Even then I could fancy the relic hunter plying his vocation, and bruing his ready handkerchief in the clotted blood, that he might preserve, exhibit, and mayhap peddle his gruesome trophy! I have lately seen in print an account of the preservation and partition of the blood-stained dress of Laura Keene."

William J. Ferguson, a prior visitor to the Petersen house, had seen the spindle bed in the photograph before. He returned to the house on the night of April 14. "I joined Mr. Petersen's son—a lad with whom I chummed; and went with him through the basement of the house to the stairs at the rear. Climbing them, we came to the floor of the room where Mr. Lincoln had been taken. It was a room formerly occupied by a Mr. Matthews, still a member of our company. I had delivered parts during the season to him and others in the room. On one of these visits I saw John Wilkes Booth lying and smoking a pipe on the same bed in which Mr. Lincoln died."

The complete story of the Ulkes and their remarkable photograph remains a mystery. In an odd twist, a few years after the assassination, Henry Ulke painted an official oil portrait of Edwin M. Stanton that hangs today in Washington at the National Portrait Gallery.

Soon other artists created assassination oil paintings, including the first, Carl Bersch, who painted the scene of Lincoln being carried across the street from Ford's to the Petersen house, to "make it the center and outstanding part of the large painting I shall make, using the sketches I made earlier in the evening, as an appropriate background. A fitting title for the picture would, I think, be 'Lincoln Borne by Loving Hands on the Fatal Night of April 14, 1865.' Altogether, it was the most tragic and impressive scene I have ever witnessed." Once in the collection of the White House, this haunting painting was transferred to the National Park Service in 1978. The morbid work, judged unsuitable for the eyes of future presidents, made its way back to Ford's Theatre, where it hung for almost thirty years until it was banished to storage.

CHAPTER FIVE

"The Body of the President Embalmed!"

Mary Lincoln had arrived at the White House about two hours ahead of her husband. She had not slept all night, but this was not a place where she could find rest. Elizabeth Dixon had accompanied her from the Petersen house. "At nine o'clock we took her home to that house so changed for her and the Doctor said she must go immediately to bed. She refused to go into any of the rooms she had previously occupied, 'not there! Oh not there' she said—and so we took her to one she had arranged for the President for a summer room to write in—I remained till eleven o'clock (twelve hours from the time I went to see her) and then left her a lonely widow, everything changed for her, since they left so happily the evening previous."

When the cortege arrived from Tenth Street, Mary did not gather herself, go downstairs, and receive the president with honors as he entered his White House for the last time. Lincoln's army tendered those honors without her. Abraham Lincoln's homecoming at 11:00 A.M. on April 15, 1865, was the most dramatic moment in the Execu-

THE PRESIDENT IS DEAD!

WAR DEPARTMENT,
Washington, April 15, 1865.

To MAJ. GEN. DIX,
Abraham Lincoln died this morning at 22 minutes after Seven o'clock.

E. M. STANTON, Sec. of War.

An early broadside announcing Lincoln's death.

tive Mansion's history since it was burned by the British in the War of 1812.

Although Mary refused to bear witness, Mrs. Dixon did: "As I started to go down the stairs I met the cortege bringing up the remains of the murdered President." Shocked by the unexpected encounter, Dixon watched as soldiers dressed in dark, Union-blue frock coats carried the flag-draped pine coffin slowly up the grand staircase.

The soldiers carried the temporary coffin into the Prince of Wales Room, also known as the Guest Room, removed the flag that draped the box, and unscrewed the lid. They lifted the body and laid it on wooden boards supported by wood trestles. They unwrapped from his corpse the bloody flag that shrouded him. At the Petersen house, his clothes—suit coat, torn shirt, pants, plus pocket contents—had been tossed in the coffin. Somebody had forgotten his boots; they were still under Willie Clarke's bed. His tie was also missing—somebody had already taken it. For the moment Lincoln lay naked on the improvised table, which looked like a carpenter's bench. He had been dead for less than five hours, and his body was still cooling.

The physicians and witnesses were waiting. Present were Dr. Joseph K. Barnes, surgeon general; Dr. Charles H. Crane, assistant surgeon general; Dr. H. M. Notson, assistant surgeon; Dr. Charles S. Taft, assistant acting surgeon; Dr. Robert King Stone, the Lincolns' family doctor; Dr. Janvier J. Woodward, assistant surgeon; and Dr. Joseph Curtis, assistant surgeon. Dr. Charles Leale had declined the invitation to watch them cut open the body of the man whose life he had tried to save. Civilian observers included Lincoln's friend and former Illinois senator Orville Hickman Browning and Benjamin Brown French.

At 9:00 A.M. French had left his home and headed to the White House. His carriage arrived at the gate not long after the remains had been taken inside. "I went immediately to the room where they were and saw them taken from the temporary coffin in which they had been brought here." French did not stay for the autopsy. Instead, at somebody's request, he went to Mary Lincoln's room. "She was in bed, Mrs. Welles being alone with her. She was in great distress, and I remained only a moment." He was already thinking of how the national capital should honor the dead president. "I then gave all the directions I could as to the preparations for the funeral."

Overcome by a severe headache, French left the White House by noon and rode to Capitol Hill. "I came through the Capitol, gave directions for clothing it in mourning . . . and then came home." French's workers used shears to cut long panels of black bunting for that purpose.

Soon after the body was laid out in the Guest Room, the doctors prepared their instruments to cut open Lincoln's body. Dr. Curtis, the assistant surgeon, described the scene. "The room . . . contained but little furniture: a large, heavily curtained bed, a sofa or two, bureaus, wardrobe, and chairs comprised all there was." He noticed that the generals and civilians in the room with him were silent or conversed quietly in whispers. He saw that at one side of the room "stretched upon a rough framework of boards covered only with sheets and towels, lay—cold and immovable—what but a few hours before was the soul of a great nation."

He recalled the surgeon general saying that "the President showed most wonderful tenacity of life, and, had not his wound been necessarily mortal, might have survived an injury to which most men would succumb."

Dr. Woodward would expose the brain. He reached into his medical kit for a scalpel, sliced through the skin at the back of the president's head, and peeled the scalp forward to expose the skull. Then he

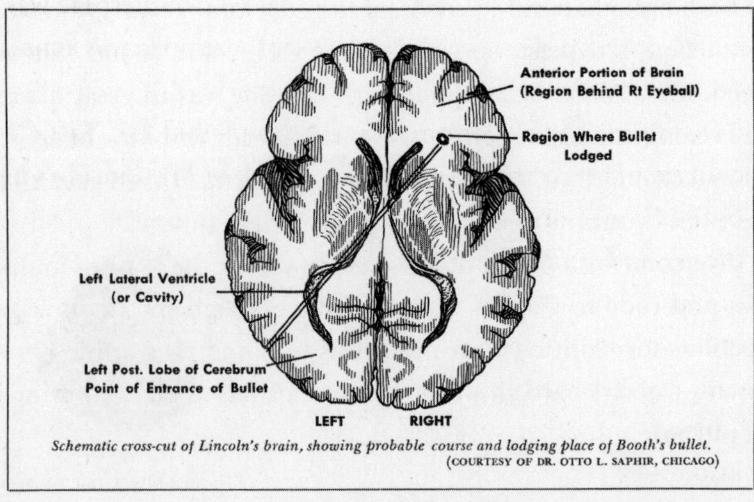

Schematic cross-cut of Lincoln's brain, showing probable course and lodging place of Booth's bullet.
(COURTESY OF DR. OTTO L. SAPHIR, CHICAGO)

THE BULLET'S FATAL PATH.

reached for the bone saw. To get to the brain, he needed to cut off the top of Lincoln's skull. Dr. Curtis described the procedure:

> Dr. Woodward and I proceeded to open the head and remove the brain down to the track of the ball. The latter had entered a little to the left of the median line at the back of the head, had passed almost directly forwards through the center of the brain and lodged. Not finding it readily, we proceeded to remove the entire brain, when, as I was lifting the matter from the cavity of the skull, suddenly the bullet dropped out through my fingers and fell, breaking the solemn silence of the room with its clatter, into an empty basin that was standing beneath. There it lay upon the white china, a little black mass no bigger than the end of my finger—dull, motionless and harmless, yet the cause of such mighty changes in the world's history as we may perhaps never realize.

During the autopsy, a man opened the door and walked into the room, breaking everyone's intense concentration. Was the intruder a curiosity seeker, or, even worse, an infernal relic hunter? He was the latter, but one authorized by the highest authority. He was a messenger from the first lady.

"During the post-mortem examination," said Dr. Taft, "Mrs. Lincoln sent him in with a request for a lock of Mr. Lincoln's hair." Dr. Stone clipped one from the region of the wound and dispatched it to her room. Taft wanted one too. "I extended my hand to him in mute appeal, and received a lock stained with blood, and other surgeons present also received one."

THE BULLET THAT ENDED LINCOLN'S LIFE.

The doctors marked the bullet for

identification—Dr. Stone scratched the initials "A.L." on it—and dropped it into a paper envelope, sealed it, and surrendered it to Secretary Stanton. It would make a prized and historic addition to the gruesome collection of wounded-tissue specimens, shattered bones, and deadly projectiles being assembled at the new U.S. Army Medical Museum. The fatal bullet from Booth's Deringer pistol became an object of fascination not just for Dr. Curtis but for the American people. It even became the subject of a bizarre allegorical print.

This lithograph, published in Chicago within a few weeks of the assassination, depicts the bullet resting beneath a powerful magnifying glass, with an eerie, all-seeing eye peering through the lens at the figure of John Wilkes Booth—imprisoned inside the bullet! In a colossal error, the artist rendered the bullet as an elongated, conical round of the type fired by Civil War rifled muskets and not as the spherical pistol ball recovered from Lincoln's brain. That inaccuracy does not disqualify the print as the most bizarre artwork created in the aftermath of the president's murder.

The bullet now recovered, and the direction of its path through Lincoln's brain confirmed, Dr. Curtis asked his superiors if he would be allowed to weigh the brain. Was it possible that an unusually large brain mass accounted for Lincoln's genius? Curtis unpacked the scale he and Dr. Woodward had brought to the White House for this purpose. Curtis would be the first of many, over the next century and a half, to speculate on the origins of Lincoln's greatness.

"Silently, in one corner of the room, I prepared the brain for weighing," Curtis remembered. "As I looked at the mass of soft gray and white substance that I was carefully washing, it was impossible to realize that it was that mere clay upon whose workings, but the day before, rested the hopes of the nation. I felt more profoundly impressed than ever with the mystery of that unknown something which may be named 'vital spark' as well as anything else, whose absence or presence makes all the immeasurable difference between an inert mass of matter owning obedience to no laws but those gov-

A bizarre print depicting Booth imprisoned inside his own bullet.

erning the physical and chemical forces of the universe, and on the other hand, a living brain by whose silent, subtle machinery a world may be ruled."

Oddly, no one at the autopsy made contemporaneous notes indicating the disposition of Lincoln's brain. Nor, as best it can be determined, did any of the doctors or witnesses ever make any oral statements regarding the fate of the brain. The Army Medical Museum would have been its natural repository. That is where the bullet, skull fragments, blood relics, instruments, probe, and more were sent. But no records survive to suggest that Lincoln's brain or blood was placed on secret deposit there.

The reading on the scale disappointed Dr. Curtis: "The weighing of the brain . . . gave approximate results only, since there had been some loss of brain substance, in consequence of the wound, during the hours of life after the shooting. But the figures, as they were, seemed to show that the brain weight was not above the ordinary for a man of Lincoln's size."

Their work done, Drs. Curtis and Woodward stepped back from the corpse and wiped their tools clean of Lincoln's blood, hair, flesh, brain matter, and bone chips. Their shirt cuffs exhibited the signs of their trade—blood spots and brain fluid stained the absorbent, white cotton fabric. They packed their instruments away in their medical kits and returned the bone saw, scalpels, and other devices to their proper, velvet-lined niches in the trays. They and the witnesses beheld the president's body. It looked ghastly. The skin was pale, the jaw slack, the eyelids slightly open, the face bruised (especially in the area surrounding the right eye socket, behind which the bullet had been lodged), the scalp peeled back, the top of the skull sawn off, and the brain, now washed clean of blood and void of Booth's profane missile, lying nearby in a basin. It was time for the embalmers to arrive. They would be responsible for repairing the damage and concealing the violations the pathologists had committed upon the president's corpse.

Secretary of the Navy Gideon Welles described what happened: "When Mr. Lincoln's body had been removed to the President's House, the embalmers proceeded to prepare it for the grave. Mr. Harry P. Cattell, in the employ of Doctors Brown and Alexander, who, three years before, had prepared so beautifully the body of little Willie Lincoln, now made as perpetual as art could effect the peculiar features of the late beloved President. The body was drained of its blood, and the parts necessary to remove decay were carefully withdrawn, and a chemical preparation injected, which soon hardened to a consistence of stone, giving the body the firmness and solid immobility of a statue."

Edwin Stanton supervised the dressing of the corpse. He went through Lincoln's wardrobe to choose the suit. Lincoln did not own an extensive collection of clothing. He was always an indifferent dresser, as many of his photographs testify. It was an old habit from his circuit-riding days as a lawyer, when he needed to travel light. He never packed many clothes on those trips. Lincoln was the kind of man who did not replace his clothes until he wore them out.

Stanton eyed the suits. There was the one the president had worn during the day of April 14, before he changed to attend the theater. If necessary, it would do. Another good suit, tailored by Brooks Brothers, lay crumpled in the temporary wood coffin on the floor. Stanton couldn't bury him in the suit he wore when he was shot. (A century later, this suit surfaced from obscurity and became a sensational collector's prize.)

There was another suit in Lincoln's closet. It was new, so the president had not had the opportunity to wear it out. This suit was one of the finest garments he had ever owned. Stanton selected it. He watched as they fitted the president with a white cotton shirt, looped the bow tie under the collar (Lincoln bought his neckties pre-tied), and dressed him in the suit. They did not put him in a coffin yet.

While the morticians embalmed and dressed Lincoln's corpse, Benjamin Brown French dined at home at 3:00 P.M. His headache was worse. He thought of his diary. He did not want to let the events of this day pass without committing them to writing in his thick, quarto-sized, leather-bound journal. But he couldn't concentrate, so he went to bed. He slept until 7:00 P.M., then rose, took tea, and opened his diary. What he wrote that night and in the days to come, in his distinctive, beautiful script, fills the pages of one of the great American journals.

Lincoln's corpse was ready for burial, but it was unclear where that would occur. Mary had the right to choose the site, but given her mental state, she was in no condition to discuss the subject only hours after her husband's death. Edwin Stanton would confer with her and Robert Lincoln later. In the meantime, whatever the final destination of the president's remains, official funeral events would have to take place in the national capital within the next few days. Stanton may have had time to supervise the dressing of Lincoln's body, but he had no time to plan and supervise a major public funeral, the biggest, no doubt, that the District of Columbia had ever seen. The secretary of war needed to delegate this responsibility, and there were several qualified candidates. Ward Hill Lamon, marshal of the District of Columbia, had known Lincoln for years, ever since their days as circuit-riding Illinois lawyers. Lamon had accompanied the president-elect on the railroad journey from Springfield to Washington in 1861 and had appointed himself the president's unofficial bodyguard. Lamon, a big, strong, barrel-chested man, was once found sleeping outside Lincoln's door at the White House clutching pistols in both hands. And it was Lamon who had organized the procession at Gettysburg on November 19, 1863, when Lincoln spoke there at the dedication of the national cemetery.

On less than a week's notice, Lamon had planned everything, devised and printed the order of march and program of events, recruited U.S. marshals from several other states to assist him, and stood on the platform with Lincoln and announced to the crowd in his bellowing voice, "Ladies and gentlemen, the president of the United States!" But Lamon was in Richmond the night of the assassination and still had not yet returned to Washington. Stanton would have to select someone else.

Benjamin Brown French was another obvious choice. The old Washington veteran had been on hand for decades of historic events, including the deaths of other presidents, and a multitude of public ceremonies and processions. French was perfect but he was needed to play another role—decorator in chief of the public buildings, especially at the U.S. Capitol, where without doubt Lincoln's corpse would lie in state.

Lincoln was, in addition to chief executive, commander in chief of the armed forces, and if Stanton wanted to entrust planning the funeral events to a military officer, he had several options. Major General Montgomery Meigs, quartermaster general of the U.S. Army, was a master planner with superb organizational skills. Stanton relied on him to supply the entire Union army with muskets, uniforms, blankets, food, and more, and to deliver those goods wherever and whenever needed. But the war was not over, and Stanton could not spare Meigs from his vital mission.

There was Brigadier General Edward D. Townsend, the brilliant assistant adjutant general of the army. Townsend had served the legendary General Winfield Scott, hero of the War of 1812 and the Mexican War, who in 1861 still held command of the army at the start of the Civil War. Upon Scott's resignation, Lincoln, who knew Townsend's qualities, offered him his choice of spots: "On reporting to the President, he asked what I desired. I replied I did not think it right to indicate for what duty I was most required, but was ready for any orders that might be given me."

Townsend hoped for a field command, but the new general in chief, George B. McClellan, said that he was too valuable in army administration. Townsend served under Lincoln's first secretary of war, Simon Cameron, and when he resigned the new secretary, Edwin M. Stanton, kept Townsend on, relying on him to run the adjutant general's office during the extended absences of its titular head, General Lorenzo Thomas. Townsend whipped the office into shape and was willing to stand up to the formidable Stanton, thereby earning his respect. He would be a good choice to plan the funeral. But already Stanton had Townsend in mind for a special duty of utmost importance, one even more critical than planning the president's funeral in the nation's capital. He held Townsend in reserve.

Stanton turned to another government department and considered George Harrington, assistant secretary of the Treasury. Harrington, fifty years old, was experienced in the ways of Washington. He had served as a delegate from the District of Columbia to the Republican National Convention of 1860 and had won the confidence of Lincoln's second secretary of the Treasury, William P. Fessenden. Lincoln, Stanton, and all the other members of the cabinet

George Harrington, the man who planned the Lincoln funeral events in Washington.

knew Harrington, and on occasions when Fessenden was absent from Washington, Lincoln had appointed him acting secretary of the Treasury. Upon Fessenden's resignation in 1865, Harrington continued to serve under the new secretary, Hugh McCulloch. Stanton believed that Harrington had the keen, quick, and thorough organizational mind essential for this assignment and chose him to take charge of all Washington events honoring the late president.

Harrington accepted the appointment, which involved more than merely taking charge of events. It was up to him to conjure how the national capital should honor its first assassinated president. Two presidents, William Henry Harrison and Zachary Taylor, had died in office, but they had expired from natural causes during peacetime, not from murder at the climax of a momentous civil war. Their more modest funerals were of limited value in planning Lincoln's. It had been only five years since Harrington and his fellow delegates had nominated Lincoln at the Chicago convention of May 1860.

Once Harrington got to work, Stanton could focus on what should be done with Lincoln's corpse after the Washington, D.C., ceremonies. Would the president be interred at the U.S. Capitol, in the underground crypt below the Great Dome, once intended as the final resting place for George Washington? Or would Mary Lincoln take the body home to Illinois, for burial in Chicago, its most important city, or in Springfield, the state capital and the Lincolns' home for twenty-four years?

On Saturday afternoon, the autopsy doctors, witnesses, and embalmers departed the Guest Room, leaving Lincoln's body alone on an undertaker's board. Now he would repose in the Executive Mansion for three days and two nights, dressed in the same splendid clothes he wore on March 4, 1865, when he rode in a carriage in a grand procession from the White House to the Capitol, where he swore to preserve, protect, and defend the Constitution of the United States. On that great day the breast pocket of his suit contained a folded sheet of paper bearing 701 words. "With malice toward none," he said from the marble steps at the East Front of the Capitol. "With charity for all," he beseeched, "let us bind up the nation's wounds." Now, six weeks later, his suit pocket was empty. Few visitors were permitted to view the body until the remains would be carried downstairs to the East Room on April 17 in preparation for opening the Executive Mansion to the public the next morning. Before then only relatives, close friends, and high officials crossed the threshold of his sanctuary and intruded upon his rest.

Mary Lincoln's confidante Elizabeth Keckly was one of them. After the president was shot, Mary had sent a messenger from the Petersen house summoning Keckly to her side. Elizabeth mistakenly rushed to the White House, but guards there denied entrance to the free black woman, and she did not gain admittance until the next day. The very sight of Lizzie, as Mary affectionately called her, soothed some of the widow's pain. They talked, and Mary told Lizzie about her terrible night at Ford's Theatre and the morning at the Petersen house. Keckly comforted her and then asked to see Abraham Lincoln.

"[Mrs. Lincoln] was nearly exhausted with grief," Keckly remembered, "and when she became a little quiet, I received permission to go into the Guest Room, where the body of the President lay in state. When I crossed the threshold . . . I could not help recalling the day on which I had seen little Willie lying in his coffin where the body of his father now lay." Three years earlier Keckly had helped wash and dress Willie's body. "I remembered how the President had wept over the pale beautiful face of his gifted boy, and now the President himself was dead." Keckly lifted the white cloth shrouding Lincoln's body. "I gazed long at the face," she said, "and turned away with tears in my eyes and a choking sensation in my throat."

Benjamin Brown French also went to the White House on the afternoon of Sunday, April 16, to confirm that all was going well in the East Room with the preparations for Lincoln's April 19 funeral. Then he went upstairs to view the embalmed corpse: "I saw the remains of the President, which are growing more and more natural . . . but for the bloodshot appearance of the cheek directly under the right eye, the face would look perfectly natural."

After spending an hour at the Executive Mansion, French visited Secretary of State Seward, who gave him a firsthand account of the savage knife attack he was lucky to have survived. French left Seward's home with Senator Solomon Foot of Vermont, and they went back to the White House. French wanted to see the corpse again: "We stood together at the side of the form of him whom, in life, we both loved so well."

Orville Hickman Browning came to the White House to view the corpse at least twice before the funeral, first for the autopsy on Saturday the fifteenth, and then on Monday the seventeenth. Browning had watched the surgeons saw off the top of his friend's cranium and remove his brain. It was bloody, ugly work. Two days later, Browning observed how the embalmer's artistry had improved Lincoln's appearance. The president, he wrote, "was looking as natural as life, and if in a quiet sleep. We all think the body should be taken to Springfield for internment, but Mrs. Lincoln is vehemently opposed to it, and wishes it to go to Chicago."

Except for these visitors and a few others—and the ever-present military honor guards standing in motionless vigil—the corpse was alone. Lincoln had finally won the solitude he craved during his presidency, when patronage seekers, influence peddlers, and lobbyists camped out near his door and harassed him so thoroughly that he felt under perpetual siege in his own office. Lincoln fought back by having a White House carpenter build a partition in the anteroom to conceal him from public view as he crossed between his second-floor office and his living quarters.

The construction of this private passage amused the great Civil War journalist George Alfred Townsend.

> It tells a long story of duns and loiterers, contract-hunters and seekers for commissions, garrulous parents on paltry errands, toadies without measure and talkers without conscience. They pressed upon him through the great door opposite his window, and hat in hand, came courtsying to his chair, with an obsequious "Mr. President!" If he dared, though the chief magistrate and commander of the army and navy, to go out the great door, these vampires leaped upon him with their Babylonian pleas, and barred his walk to his hearthside. He could not insult them since it was not in his nature, and perhaps

many of them had really urgent errands. So he called up the carpenter and ordered a strategic route cut from his office to his hearth, and perhaps told of it after with much merriment.

Now that traffic had ceased, and for the first time in four years, the human jackals did not skulk about the second floor, staking out his office. Once, when sick with smallpox, Lincoln had joked, "Now I have something I can give to everyone!"

The only sounds now were ones Lincoln would have recalled from childhood—wood saws cutting, hammers pounding nails, carpenters at work. Workmen in the East Room were building the catafalque upon which his elaborate coffin, not yet finished, would rest during the public viewing and state funeral. These sounds were the familiar music of his youth, made by his carpenter father, Thomas. It was the echo of his own labor too, when he had a rail-splitting axe placed in his hands at the age of nine. But the noise frightened Mary Lincoln; the hammer strikes reminded her of gunshots.

Strangely, not once during the days and nights that the president's corpse lay in seclusion at the White House did Mary make a private visit to her husband. Her last nightmare vision of him, bleeding, gasping, mortally wounded, dying in the overcrowded, stuffy little back bedroom of the Petersen house, had traumatized her and she could not bear to walk the short distance from her bedchamber to the Guest Room and look upon his face now. History does not record whether her son Robert defied her morbid imprisonment of Tad in her frightening mourning chamber and whether he took his little brother to the Guest Room to view their father, just as, three years before, Abraham had carried Tad from his bed to view his brother Willie in death.

Elizabeth Keckly, one of the few people allowed into Mary Lincoln's room, witnessed her tortured paroxysms. "I shall never forget the scene—the wails of a broken heart, the unearthly shrieks, the terrible convulsions, the wild, tempestuous outbursts of grief from the soul." Keckly worried about Tad. "[His] grief at his father's death was as great as the grief of his mother, but her terrible outbursts awed the boy into silence. Sometimes he would throw his arms around her neck, and exclaim, between his broken sobs, 'Don't cry so, Momma! Don't cry, or you will make me cry too! You will break my heart.' "

Outside this room, away from the Executive Mansion, the nation was in upheaval. The assassin John Wilkes Booth had escaped, and Stanton was coordinating an unprecedented manhunt to capture him. Secretary of State Seward and his son Fred were fighting for their lives after Lewis Powell's botched assassination attempt. Stanton suspected that numerous conspirators, their plans still secret and their strength yet unknown, might still lurk in Washington. Perhaps some of Booth's conspirators still at large planned to commit additional bloody crimes—like the murder of Lincoln's entire cabinet. As a precaution, Stanton assigned an around-the-clock military guard to every one of them.

Jefferson Davis was still on the run. Stanton worried that the "rebel chief," who was not satisfied to escape Richmond with his life, had ignored General Lee's surrender at Appomattox and, from his mobile command post, was attempting to rally the South to fight on and continue the Civil War. Confederate armies were still in the field and some of its ships still at sea. Hurriedly published newspaper extras shouted the latest news several times a day. Many stories suggested that Lincoln's murder was part of Davis's plan to reverse the outcome of the war.

The newspapers also reported what had been done to Lincoln's corpse. "The Body of the President Embalmed!" shouted a headline in one broadside extra. The number of deaths during the Civil War had advanced the art and social acceptance of embalming. Once a novelty viewed with distaste and even suspicion, the practice had become commonplace when the broken bodies of so many fallen soldiers were shipped from distant battlefields back home to waiting parents and widows.

On Sunday morning, April 16, "Black Easter," ministers across the land mounted their pulpits and, within a few hours, began to transform Abraham Lincoln from a mortal man into a secular saint. While the preachers were delivering their sermons that day, George Harrington began organizing the grandest funeral ceremonies in American history. He took pen, paper, and ink and wrote out in longhand his proposal for honoring the first American president slain by an assassin. The document was brief and only a draft. But it would, over the next three days, set in motion the intertwined actions and coordinated movements of more than one hundred thousand men. In an inspired moment, Harrington had dreamed up a grand idea out of thin air and then captured it on paper. This was his plan:

> Proposed arrangements for the Funeral and disposition of the Remains of the late President, submitted for approval.
>
> The Executive Mansion, under proper police and guards, to be thrown open during Tuesday, the 18th . . . for the public to show their respect,—the remains to be in the East Room, under a guard of commissioned [Harrington originally wrote "competent" but struck out that word and replaced it with "commissioned"] Officers of the Army.
>
> On Wednesday, the procession to form at 11 o'clock, the religious ceremonies to commence at 12, and the procession to move at 2 P.M.
>
> The remains to be escorted to the Capitol, and there deposited in the Rotunda, to remain under a suitable guard, to be provided by the proper military authorities.
>
> The delegation especially appointed from Illinois to receive the remains and escort them thither, to be called the "Body Guard," to have them in official charge after they shall have been deposited in the Capitol:

> The remains to be taken to the depot on Thursday morning, by military escort, a guard of honor, consisting of such Senators and Members of the House of Representatives as may be designated for that purpose by those bodies respectively, and also such other civilians as the Cabinet may determine to accompany the remains to their final resting place. The whole to be accompanied by such military escort as the proper authorities may designate.

In five short paragraphs, Harrington had his template—even if the arrangements raised as many questions as they answered. For how many hours should the White House be kept open for the public to view the remains? How many people per hour could squeeze through the doors—and how many thousands more would try? Who should receive invitations to the funeral? The East Room was the biggest chamber in the White House, but it could never hold everyone who would demand the right to attend. Without even calculating the dimensions and square footage, Harrington knew the room could hold fewer than a thousand people at a standing reception, and the funeral guests would be seated, thus consuming additional, scarce space. And where would he get all those chairs? The entire Executive Mansion did not contain enough furniture to seat hundreds of people. And when arranging the chairs, Harrington would have to be careful to reserve enough space for the catafalque, and for aisles. That was just for the White House events. After the funeral, who would march in the procession to the Capitol? How would this procession be organized? Who would keep order in the streets? George Harrington needed help.

It was one thing to sketch an outline of the funeral events on a piece of paper but quite another to fill in all the details and then execute them. Harrington knew that Lincoln's funeral ceremonies would be the largest and most elaborate series of public events ever held in the nation's capital, and possibly the entire nation. He could not possibly organize all of them himself—a public viewing at the White House on April 18; a private White House funeral attended by hundreds of dignitaries on April 19, followed immediately by a grand, synchronized, and incomparable procession from the White House to the U.S. Capitol; a lying in state and public viewing in the Capitol rotunda under the Great Dome on April 20; and the departure of the president's remains from Washington on April 21. Only one institution in the country possessed the men, command structure, and logistical experience to conduct such an event—the U.S. Army.

That afternoon, Harrington called a crucial meeting at the Treasury Department for 5:00 P.M., and he summoned by messenger several of the most important army officers in Washington, including Major General and Chief of Staff Henry W. Halleck, Major General and commander of the military district of Washington C. C. Augur, and Assistant Adjutant General and skilled War Department administrator William A. Nichols. Harrington also invited Benjamin Brown French to attend.

"I had agreed to meet Assist. Secretary Harrington at the Treasury Dept. at 5, to aid in making the programme of Arrangements for the funeral," French recalled, "so I remained at the President's until that hour, then went to the Treasury Dept."

One by one, messengers arrived at Harrington's office bearing responses. Among the acceptances were those from W. A. Nichols, assistant adjutant general: "I have the honor to acknowledge the receipt of your note of this date, stating that the Sec'y of War had designated me as one, on part of the Army, to confer in relation to the funeral ceremonies of the late President. As requested I will be present at the meeting fixed at the hour of 5 o'clk P.M. to-day"; from H. W. Halleck, army chief of staff: "I was notified by the Secy of War to meet you at 7 o.c. this evening & so wrote to Genl Augur, but will meet you as soon after *5* as I can"; and from C. C. Augur, commander of the military district of Washington: "I have received your note, and will be at the place you indicate at 5. P.M. today."

When Harrington's chosen men convened at the Treasury Department next door to the White House, the footsteps of their heavy boots echoed through the marble-paved halls. They had much work to do and little time. Gathered around Harrington's desk, they had just sixty-eight hours to plan Abraham Lincoln's state funeral.

They met for an hour, adjourned at 6:00 P.M., and agreed to reconvene in one hour. "[We] agreed," French wrote in his diary, "to return at 7 to meet with several Senators, Members of the House & Military officers." When the commissioner of public buildings returned he found, among others, two assistant secretaries of the Treasury, George Harrington and Maunsell B. Field (who had been at the Petersen house); Senator Solomon Foot of Vermont; Richard Yates, former Illinois congressman, Civil War governor, and now U.S. senator; former congressman Isaac N. Arnold from Illinois; Governor Richard J. Oglesby; Major Generals Henry W. Halleck and C. C. Augur; Brigadier General George W. Nichols from the adjutant general's office; Admiral William B. Shubrick; and Lawrence A. Gobright, longtime Associated Press correspondent in Washington. They spent another hour talking about the arrangements and agreed to meet again the next day at 2:00 P.M.

Harrington appeared strained under the burden. Before Easter evening was over, he wrote a letter to his patron, former Treasury secretary William Pitt Fessenden, updating him on various events but ending by saying, "What shadows we are and what shadows we pursue . . . the whole charge of the funeral fixed for Wednesday has been put on me. Heavens I have enough to do without this."

Although Harrington complained privately about his duties, he knew, at least, that they would end once the Washington ceremonies concluded. As soon as Lincoln's corpse was ready to depart the national capital, his work would be done. Stanton had taken upon himself the responsibility for the next stage of the president's journey.

The next day, Monday, April 17, Harrington was overwhelmed

by letters, telegrams, and personal visitors who hounded him and beseeched him, seeking advantage. Some sought tickets to the funeral, others the right to march in the procession. Some wanted a license to sell mourning goods to the government, while others alerted him to special deliveries of flowers and asked him to confirm their arrival. Some supplicants did not wait for invitations and simply announced that they were coming.

The War Department sent over a document laying out the military's role in the procession, but the draft was blank where the civic portion of the parade would be described. Ward Hill Lamon would organize and lead that, and Harrington had one day to get the information to Nichols so the War Department could publish a printed handbill with the order of march for both the military and civic processions.

Harrington lamented that so much was still left undone: "Nothing has been done to provide for the admission of persons who are to be at the President's House, and to have a right to places in the carriages. Of course those who have cards to the Green & East Rooms will pass but it is important that all of them can get into the carriages—Nor has any arrangement been made as to the *number of carriages* to be admitted into the President's grounds—nor for the admission of the delegations from Illinois and Kentucky. Who has charge of the carriages for the procession?"

On April 17, Stanton requested an interview with Mary and Robert Lincoln to ascertain the family's wishes for the final disposition of the remains. Would they be laid to rest in Washington, Illinois, or some other place? Some federal officials, including Benjamin Brown French, argued in favor of entombment in the U.S. Capitol. From Kentucky came an urgent telegram imploring consideration of Lincoln's birthplace as a suitable final resting place. Chicago, where Lincoln had practiced law in the federal courts, and where he received the Republican nomination for the presidency, put in a bid. The Illinois congressional delegation, acting via telegraph with officials back home, lobbied hard for burial in Springfield. Some of them, without the Lincoln family's consent, had already begun an extravagant scheme to purchase an entire city block and erect a stupendous monument on the site.

Mary was appalled when she discovered what she thought was a hometown conspiracy to hijack the martyr's remains and wrest control of her husband's body from her. She threatened to thwart her former neighbors' grandiose plans, and emotional telegrams went back and forth between Springfield and Washington. Influential Illinoisans in the national capital, including Orville Hickman Browning, sought to lobby Mary in person, but she refused to receive them at the White House. President Lincoln would be buried wherever she, and no one else, designated. Perhaps, she hinted, it might be Washington. Or perhaps Chicago. Or maybe somewhere else.

Stanton had to find out. If Lincoln's body was to travel to some distant place, it would be the War Department's and the United States Military Railroad's job to transport him there. Such a journey would take time to plan, and the Washington funeral was just two days away. Mary and Robert could feud with Springfield all they wanted; Stanton need not involve himself in that dispute. He only needed to know where he had to send the train. The Lincolns agreed to receive Stanton and told him they had decided on Illinois. And it would be Springfield, not Chicago.

Now the secretary of war could plan the route and devise the timetables. The train *could* proceed directly to Illinois on the shortest and most direct route, stopping along the way only to replenish water for the steam engine and fuel for the fire. But the most efficient route might not be the most desirable one. Lincoln had established a precedent four years earlier when he journeyed east as president-elect. Instead of a hurried run to Washington, D.C., he took a circuitous route through several of the major Northern states that had elected him so that he could see the American people, and they could see him. Lincoln hoped to reassure the country, sustain support for the Union and the Constitution, and avoid civil war.

His train stopped many times. He gave impromptu, unscripted speeches; greeted delegations of important officials; mingled with ordinary citizens; accepted tributes and well wishes; and participated in public ceremonies. Lincoln presented himself to his fellow citizens as a man of the people elevated temporarily to high office. For Lincoln, his inaugural train symbolized neither personal triumph nor glory, but the simplicity and integrity of the republican form of government established by the Constitution and laws of the United States. This journey represented a living bond between Lincoln and the American people.

Now he was dead. In their grief, Americans had not forgotten the inaugural train of 1861. Telegrams began to pour into the War Department from the cities and towns that had wished him Godspeed on his journey east four years ago, beseeching Stanton to send him back to them. Once the news spread that Lincoln would make the long westward journey home to Illinois, a groundswell of public opinion clamored for the government to re-create Lincoln's inaugural trip in reverse. Edwin Stanton liked the idea. The assassination of President Lincoln was a national tragedy. But the American people could not come from all over the country and converge on Washington to view the president's body, attend the funeral, or march in the procession. Why couldn't Abraham Lincoln go to them?

It was possible. It would require a special train fitted out properly to transport the body, a military escort to guard Lincoln's corpse around the clock to ensure that the remains were treated with the utmost dignity, coordination between the military railroad and the major commercial lines, cooperation between the War Department and state and local governments, and the resources and will to do it. Stanton believed it could be done. There was only one obstacle—the president's grieving, mercurial, and unpredictable widow. Executing this unprecedented plan would be impossible without her explicit consent.

Stanton broached the delicate subject with the first lady. Might she consider assuaging the American people's profound sadness by consenting to an extended route that would take her husband through the great cities of the Union? From Washington, the War Department could divert the train north through Maryland, Pennsylvania, New Jersey, and New York State, then make the great turn west, passing through Ohio and Indiana and into northern Illinois, then make a final turn south from Chicago, down through the prairies and home to Springfield. This route would take many days to travel, longer than a fast run to Illinois. The exact duration of the extended trip would depend upon the number of times the train stopped for water, fuel, and public ceremonies along the way. Stanton promised that if she said yes, he and his aides would handle all the details. It would be his final service to the president who once called him his "Mars," his god of war.

There was one more thing. The people wanted to see their Father Abraham, not just his closed coffin. They wanted to look upon his face. That meant an open casket. Mary had consented to an open coffin at the White House and the U.S. Capitol. But an open coffin at multiple ceremonies, all the way from Washington to Springfield, a distance of more than 1,600 miles? In warm weather, without refrigeration, it would test the limits of the embalmer's art. Mary thought the idea seemed morbid and ghoulish, but a grand, national funeral pageant that affirmed her husband's greatness appealed to her. She consented.

This epic train journey symbolized the importance of railroads in Abraham Lincoln's life. From early in his political career, Lincoln believed the government should invest in "internal improvements" to advance settlement and commerce. In Illinois in the 1830s and early 1840s, that meant navigable waterways. His youthful experiences on the Sangamon River, of floating a flatboat down the Mississippi River to New Orleans, and of living in a small river town, New Salem, created in him an enduring fascination with water transportation. Indeed, he even patented a device to raise trapped vessels over shoals. Later, after he became a lawyer, Lincoln represented the Illinois Central and other railroads in a number of cases, earning substantial fees. In one dispute, the Effie Afton case, he had to choose between water and rail. A river vessel had struck and damaged a railroad bridge. Each side blamed the other for the accident. Lincoln represented the railroad. Trains were the future. He knew they were the key to conquering the American continent.

Still, years later, Lincoln retained his sentimental affection for waterways. When Vicksburg and Port Hudson fell to the Union, he rejoiced that "the Father of Waters again goes unvexed to the sea." He enjoyed riding aboard warships and steam-powered paddleboats—joking to sailors that he was a "fresh water man" with little firsthand knowledge of the sea—and on the last day of his life he inspected ironclads at the Washington Navy Yard. And though in the White House he dreamed of mysterious journeys by sea, Lincoln had made the most important journeys of his life by railroad: to Washington after his election to Congress; to the federal courts in Chicago; east for political speeches in the 1850s; to New York City in 1860 to speak at Cooper Union; to Washington again as president-elect in 1861; to the battlefields of Antietam in 1862 and Gettysburg in 1863. And now, home to Illinois.

At the Treasury Department, George Harrington began adding up the number of people from the government departments, the military, civil organizations, and diplomatic missions who should receive an invitation to the White House state funeral. He divided potential guests by category, then tallied a raw count—630 people. He would worry about the individual names later.

Officials from various cities contacted Harrington and tried to influence the route of the train, or at least obtain White House funeral tickets. In an April 17 letter from the collector's office at the U.S. Custom House in Philadelphia, one official reminded Harrington that the president had in that city once vowed to sacrifice his life for liberty. Now he *must* come back: "It is the general desire of the Citizens of Philadelphia that the remains of President Lincoln should pass through this city and remain a day in 'State' in Independence Hall, on its way to Illinois. It was in this city and at Independence Hall that he raised the flag of the Union with his own hands, and expressed his willingness to be assassinated on the spot rather than sacrifice the principles of Liberty on which he conceived the government to be based. I trust that the wishes of our people will be gratified. Very Truly Yours Wm. B. Thomas."

Unsurprisingly, New Yorkers proved quite assertive in announcing their participation. Typical was a telegram dispatched from New York to Harrington on April 17: "A committee of thirteen members of the Union League club on behalf thereof will attend the funeral of the late President. The committee will be in Washington at Willard's Hotel tomorrow Tuesday by the morning train from this city. / Otis D. Swan / Secty of Union League Club." And just in case Harrington did not get the message, the Union League dispatched another one to another assistant secretary of the Treasury, Simeon Draper: "You will oblige the Union League Club by notifying Assistant Secretary Harrington that a committee of thirteen members of the club on the behalf thereof will attend the funeral of President Lincoln at Washington Wednesday / Otis D. Swan / Secty of Union League Club."

Attorney General James Speed forwarded to Harrington a silly telegram that arrived at the Department of Justice from Louisville on April 17: "A wreath of rare flowers for the bier of our loved & lamented President is sent by Express by the German Gymnastic Association of this city / [Signed] Phillip Speed." It was absurd, as if the harried assistant secretary had any time to devote to something so trivial as the safe delivery of a particular arrangement of exotic flowers.

The letters and telegrams kept coming: "As chairman of the Committee of colored citizens of Washington, who desire to participate in the funeral ceremonies of our late President and friend Abraham Lincoln I have to solicit the favor of being placed in such a position in the line of procession as you may assign. Hoping an immediate answer I am sir your humble servants, James Wormley, Chairman / G. Snowdin / H. Harris / Committee."

One correspondent sought to exploit Lincoln's death for commercial gain, urging Harrington to purchase mourning goods from a recommended source: "Washington, April 17 1865. / Hon. George Harrington, Assistant Sec. of Treas. / My dear sir:/ Allow me to introduce my friend Wm. S. Mitchell Esq. a merchant of this city who is desirous of furnishing articles connected with the funeral ceremonies. He is an honorable gentleman, and the best guarantee of his patriotism is the fact that he is a cherished friend of President Johnson. / I have the honor to be / your humble & obd servt / Daniel R. Goodloe." The government would spend huge sums on funeral goods, but in the collection of voluminous bills and receipts compiled after the Washington events, the name of the enterprising William S. Mitchell cannot be found.

By nightfall of the seventeenth, Harrington still labored on the details. As soldiers prepared to carry Lincoln down the staircase to lie in the East Room, the War Department had still not finished organizing the funeral procession from the White House to the Capitol. The procession would begin in less than forty hours, and the final order for the line of march had not been printed yet. Officials were still making last-minute changes. Assistant Adjutant General Nichols tried to locate Harrington that night by writing to Maunsell Field, one of the Petersen house visitors, at the Treasury Department: "The Hon. Mr. Harrington directed the publishing of the order of the Funeral Ceremonies. If he is in the Dept. please ask him to cause the names of Messrs George Ashmun & Simon Cameron to be inserted with the names of the Pall bearers—if not in—please request the

Chronicle & Intelligencer to insert their names on the order under the caption of 'Civilians.'"

While Harrington worked and planned, Abraham Lincoln spent his last night in the White House. He had lived there four years, one month, and nine days. He had reposed there in death for four nights.

CHAPTER SIX

"We Shall See and Know Our Friends in Heaven"

Two days after the president died, the coffin was ready. Soldiers carried it to the second-floor Guest Room and placed it on the floor. They approached the president and lifted him from the table where he had lain since Saturday afternoon. The soldiers carried him to the coffin—it looked too small. The casket appeared no taller than the president. Lincoln was six feet, four inches tall, and the coffin was just two inches taller. They lowered his body into the casket, and it was definitely a snug fit. If they had tried to bury him in his boots, the body would have been too tall. The soldiers lifted the coffin and carried it down the stairs. Gaslights illuminated the silent, eerie journey. Noah Brooks described the scene: "On the night of the seventeenth the remains of Lincoln were laid in the casket prepared for their reception, and were taken from the large guest-chamber of the house to the famous East Room, where so many brilliant receptions and so many important public events had been witnessed; and there they lay in state until the day of the funeral."

They carried the coffin to the center of the room and rested it upon the catafalque. It was magnificent, more impressive than any coffin Abraham Lincoln had ever seen—finer than the crude one he helped build for his mother when he was a little boy, finer than the simple one that buried the hopes and body of young Ann Rutledge, and finer than the child-size coffins for his sons Eddie and Willie.

In life Abraham had eschewed his wife Mary's love of frills and finery. He would have never chosen such a stately and expensive coffin for himself. It had cost almost as much as he paid for his house at Eighth and Jackson streets in Springfield. He would have preferred the pine box they put him in at the Petersen house.

And the decorations. Lincoln had always laughed at Mary's obsession with decorating the White House—as had the newspapers and official Washington. But no one who entered the East Room over the next two days mocked its lavish vestments of death. When the public and press saw it, they were so impressed they named it the "Temple of Death." Lincoln, claimed one of his friends, had foreseen this tableau in one of his prophetic dreams.

Ward Hill Lamon recalled a small gathering at the White House a few days before the assassination where only he, the president, Mary Lincoln, and two or three other people were present. Lamon observed that Lincoln was in a "melancholy, meditative mood, and had been for some time." Mary commented on his demeanor. Then the president spoke:

"It seems strange how much there is in the Bible about dreams... If we believe in the Bible, we must accept the fact that in the old days God and his angels came to men in their sleep and made themselves known in dreams. Nowadays dreams are regarded as very foolish, and are seldom told, except by old women and by young men and maidens in love."

Mary asked her husband if he believed in dreams. "I can't say that I do, but I had one the other night which has haunted me ever since . . . somehow the thing has gotten possession of me, and, like Banquo's ghost, it will not down." Lincoln then narrated his troublesome dream:

> About ten days ago, I retired very late. I had been up waiting for important dispatches from the front. I could not have been long in bed when I fell into a slumber, for I was weary. I soon began to dream. There seemed to be a death-like stillness about me. Then I heard subdued sobs, as if a number of people were weeping. I thought I left my bed and wandered downstairs. There the silence was broken by the same pitiful sobbing, but the mourners were invisible. I went from room to room; no living person was in sight, but the same mournful sounds of distress met me as I passed along. It was light in all the rooms; every object was familiar to me; but where were all the people who were grieving as if their hearts would break? I was puzzled and alarmed. What could be the meaning of all this? Determined to find the cause of a state of things so mysterious and shocking, I kept on until I arrived at the East Room, which I entered. There I met with a sickening surprise. Before me was a catafalque, on which rested a corpse wrapped in funeral vestments. Around it were soldiers who were acting as guards; and there was a throng of people, some gazing mournfully upon the corpse, whose face was covered, others weeping pitifully. "Who is dead in the White House?" I demanded of one of the soldiers. "The President," was his answer; "he was killed by an assassin!" Then came a loud burst of grief from the crowd, which awoke me from my dream. I slept no more that night.

Mary Lincoln recoiled. "That is horrid! I wish you had not told it. I am glad that I don't believe in dreams, or I should be in terror from this time forth."

"Well," replied the president, "it is only a dream, Mary. Let us say no more about it, and try to forget it."

Some accused Lincoln's old friend Lamon of embellishing, even of concocting the dream. Its Lincolnesque qualities cannot be denied, but Lamon did not write about it until a number of years after Lincoln's death. Whether or not Lincoln had foreseen his own coffin lying in state in the East Room, he was haunted by other coffins he had seen there, and other places in wartime Washington. And he had carried with him to Washington the memories of other coffins and funerals of long ago.

In May 1861, Elmer Ellsworth, a flamboyant twenty-four-year-old friend of Lincoln's, was shot and killed after he hauled down a Confederate flag at a hotel in Alexandria, Virginia. Ellsworth had worked in Lincoln's Illinois law office, delivered exciting political speeches to advance Lincoln's career, and commanded a famous quasimilitary unit, Ellsworth's Zouaves, a drill team that had thrilled spectators with its exotic costumes and precision choreography. After Lincoln's nomination in May 1860, people from all over America—including autograph hounds—had inundated Springfield with letters. Lincoln tried to comply with the requests, and he asked Ellsworth to draft a number of replies in his own hand for Lincoln's signature. Dozens of letters from that period survive, each bearing the identical message in Ellsworth's neat script—"It gives me pleasure to comply with your request for my autograph"—and each then signed "A. Lincoln" by the nominee himself. Lincoln had grown fond of his enthusiastic, impetuous protégé, and invited him to travel aboard the special train that took the president-elect to Washington in February 1861.

In May 1861, Colonel Ellsworth's Zouaves were sworn into military service while Lincoln watched, and on May 24 Ellsworth and his troops crossed the Potomac River to take possession of Alexandria, Virginia. For days a defiant rebel flag, visible from the White House, had flown over the town. After landing, Ellsworth led his men in the direction of the telegraph office, but he could not resist stopping at Marshall House, the hotel upon which the offending flag waved atop the roof. The colonel burst inside, climbed the stairs, and hauled down the flag. As he descended to the lobby, the proprietor, James W. Jackson, fired a shotgun blast at his chest, killing him instantly. Ellsworth was still clutching the Confederate flag that had cost him his life. In vengeance, one of his Zouaves killed Jackson on the spot. To this day, a bronze plaque marks the site on Alexandria's King Street, not in memory of the slain Union officer, but in honor of the secessionist hotel proprietor who was "killed by federal soldiers while defending his property and personal rights." The plaque goes on to honor Jackson as a "martyr to the cause of Southern Independence."

Ellsworth's corpse was brought to the Washington Navy Yard. Someone would have to go to the White House to inform the president. That duty fell to navy captain Gustavus Fox. He arrived at the Executive Mansion, where he went upstairs to Lincoln's second-floor office. The news staggered the president. Fox left him alone. Lincoln walked to the window. He did not notice when Senator Henry Wilson and a reporter from the *New York Herald* entered his office. They approached him from behind, but he did not move. He stared through the window, his eyes fixed on the Potomac River and beyond. Startled, Lincoln made an abrupt turn and faced Wilson and the newsman. The president had thought he was alone.

"Excuse me," he said, "but I cannot talk." Then he burst into tears and buried his face in a handkerchief. The visitors retreated in silence. Lincoln sat down, composed himself, and spoke. "I will make no apology, gentlemen, for my weakness; but I knew poor Ellsworth well, and held him in great regard. Just as you entered the room, Captain Fox left me, after giving me the painful details of Ellsworth's unfortunate death. The event was so unexpected, and the recital so touching, that it quite unmanned me."

Lincoln recalled his young friend's impetuosity. "Poor fellow! It was undoubtedly an act of rashness, but it only shows the heroic spirit that animates our soldiers . . . in this righteous cause of ours. Yet who can restrain their grief to see them fall in such a way as this; not by fortunes of war, but by the hand of an assassin."

He wanted to see the body. Abraham and Mary Lincoln rode in their carriage east from the White House, down Pennsylvania Avenue, past the Great Dome, and south to the navy yard. For a long time, the president gazed at Ellsworth's handsome face, unmarred by the shotgun pellets.

"My boy! My boy!" pleaded Lincoln to the dead man. "Was it necessary that this sacrifice be made?" The president wept.

The bloody frock coat, its breast shredded by the fatal blast, would be preserved as a relic and displayed at Union patriotic fairs. Soon the body would be sent north by train to Ellsworth's parents and fiancée in New York. But not before, the president decided, he could honor his young martyred friend. Lincoln ordered that Ellsworth's corpse be brought to the White House. There, the next day, at noon on May 25, the president presided over the East Room funeral. After the service, Lincoln rode in the procession that carried Ellsworth's coffin to the railroad station for the 2:00 P.M. train to New York. Mary Lincoln kept the Confederate flag he had clutched in death. That afternoon the president wrote a letter to Ellsworth's parents.

My dear Sir and Madam,

In the untimely loss of your noble son, our affliction here, is scarcely less than your own. So much of promised usefulness to one's country, and of bright hopes for one's self and friends, have rarely been so suddenly dashed, as in his fall. In size, in years, and in youthful appearance, a boy only, his power to command men, was surpassingly great. This power, combined with a fine intellect, an indomitable energy, and a taste altogether military, constituted in him, as seemed to me, the best natural talent, in that department, I ever knew. And yet he was singularly modest and deferential in social intercourse. My acquaintance with him began less than two years ago; yet through the latter half of the intervening period, it was as intimate as the disparity of our ages, and my engrossing engagements, would permit. To me, he appeared to have no indulgences or pastimes; and I never heard him utter a profane, or intemperate word. What was conclusive of his good heart, he never forgot his parents. The honors he labored for so laudably, and, in the sad end, so gallantly gave his life, he meant for them, no less than for himself.

In the hope that it may be no intrusion upon the sacredness of your sorrow, I have ventured to address you this tribute to the memory of my young friend, and your brave and early fallen child.

May God give you that consolation which is beyond all earthly power. Sincerely your friend in a common affliction.

A. Lincoln

This was not the last condolence letter Lincoln wrote during the war, nor was this the last funeral he witnessed in his White House. As word of Ellsworth's death spread across the country, he became a popular hero celebrated in prints, sheet music, badges, and, in the most unusual mourning relic of the Civil War, an imposing and elaborate ceramic pitcher decorated with painted bas-relief panels depicting his death and the slaying of his murderer. Noah Brooks said of the Ellsworth craze: "The death of Ellsworth, needless though it may have been, caused a profound sensation throughout the country, where he was well known. He was among the very first martyrs of the war, as he had been one of the first volunteers. Lincoln was overwhelmed with sorrow . . . and even in the midst of his increasing cares, he found time to sit alone and in grief-stricken meditation by the bier of the dead young soldier of whose career he had cherished such great hopes."

The Civil War had opened with a funeral and now it would close with one. Indeed, a member of the Zouaves wrote that "Colonel

Ellsworth was the war's first conspicuous victim; Lincoln himself the last."

In October 1861, five months later, Lincoln suffered another personal loss. Edward D. Baker, an old friend from the Illinois political scene in the 1830s and 1840s, a former congressman who had moved to California in the 1850s, was now a U.S. senator from Oregon. Lincoln had named his firstborn son, Eddie, who died in 1850 when he was three years old, after him. Baker was in Washington in March 1861 for Lincoln's inaugural ball and could have served out the war in the safety of the halls of Congress, but he wanted to see action, so Lincoln offered him a commission as a brigadier general of volunteers. Such a high rank would require that Baker resign from the Senate, so he asked the president to make him a colonel, allowing him to keep his seat.

On October 21, 1861, during the Battle of Ball's Bluff, a confusing and embarrassing Union disaster fought not far from Washington near Leesburg, Virginia, Baker was killed. On the day before his death, on a beautiful fall afternoon, he had been idling with Lincoln on the White House lawn. The next day, when the president visited General McClellan's headquarters, the army commander told him that Baker was dead. Lincoln reeled, and when he left McClellan's office, he almost fell into the street.

When he returned to the White House, he gave orders that he would receive no visitors and he was unable to sleep that night. On October 24, Abraham and Mary Lincoln attended Baker's funeral at a private home in Washington. After the funeral Lincoln joined the procession to Congressional Cemetery on Capitol Hill. In 1848, when Lincoln was a congressman, he helped organize the funeral there for former president and member of the House of Representatives John Quincy Adams, and had helped escort Adams's remains to Congressional. The day after Baker's funeral Lincoln canceled a cabinet meeting, and that evening he received the colonel's father, son, and nephew.

By now death in the war was no longer a distant, abstract thing to Lincoln. It reached into his own house, as death had many times before in his boyhood, youth, and manhood in Illinois. First his mother, then his sister, then Ann Rutledge, then his little son Eddie, and the others. Baker's death hit him hard. Yes, he had grieved for the young Ellsworth, but he had known the youth for less than two years. Baker was different—their friendship went back more than a quarter of a century. Twice in the first year of the war Lincoln had bid farewell to two friends.

Willie Lincoln, only ten years old, also bid Baker good-bye, by composing for him what can only be called a death poem, which was published in Washington's *National Republican* newspaper. Willie had inherited many of his father's traits, and in this composition, he revealed a glimmer of Abraham Lincoln's lifelong fascination with poems of loss and death. Indeed, Lincoln not only read and recited such poetry, he wrote it. Willie's melancholy poem foreshadowed a time when he would be not the author but rather the subject of sad poems.

In February 1862, four months after the death of Edward Baker, Lincoln suffered the most painful loss in his life. It was the single event, more than any other, that crushed his spirit and killed the joy inside him. William Wallace Lincoln, age eleven, was the president's favorite son. Eddie had died too young and too long ago for Abraham to envision the kind of young man he might have become. Abraham and his oldest boy, Robert, were not close. Robert did not look like his father, and he was in temperament more like his mother's family, the Todds.

Lincoln had provided Robert all the opportunities that life had denied him. This son never had to work a hard day of manual labor in his life, and the president had him educated at Exeter and Harvard. But Robert was becoming a snob who seemed at times embarrassed by the crude vestiges of Abraham's frontier background and lack of education. Tad was a lovable, impulsive, impish, and undisciplined little boy who suffered from a speech impediment and who had inherited his mother's mercurial and selfish nature.

William was Lincoln's true heir. Tall and thin, he resembled his father in posture and physical gestures. Intellectual, analytical, and thoughtful, his mind worked in ways that reminded Lincoln of himself—and this pleased him. Willie was his father's true companion in the White House. The president was not alone in his admiration. Willie's maturity, splendid manners, and winning personality impressed all who met him, including Lincoln's cabinet. He was a favorite of many White House regulars. Lincoln loved no one more.

In February 1862, Tad and Willie fell ill with a fever, probably contracted from contaminated water that supplied the White House. Their condition worsened, and the president watched over them with a keen eye. On the night of February 5, he left a White House reception and went upstairs to their sickbeds. His boys were not improving, and during the next two weeks, they became grievously ill. The *Evening Star* began publishing daily reports on their condition.

> February 18: "NO RECEPTION TO-NIGHT—The continued indisposition of the President's children, one of whom, Willie, we regret to say, is extremely ill, will prevent the usual Tuesday night's reception at the Executive Mansion from taking place."
>
> February 19: The paper reports that Willie continues critically ill, but that Tad has not yet been dangerously sick. "Everything that skillful physicians—Drs. Hall and Stone—and ceaseless care can do for the little sufferers, is being done."
>
> February 20: "BETTER.—We are glad to say that the President's second son—Willie—who has been so dangerously ill seems better to-day."

The last report proved wrong. Willie was delirious, and he died the afternoon of February 20, at 5:00 P.M. Lincoln cried out to his secretary, John G. Nicolay: "Well, Nicolay, my boy is gone—he is actually gone!"

On the morning of February 21, all the members of the cabinet called upon the president and later that day signed a joint letter addressed to the Senate and House of Representatives asking Congress to cancel the annual Washington's birthday illumination of the public buildings, scheduled for the next night. William H. Seward, Salmon P. Chase, Edwin M. Stanton, Gideon Welles, Edward Bates, and Montgomery Blair wrote that the president "had been plunged into affliction" by his son's death.

On February 22, in a story headlined "Little Willie Lincoln," the *Evening Star* reported the sad details of the boy's death. After Willie was embalmed, Lincoln viewed his son's body in the Green Room.

That same day in Richmond, which had replaced Montgomery, Alabama, as the new Confederate capital, Willie Lincoln's death did not postpone the February 22 inauguration of Jefferson Davis, which was scheduled specifically on George Washington's birthday. It was a glorious and auspicious day for Davis. The Confederate president saw himself not as a rebel or a traitor but as the true inheritor of the legacy of George Washington and the revolutionary generation. He believed it was the Southern Confederacy, not the federal Union, that upheld the spirit of 1776. That evening, Davis was feted at a wonderful party in the White House of the Confederacy.

In Washington, at Lincoln's White House, an opportunistic office-seeker made the mistake of intruding upon Lincoln's anguish to request a petty postmaster's position. Like George Washington, Lincoln had made it a lifelong habit to control his temper, and only rarely did he show anger in public. But if pushed too far, Lincoln would on occasion explode. This was one of those moments.

"When you came to the door here, didn't you see the crepe on it?" Lincoln demanded. "Didn't you realize that meant somebody must be lying dead in this house?"

"Yes, Mr. Lincoln, I did. But what I wanted to see you about was very important."

"That crepe is hanging there for my son; his dead body at this moment is lying unburied in this house, and you come here, push yourself in with such a request! Couldn't you at least have the decency to wait until after we had buried him?"

The president asked his old friend Orville Hickman Browning to be in charge of the funeral arrangements and burial. Browning rode in a carriage to Georgetown's Oak Hill Cemetery to inspect the family tomb that the clerk of the United States Supreme Court, William T. Carroll, had offered as Willie's temporary resting place until the president could take him home to Illinois. It was also Oak Hill where Jefferson Davis had buried his son Samuel Emory Davis, who died of illness on June 13, 1854, when he was less than two years old.

On February 23, friends and family viewed Willie's body at the White House. On February 24, the day of Willie's funeral, the government offices were closed, as if an important man of state had died. All official Washington knew the boy. Members of the cabinet, foreign ministers, members of Congress, military officers, and other important Washingtonians attended his funeral.

The *Evening Star* published a heartbreaking description of the scene:

> His remains were placed in the Green room at the Executive mansion, where this morning a great many friends of the family called to take a last look at the little favorite, who had endeared himself to all guests of the family. The body was clothed in the usual every-day attire of youths of his age, consisting of pants and jacket with white stockings and low shoes—the white collar and wristbands being turned over the black cloth of the jacket.

The countenance wore a natural and placid look, the only signs of death being a slight discoloration of the features.

The body lay in the lower section of a metallic case, the sides of which were covered by the winding sheet of white crape. The deceased held in the right hand a boquet composed of a superb camellia, around which were grouped azalias and sprigs of mignionette. This, when the case is closed, is to be reserved for the bereaved mother. On the breast of the deceased, was a beautiful wreath of the flowers, already named, interspersed with ivy leaves and other evergreens; near the feet was another wreath of the same kind, while azalias and sprigs of mignionette were disposed about the body.

The metallic case is very plain, and is an imitation of rosewood. On the upper section is a square silver plate, bearing, in plain characters, the simple inscription:

William Wallace Lincoln.
Born December 21st, 1850.
Died February 20th, 1862.

The mirrors in the East Room, the Green Room, and all the other reception rooms were covered with mourning drapery, the frames wrapped with black and the glass concealed by white crepe. It was impossible to see a reflection. It was Lincoln's wish that Willie's body remain in the Green Room and not be moved to the East Room for the funeral service, which was conducted by Rev. Dr. Gurley.

Gurley described Willie as "a child of bright intelligence and peculiar promise." The minister listed his qualities: "His mind was active, inquisitive, and conscientious; his disposition was amiable and affectionate; his impulses were kind and generous; and his words and manners were gentle and attractive." Everyone who knew the boy, Gurley continued, loved him: "It is easy to see how a child, thus endowed, would, in the course of eleven years, entwine himself around the hearts of those who knew him best."

The president, who could usually speak with pride about his ability to master his emotions, could not contain himself. Willie, he said, "was too good for this earth . . . but then we loved him so. It is hard, hard to have him die!" Willie's death seemed to summon forth his father's accumulated, buried pain from a lifetime's worth of losses. "This is the hardest trial of my life," Lincoln moaned. "Why is it? Oh, why is it?" He was coming apart. No one in Washington had known Lincoln during the old New Salem days three decades ago. If any friends from that ghost town of Lincoln's long-lost past had been present at Willie's funeral, they would have recognized the familiar signs that made them fear for Lincoln's mind and life thirty years ago, after the death of Ann Rutledge.

Most of the guests in the East Room joined the procession to Georgetown. At Oak Hill, Willie's coffin was carried into the small chapel, where the Rev. Dr. Gurley performed a brief service. After the funeral guests went home, Willie's casket was hidden below the floorboards of the chapel in a subterranean storage pit until graveyard workers carried him to the Carroll vault.

Lincoln prayed that Tad, still sick, would be spared. On February 26, the *Evening Star* reported that he would live: "We are glad to learn that the youngest son of the President is still improving in health, and is now considered entirely out of danger from the disease which prostrated him." The *Star* went beyond reporting of the facts, and in an editorial beseeched its readers to consider the president: "Death has invaded the home of our Chief Magistrate, 'whose heart is torn.' Let the people stop to shed a tear with the President, who has so nobly earned their regard."

In the days ahead Abraham and Mary mourned Willie in different ways. Mary sought relief in the world of dreams and spirits. "He comes to me every night," she swore to her sister Emilie Todd Helm. "He comes to me . . . and stands at the foot of my bed with the same sweet, adorable smile he has always had; he does not always come alone; little Eddie is sometimes with him and twice he has come with our brother Alec, he tells me he loves his Uncle Alec and is with him most of the time. You cannot dream of the comfort this gives me. When I thought of my little son in immensity, alone, without his mother to direct him, no one to hold his little hand in loving guidance, it nearly broke my heart." Soon Mary would call upon spiritualists and mediums to cross over to Willie's realm. Mary banished from her sight all earthly reminders of her dead son. She disposed of his toys and forbade his playmates to return to the White House to play with Tad. The sight of them, she said, upset her too much.

No ghosts came to Lincoln's bedchamber. Willie had died on a Thursday, and for several weeks, the president locked himself in his office every Thursday for a time to mourn and to conjure up memories of his son. No one dared intrude upon these reveries. And at night he dreamed of his lost boy.

Lincoln loved to read passages from literature aloud to his friends. One day in May, he recited lines from Shakespeare's *King John*. "And, Father Cardinal, I have heard you say / That we shall see and know our friends in heaven. / If that be true, I shall see my boy again." Then he wept uncontrollably.

Death also visited Jefferson Davis's White House. On the afternoon of Saturday, April 30, 1864, an officer walking near the Confederate White House saw a crying young girl run out of the mansion and yank violently on the bell cord of the house next door. Then another girl and a boy fled the White House. A black female servant who followed them told the officer that one of the Davis children was badly hurt. The officer ran inside and found a male servant holding in his arms a little boy, "insensible and almost dead." It was five-year-old Joseph Evan Davis. His brother, Jeff Jr., was kneeling beside him, trying to make him speak. "I have said all the prayers I know," said Jeff, "but God will not wake Joe." Jefferson and Varina were not home.

Joseph had fallen fifteen feet from a porch. He was found lying on the brick pavement, unconscious, with a broken left thigh and a severely contused forehead. His chest evidenced signs of internal injuries. The officer sent for a doctor and then began to rub the boy with camphor and brandy, and applied mustard on his feet and wrists. The child, he observed, "had beautiful black eyes and hair, and was a very handsome boy." The treatment, wrote the officer in a letter a few days after the event, seemed to work: "In a short time he began to breathe better, and opened his eyes, and we all thought he was reviving, but it was the last bright gleaming of the wick in the socket before the light is extinguished for ever."

Messengers summoned the president and Varina. When she saw Joseph, she "relieved herself in a flood of tears and wild lamentations." Jefferson kneeled beside his son, squeezed his hands, and watched him die. The Confederate officer, whose name remains unknown to this day, described the president's appearance: "Such a look of petrified, unutterable anguish I never saw. His pale, intellectual face . . . seemed suddenly ready to burst with unspeakable grief, and thus transfixed into a stony rigidity." Almost thirty years earlier, watching Knox Taylor die had driven him into his "great seclusion." He could not indulge in private grief now. His struggling nation needed him. Davis mastered his emotions in public, but his face could not hide them. "When I recall the picture of our poor president," wrote the officer, "grief-stricken, speechless, tearless and crushed, I can scarcely refrain from tears myself."

That night family friends and Confederate officials called at the mansion, but Jefferson Davis refused to come downstairs. Above their heads, guests could hear his creaking footsteps on the floorboards as he paced through the night. Mary Chestnut remembered "the tramp of Mr. Davis's step as he walked up and down the room above—not another sound. The whole house [was] as silent as death." The funeral at St. Paul's Church, reported the newspapers, drew the largest crowd of any public event in Richmond since the beginning of the war. Hundreds of children packed the pews, each carrying a green bough or flowers to lay upon Joe's grave. Later, Davis had the porch torn down.

In December 1862, Lincoln received word that Lieutenant Colonel William McCullough, the former clerk of the McLean County Circuit Court in Bloomington, Illinois, had been killed in action on December 5, and that his teenage daughter was overcome with grief. On December 13, in Fredericksburg, Virginia, the Army of the Potomac suffered terrible casualties in a series of futile infantry charges against Confederate troops sheltered behind stone walls. It was an ill-conceived, costly, senseless, and even shameful performance by General Ambrose Burnside. Two days before Christmas, on a day Lincoln might have taken Willie—gone ten months now—to his favorite toy store on New York Avenue, and while Mary worked downstairs with the White House staff making final arrangements for serving Christmas Day dinner to wounded soldiers, the president thought of another child and wrote a condolence letter to Fanny McCullough.

Lincoln sat at the big table in his second-floor office, reached for an eight-by-ten-inch sheet of lined paper bearing the engraved letterhead "Executive Mansion," and began to write. What came from his pen was more than a polite and perfunctory note. In one of the most moving and revealing letters he ever wrote, Lincoln set down his hard-earned knowledge of life and death for an inexperienced girl. It was as if Lincoln had composed the letter not to one sad girl but to the American people.

Washington,
December 23, 1862

Dear Fanny

It is with deep regret that I learn of the death of your kind and brave Father; and, especially, that it is affecting your young heart beyond what is common in such cases. In this sad world of ours, sorrow comes to all; and, to the young, it comes with bitterest agony, because it takes them unawares. The older have learned to ever expect it. I am anxious to afford some alleviation of your present distress. Perfect relief is not possible, except with time. You can not now realize that you will ever feel better. Is not this so? And yet it is a mistake. You are sure to be happy again. To know this, which is certainly true, will make you some less miserable now. I have had experience enough to know what I say; and you need only to believe it, to feel better at once. The memory of your dear Father, instead of an agony, will yet be a sad sweet feeling in your heart, of a purer and holier sort than you have known before.

Please present my kind regards to your afflicted mother.

Your sincere friend,
A. Lincoln

Two days later, on their first Christmas without Willie, Abraham and Mary Lincoln, remembering their lost boy, and recalling Fanny McCullough and all the men who fell at Fredericksburg, and perhaps all the fallen men from all the battles, left the White House on Christmas Day and rode in their carriage from hospital to hospital, visiting wounded soldiers.

There was more grief to come. No wartime funeral in Washington had prepared the population—or the president—for the sensational catastrophe of June 17, 1864. It happened while Lincoln was returning to the capital aboard a special 8:00 A.M. train from Philadelphia, where the day before he had attended the Great Central Fair to benefit the U.S. Sanitary Commission, an organization that aided wounded soldiers.

On the morning of the seventeenth, as Lincoln's train steamed south to Washington, more than one hundred young women were at work in the so-called "laboratory" of the U.S. Arsenal, making small-arms ammunition. The room was filled with unstable, combustible black powder. Outside the building, someone had set out several pans of fireworks to dry in the sunlight. At ten minutes before noon, a pan of fireworks ignited and cast a spark through an open window into the laboratory.

The president, a lifelong newspaper addict, must have read in the afternoon editions of the *Evening Star* what happened next: "After the powder on the benches caught, the fire spread down rapidly, blinding the girls and setting fire to their clothes. Many of them ran to the windows wrapped in flames, and on their way communicated the fire to the dresses of others."

The fire, followed by a terrific explosion, caused male workers on the grounds to sprint to the laboratory from all directions. Some of the men wrapped the fleeing, burning girls in tarpaulins to extinguish the flames. Other men gathered the girls up in their arms and ran for the river: "One young lady ran out of the building with her dress all in flames, and was at once seized by a gentleman, who, in order to save her, plunged her into the river. He, however, burned his arms and hands badly in the effort. Three others, also in flames, started to run up the hill, the upper part of their clothing was torn off by two gentlemen near by, and who thus probably saved the girls from a horrific death, but in the effort, they too were badly injured."

Desperate arsenal workers searched the debris for survivors. They knew these girls and had flirted with some of them. In an undated photograph taken some time before the explosion, a group of the women, dressed in bright, pretty hoopskirts and joined by several of the men, posed on the front porch of the laboratory. Now, in the ruins, they found only the dead. "The bodies were in such a condition that it was found necessary to place boards under each one in order to remove them from the ruins . . . they were carried out and placed upon the ground." Unsupported, the burned corpses would have crumbled and broken into pieces. The "charred remains of those who had perished," the *Evening Star* reported, "were laid upon the ground and covered over with canvass."

The *Star*'s reporter rushed back to newspaper row to file his story in time to make the 2:30 P.M. edition: "When our reporter left the scene of the disaster nineteen bodies had been taken from the ruins, but they were so completely burnt to a crisp that recognition was impossible." The survivors were "frightfully" wounded.

A little after 4:00 P.M., the coroner arrived to examine the dead. "The canvas covering the remains was then removed, and the most terrible sight presented itself to the view of those standing around. The charred remains of seventeen dead bodies lay scattered about, some in boxes, some on pieces of boards, and some in large tin pans, they having been removed from the ruins in these receptacles. In nearly every case only the trunk of the body remained, the arms and legs being missing or detached. A singular feature of the sad spectacle was that presented by a number of bodies nearly burned to a cinder being caged, as it were, in the wire of their hooped skirts . . . Many of the bodies seem to have been crisped quite bloodless."

The scene was like a battlefield field hospital littered with the grisly evidence of amputations. "In a box was collected together a large number of feet, hands, arms and legs, and portions of the bones of the head, which it would be impossible to recognize."

One woman was identified by her boots. Another still wore a fragment of blouse or skirt, and "her remains were subsequently recognized by a portion of dress which remained upon her unconsumed. The whole top of her head was, however, gone, and the brain was visible; and but for the fragment of dress it would have been impossible to recognize her."

The youth of the victims—one was just thirteen years old—and the horrific nature of their hideous injuries shocked the city. "Seventeen Young Women Blown to Atoms" said the headline of the *Daily Morning Chronicle* the next day.

The funeral service, an outdoor ceremony to be held on the site of the tragedy, was scheduled for Sunday, June 19. The arsenal's master carpenter needed time to make proper coffins. He also built a wood pavilion measuring twenty by fifteen feet and standing three feet off the ground. Upon it fifteen coffins lay side by side. Twenty-five thousand people, including President Lincoln—described by the press as "mourner in chief"—and Secretary of War Stanton, assembled on the arsenal grounds. After the service, the burial procession left the arsenal at 3:00 P.M., moved up Four and a Half Street, and then along Pennsylvania Avenue to the Congressional Cemetery. Lincoln's carriage followed the hearses. He had come this way before, first for John Quincy Adams and then for Colonel Edward Baker.

At Fourth Street, the small funeral procession of thirteen-year-old Sallie McElfresh joined the main procession. "Her body," reported the *Daily Morning Chronicle,* "was encased in a splendid coffin, decorated with wreaths, which was carried in a beautiful modern child's hearse." Lincoln could not have avoided seeing Sallie's tiny coffin.

At the cemetery, two large burial pits—each one six feet long, fifteen feet wide, and five and a half feet deep—had been dug six feet apart on the west side of the cemetery. The dead had been divided into two groups, the known and the unknown, and they would be buried that way. Male employees of the arsenal handled the ropes and lowered each coffin, one at a time, into its grave. The crowd was dense, and as it pressed forward many women had their dresses torn in the scrum. Police held the throng at bay to allow the families to approach the pits. There, reported the *Evening Star,* "was another scene of anguish—the relatives, or many of them, giving way to loud cries, and hanging over the chasm, calling the deceased by their names."

The ministers read services for the dead, and the crowd repeated the chant "Farewell, sisters, farewell." Standing nearby, Lincoln did not speak publicly that day. It was the biggest funeral he had ever seen. Yes, seven months earlier he had spoken at the dedication of the new national military cemetery at Gettysburg, a battlefield where thousands had perished, but that was not a funeral, and the men he honored there had been long buried. The arsenal tragedy was fresh, its wounds raw. Not one of the Washington papers commented on Lincoln's demeanor at Congressional Cemetery or described how he reacted when the girls were lowered into the ground. Surviving accounts do no more than note his presence. Later, as best can be told, Lincoln never spoke or wrote of what he saw this day.

That evening the president, accompanied by his secretary John Hay, went to Ford's Theatre to attend a concert of sacred music. Abraham Lincoln often went to the theater when he wanted to forget.

While Lincoln's body lay in the East Room on the night of April 17, and while thousands mourned and prepared for the next day's public viewing, elsewhere in Washington one man gloated over his harvest of Lincoln blood relics. Mose Sandford, one of the men at the War Department hardware workshop who had built Lincoln's temporary pine-box coffin to transport his body from the Petersen house to the White House on the morning of April 15, wrote a letter to a friend, describing how he plundered Lincoln's possessions from the temporary Petersen house coffin. "I found one of the sleeves of his shirt one of his sleeve buttons," he wrote, "black enameled trimmed with gold and the letter 'L' on the out side with 'A.L.' underneath that I sent to the Sect of War. The Bosom of his shirt was the next thing which met my eye as it had considerable blood upon it so I just confiscated the whole of it." He even took the screws that had held down the box's lid.

On April 17 Jefferson Davis, on the way to Charlotte, spent the night in Salisbury, North Carolina. Seventy-two hours had elapsed since

Lincoln was assassinated, and still Davis had no news of the events in Washington.

Nor did he know that on this night, and the next morning, Union general William T. Sherman contemplated what should be done about Davis's future. On the seventeenth, Sherman met with most of his generals to discuss Confederate general Joe Johnston's army in North Carolina and to analyze the meeting Sherman had attended the day before with Johnston at the Bennett house to discuss that army's possible surrender. But Sherman and his staff also talked about the Confederate president.

"We discussed . . . whether, if Johnston made a point of it, I should assent to the escape from the country of Jeff. Davis and his fugitive cabinet; and some one of my general officers, either Logan or Blair, insisted that, if asked for, we should even provide a vessel to carry them to Nassau from Charleston."

Like Abraham Lincoln, Sherman would not have been disappointed if Jefferson Davis escaped the Union's pursuit and fled the country.

In Salisbury, Davis received a letter signed by several Confederate officers begging his permission to disband their command and send their men home. They wanted to quit the war. If Davis agreed, news of it would spread like a contagion and infect the whole army. Soon every man would want to go home, and the South would lose the war. The Confederate president replied: "Our necessities exclude the idea of disbanding any portion of the force which remains to us and constitutes our best hope of recovery from the reverses and disasters to which you refer. The considerations which move you to the request are such, if generally acted on, would reduce the Confederate power to the force which each State might raise for its own protection. On the many battle-fields within the limits of your State the sons of other States have freely bled . . ."

Didn't these men know that Davis also worried about his own wife and children? Moreover, the Confederacy's survival was at stake.

He continued writing. "My personal experience enables me fully to sympathize with your anxieties for your homes and for your families, but I hope I have said enough to satisfy you that I cannot consistently comply with your request, and that you will agree that duty to the country must take precedence of any personal desire."

Davis's morale remained high. Burton Harrison witnessed it firsthand: "During all this march Mr. Davis was singularly equable and cheerful; he seemed to have had a great load taken from his mind, to feel relieved of responsibilities, and his conversation was bright and agreeable . . . He talked of men and of books, particularly of Walter Scott and Byron; of horses and dogs and sports; of the woods and the fields; of trees and many plants; of roads, and how to make them; of the habits of birds, and of a variety of other topics. His familiarity with, and correct taste in, the English literature of the last generation, his varied experiences in life, his habits of close observation, and his extraordinary memory, made him a charming companion when disposed to talk."

Although they had evacuated Richmond more than two weeks earlier, Harrison observed that Davis's entourage shared his optimism: "Indeed . . . we were all in good spirits under adverse circumstances."

On the morning of April 18, the White House gates opened to admit the throng that had waited all night to file into the East Room to view the president's remains. Upstairs, Mary Lincoln and Tad remained in seclusion in her room. He would have liked to have seen the people who came to honor his father. He would, perhaps, have found more comfort in the consoling company of these loving strangers than in the secluded and unwholesome bedchamber of his unstable mother.

For the past three days the newspapers had been saturated with accounts of the president's assassination and death. Today was the people's first chance to come face-to-face with his corpse. While the public viewing was under way, as thousands of people walked past the coffin, with the White House funeral less than twenty-four hours away, George Harrington was trying to locate Bishop Matthew Simpson, who was in Philadelphia, to let him know he was expected to speak tomorrow at the president's funeral.

The Philadelphia Telegraph Office responded to Harrington's telegraph: "Bishop Simpson was not at home and his daughter says she cannot answer it. She says he is going to Washington tonight? Respectfully / H.B. Berry / Manager / American Telegraph Office."

Eventually the divine's family dispatched a telegram from Philadelphia to Harrington: "Bishop Simpson is absent from home he will be in Washington City tomorrow morning. E M Simpson." Then another telegram arrived, this one from Simpson: "Just received your invitation. Am willing to assist. What part of the services am I expected to take. M. Simpson." If his train was not delayed, he would arrive the next day, just in time. He would prepare his text through the night and during his train ride. There would be no time once he arrived in Washington.

By this point, Harrington was becoming overwhelmed by a last-minute deluge of requests for funeral tickets, press passes to the White House, and permission to march in the procession. For every request Harrington disposed of, another came in the door.

Not all of Harrington's correspondents demanded special favors. Some offered helpful advice. "Pardon me for suggesting that as few carriages as possible ought to be allowed in the funeral cortege of the President. There are one hundred thousand aching hearts that will be following his remains to the grave. This cannot be done if long lines of vehicles occupy the space, without adding to the volume of humanity desirous of participating." The anonymous letter was signed "Affectionately." The same writer also sent a note to Stanton on April 18 suggesting that streetcar noise might disturb the next day's funeral events: "The running of cars upon the street railroads, between 17th Street and the Congressional Cemetery, should cease tomorrow between 11 A.M., and 4 P.M. The rolling of cars, and the jingle of bells will contrast strangely with the solemnity of those sacred hours."

As the final visitors filed past the coffin, carpenters loitered nearby, impatient to start work the moment the last citizen exited the White House and the doors were shut and locked behind them. If the public had its way, the viewing would have continued through the night. Thousands of people were turned away so the crews could begin preparing the East Room for the funeral. Disappointed mourners would have one more chance to view the remains, after they were transferred to the Capitol.

Harrington had come up with an ingenious solution to the seating dilemma. He would not seat the guests in chairs at all. He had calculated that, allowing for the space required for the catafalque and the aisles, it was impossible to squeeze six hundred chairs into the East Room. He decided that only a few of the most important guests, including the Lincoln family, needed to have individual chairs. But if he built risers, or bleachers, for the rest, he could pack slightly more than six hundred people into the East Room, the minimum number of important guests he needed to seat. The White House was abuzz with activity—men carried stacks of fresh lumber into the East Room, where carpenters sawed, hammered, and nailed them into bleachers.

On the morning of April 19, cities across the North prepared to hold memorial services at the same time as the Washington funeral. Military posts across the nation marked the hour. At the White House, journalist George Alfred Townsend was among the first guests to enter the East Room that morning. As one of the most celebrated members of the press, he was allowed to approach Lincoln's corpse. His account of the event invited his readers to do the same:

Approach and look at the dead man. Death has fastened into his frozen face all the character and idiosyncrasy of life. He has not changed one line of his grave, grotesque countenance, nor smoothed out a single feature. The hue is rather bloodless and leaden; but he was always sallow. The dark eyebrows seem abruptly arched; the beard, which will grow no more, is shaved close, save for the tuft at the short small chin. The mouth is shut, and like that of one who has put the foot down firm, and so are the eyes, which look as calm as slumber. The collar is short and awkward, turned over the stiff elastic cravat, and whatever energy or humor or tender gravity marked the living face is hardened into its pulseless outline. No corpse in the world is better prepared according to its appearances. The white satin around it reflects sufficient light upon the face to show that death is really there; but there are sweet roses and early magnolias, and the balmiest of lilies strewn around, as if the flowers had begun to bloom even in his coffin . . .

Three years ago, when little Willie Lincoln died, Doctors Brown and Alexander, the embalmers or injectors, prepared his body so handsomely that the President had it twice disinterred to look upon it. The same men, in the same way, have made perpetual these beloved lineaments. There is now no blood in the body; it was drained by the jugular vein and sacredly preserved, and through a cutting on the inside of the thigh the empty blood-vessels were charged with a chemical preparation which soon hardened to the consistence of stone. The long and bony body is now hard and stiff, so that beyond its present position it cannot be moved any more than the arms or legs of a statue. It has undergone many changes. The scalp has been removed, the brain taken out, and the chest opened and the blood emptied. All that we see of Abraham Lincoln, so cunningly calculated in this splendid coffin, is a mere shell, an effigy, a sculpture. He lies in sleep, but it is the sleep of marble.

All that made this flesh vital, sentient, and affectionate is gone forever.

The morning of the funeral, requests were still being made by politicians and leading citizens hopeful that the funeral train would pass through their locale. C. W. Chapin, president of the Western Railroad Corporation, sent an urgent telegram to Massachusetts congressman George Ashmun, one of the last men to see Lincoln alive at the White House the evening of April 14, pleading with him to use his influence to divert the funeral train to New England: "In no portion of our common country do the people mourn in deeper grief than in New England," he wrote. "This slight divergence will take in the route the capital of Connecticut and also important points in Massachusetts."

To thwart gate-crashers, funeral guests were not allowed direct entry to the Executive Mansion. Instead, guards directed the bearers of the six hundred coveted tickets, printed on heavy card stock, next door, to the Treasury Department. From there they crossed a narrow,

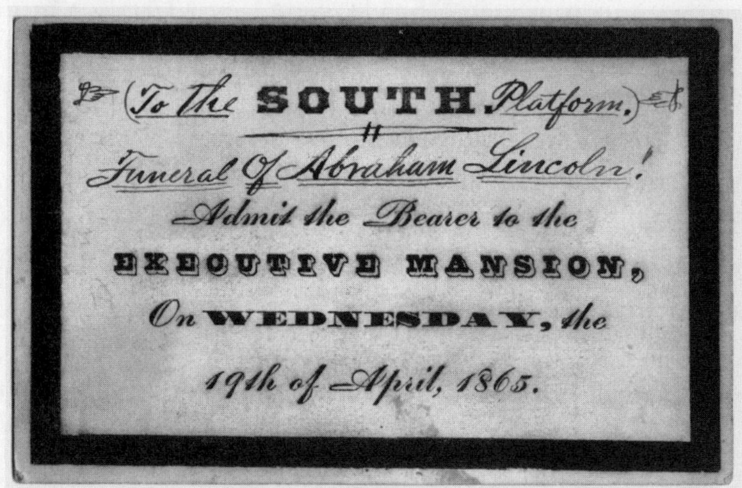
(To The SOUTH Platform.)
Funeral Of Abraham Lincoln!
Admit the Bearer to the
EXECUTIVE MANSION,
On WEDNESDAY, the
19th of April, 1865.

One of the few surviving invitations to Lincoln's White House funeral.

elevated wooden footbridge, constructed just for the occasion, that led into the White House. A funeral pass became the hottest ticket in town, more precious than tickets to Lincoln's first and second inaugural ceremonies or balls, and more desirable than an invitation to one of Mary Lincoln's White House levees. The only ticket to surpass the rarity of a funeral invitation had not been printed yet and would be for an event not yet scheduled—the July 7, 1865, execution of Booth's coconspirators, for which only two hundred tickets would be issued.

Just hours before the funeral, George Harrington was still receiving last-minute ticket requests: "Surgeon General's Office / Washington City, D.C. / April 19. / Dear Sir / Please send me by [messenger] tickets for myself & Col. Crane my executive Officer. JW Barnes, Surgeon General." Harrington could not deny tickets to the chief medical officer of the U.S. Army and senior doctor at the Petersen house, or to his deputy.

As the guests entered the Executive Mansion, none of them knew what to expect. The East Room overwhelmed them with its decorations and flowers and the catafalque. It was an unprecedented scene. Two presidents had died in office, William Henry Harrison in 1841 and Zachary Taylor in 1850, but their funerals were not as grand or elaborate as this. No president had been so honored in death, not even George Washington, who, after modest services, rested in a simple tomb at Mount Vernon.

The scene lives on only in the written accounts of those who were there, and in artists' sketches and newspaper woodcuts. Somebody should have taken a photograph. Sadly, no photographs were made in the East Room before or during the funeral. It could have been done. Alexander Gardner had photographed more complex scenes than this, including the second inaugural, where he took crystal-clear close-ups of the East Front platform and one of his operators had managed to take a long view of the Great Dome while Lincoln was reading his address. Edwin Stanton failed to invite Gardner or his rival Mathew Brady to preserve for history the majesty of Abraham Lincoln's funeral.

At exactly ten minutes past noon, a man arose from his chair, approached the coffin, and in a solitary voice broke the hush. The Reverend Mr. Hall intoned the solemn opening words of the Episcopal burial service: "I am the resurrection and the life, saith the Lord; he that believeth in me, though he were dead, yet shall he live; and whosoever liveth and believeth in me shall never die."

Then Bishop Matthew Simpson, who had arrived on time, spoke. He was followed by Lincoln's own minister, the Reverend Dr. Gurley, who delivered his sermon.

> It was a cruel, cruel hand, that dark hand of the assassin, which smote our honored, wise, and noble President, and filled the land with sorrow. But above and beyond that hand, there is another, which we must see and acknowledge. It is the chastening hand of a wise and faithful Father. He gives us this bitter cup, and the cup that our Father has given us shall we not drink it?
>
> He is dead! But the God whom he trusted lives and He can guide and strengthen his successor as He guided and strengthened him. He is dead! But the memory of his virtues; of his wise and patriotic counsels and labors; of his calm and steady faith in God, lives as precious, and will be a power for good in the country quite down to the end of time. He is dead! But the cause he so ardently loved . . . That cause survives his fall and will survive it . . . though the friends of liberty die, liberty itself is immortal. There is no assassin strong enough and no weapon deadly enough to quench its inexhaustible life . . . This is our confidence and this is our consolation, as we weep and mourn today; though our President is slain, our

beloved country is saved; and so we sing of mercy as well as of judgment. Tears of gratitude mingle with those of sorrow, while there is also the dawning of a brighter, happier day upon our stricken and weary land.

Mourning ribbon worn in Washington by Post Office workers for the April 19, 1865, funeral procession.

While the three ministers held forth for almost two hours, more than a hundred thousand people waited outside the White House for the funeral services inside to end. In the driveway of the Executive Mansion, six white horses were harnessed to the magnificent hearse that awaited their passenger. Nearby, more than fifty thousand marchers and riders had assembled in the sequence assigned to them by the War Department's printed order of procession. Another fifty thousand people lined Pennsylvania Avenue between the Treasury building and the Capitol. According to the *New York Times*, "the throng of spectators was . . . by far the greatest that ever filled the streets of the city."

Most of the people outside the White House wore symbols of mourning: black badges containing small photographs of Lincoln, white silk ribbons bordered in black and bearing his image, small American flags with sentiments of grief printed in black letters and superimposed over the stripes, or just simple strips of black crepe wrapped around coat sleeves. Some mourners had arrived by sunrise to stake out the best viewing positions. By 10:00 A.M. there were no more places left to stand on Pennsylvania Avenue.

The hearse that carried Lincoln's body down Pennsylvania Avenue.

Faces filled every window, and children and young men climbed lampposts and trees for a better view. The city was so crowded with out-of-towners that the hotels were filled and many people had to sleep along the streets or in public parks.

By the time the White House funeral services ended and the procession to the Capitol got under way, the people had been waiting for hours. It was a beautiful day. Four years earlier, on inauguration day, March 4, 1861, General Winfield Scott had placed snipers on rooftops overlooking Pennsylvania Avenue to protect president-elect Lincoln, who had received many death threats from boastful would-be assassins. No marksmen were needed today. This afternoon the only men on rooftops were spectators seeking a clear view of Lincoln's hearse.

At 2:00 p.m., soldiers in the East Room surrounded the coffin, lifted it from the catafalque, and carried Abraham Lincoln out of the White House for the last time. They placed the coffin in the hearse. Funeral guests designated to join the procession took their places in the line of march. Everything was ready. Soon, for the first time since the morning of April 15, when the soldiers took him home, Lincoln was on the move again. This second procession on April 19 dwarfed the simple one that had escorted him from the Petersen house to the White House.

Cannon fire announced the start of the procession: Minute guns boomed near St. John's Church, City Hall, and the Capitol. At churches and firehouses, every bell in Washington tolled. They tolled too in Georgetown and Alexandria. In later years, witnesses recalled the sound of the day as much as the sight of it. Tad Lincoln emerged from seclusion to join the procession, and he and his brother Robert rode in a carriage behind the hearse. The mile-and-a-half trip between the White House and the Capitol took Abraham Lincoln past familiar places—the Willard Hotel, where he had spent his first night in Washington as president-elect; Mathew Brady's studio, where he had gone to pose for many photographs. His hearse also carried him past the National Hotel, where he had once spoken from the balcony to a regiment of Union soldiers, and where his assassin had spent his last night on the eve of the murder. The procession was immense and included every imaginable category of marcher: military officers from the army, navy, and Marine Corps; enlisted men; civil officials; judges; diplomats; and doctors—"physicians to the deceased," read the printed program. One corps of marchers suggested the cost of the war: wounded and bandaged veterans, many of them amputees missing arms or legs, many on crutches. The procession was immense and took two hours to pass a given point. Indeed, when the front of the column reached the Capitol, the rear had still not cleared the Treasury Department.

George Alfred Townsend described the arrival of the remains at the Capitol:

"The cortege passed to the left [north] side of the Capitol, and entering the great gates, passed the grand stairway, opposite the splendid dome, where the coffin was disengaged and carried up the ascent. It was posted under the bright concave, now streaked with mournful trappings, and left in state, watched by guards of officers with drawn swords. This was a wonderful spectacle, the man most beloved and honored in the ark of the republic . . . Here the prayers and addresses of the noon were rehearsed and the solemn burial service read."

When the soldiers carried the flag-draped coffin up the stairs, they walked past the very spot where Lincoln had delivered his second inaugural address. That day began stormy, but as Lincoln began to speak, the sun burst through the clouds. On the day of his funeral it was beautiful and clear. The crowds watched in silence as the soldiers carried the coffin inside and laid it upon a catafalque in the center of the rotunda. Many of the people waiting to enter the Capitol to view the corpse had been there six weeks earlier and had heard Lincoln deliver his inaugural address. They marveled at this terrible reversal of fortune.

CHAPTER SEVEN

"The Cause Is Not Yet Dead"

When Jefferson Davis rode into Charlotte on April 19, its citizens were not happy to see him. North Carolina had sent more men into action in the battle at Gettysburg than any Confederate state but Virginia—they suffered heavy casualties in Pickett's Charge—and the people of Charlotte no longer felt the enthusiasm for the Confederacy that the state's valiant sons had demonstrated at Gettysburg two years earlier. Only one man, Lewis F. Bates, a transplanted Yankee, would allow Davis to set foot in his home.

An officer explained to Burton Harrison the reason for this embarrassing lack of Southern hospitality. While the people were willing to offer shelter to Davis's entourage, they were afraid that anyone who offered refuge to the president would later have his house burned down by Union cavalry raiders.

Harrison was dubious of Davis's would-be host, but "there seemed to be nothing to do but to go to the one domicile offered. It was on the main street of the town, and was occupied by Mr. Bates, a man said to be of northern birth, a bachelor of convivial habits, the local agent of the Southern Express Company, apparently living alone with his negro servants, and keeping with them a sort of 'open house,' where a broad, well equipped sideboard was the most conspicuous feature of the situation—not at all a seemly place for Mr. Davis." Davis would come to regret his stay with Mr. Bates.

Not long after he arrived in Charlotte, Davis gave a speech to an audience that included a number of Confederate soldiers in the city:

> My friends, I thank you for this evidence of your affection. If I had come as the bearer of glad tidings, if I had come to announce success at the head of a triumphant army, this is nothing more than I would have expected; but coming as I do, to tell you of a very great disaster; coming, as I do, to tell you that our national affairs have reached a very low point of depression; coming, I may say, a refugee from the capital of the country, this demonstration of your love fills me with feelings too deep for utterance. This has been a war of the people for the people, and I have been simply their executive; and if they desire to continue the struggle, I am still ready and willing to devote myself to their cause. True, General Lee's army has surrendered, but the men are still alive, the cause is not yet dead; and only show by your determination and fortitude that you are willing to suffer yet longer, and we may still hope for success. In reviewing my administration of the past four years, I am conscious of having committed errors, and very grave ones; but in all that I have done, in that I have tried to do, I can lay my hand upon my heart and appeal to God that I have had but one purpose to serve, but one mission to fulfill, the preservation of the true principles of constitutional freedom, which are as dear to me to-day as they were four years ago. I have nothing to abate or take back; if they were right then, they are right now, and no misfortune to our arms can change right into wrong. Again I thank you.

At the conclusion of the speech, somebody handed Davis a telegram just received from John C. Breckinridge. Davis read the words in silence: "President Lincoln was assassinated in the theatre in Washington on the night of the 11th inst. Seward's house was entered on the same night and he was repeatedly stabbed and is probably mortally wounded." Breckinridge had the date wrong, and Seward had survived. But Abraham Lincoln was dead.

John Reagan was there when Davis received the news: "At Charlotte . . . we received the melancholy news of the assassination of President Lincoln. [Davis] and members of the Cabinet, with one accord, greatly regretted the occurrence. We felt that his death was most unfortunate for the people of the Confederacy, because we believed that it would intensify the feeling of hostility in the Northern States against us, and because we believed we could expect better terms from Lincoln than from Johnson, who had shown a marked hostility to us, and was especially unfriendly to President Davis."

Stephen Mallory, Davis's secretary of the navy, was not present when Davis received the message, but they spoke about it a few minutes later. Mallory told Davis he did not believe it. The president said it sounded like a canard, but in revolutionary times events no less startling occurred constantly. Then Mallory expressed to Davis his "conviction of Mr. Lincoln's moderation, his sense of justice, and [Mallory's] apprehension that the South would be accused of instigating his death."

Davis replied, Mallory wrote, in a sad voice: "I certainly have no special regard for Mr. Lincoln; but there are a great many men of whose end I would much rather have heard than his. I fear it will be disastrous to our people, and I regret it deeply."

The myth took hold that Davis rejoiced at the news. Several weeks later, at the trial of Booth's accomplices, Lewis Bates swore under oath that Davis read Breckinridge's telegram aloud and then announced to the crowd: "If it were to be done, it were better it were well done." But it was a lie. Then Bates said that when Breckenridge and Davis met in his house a day or two later, and Breckenridge expressed regret at Lincoln's death, Davis disagreed: "Well, General, I don't know, if it were to be done at all, it were better that it were well done; and if the same had been done to Andy Johnson, the beast, and to Secretary Stanton, the job would then be complete." Bates's falsehoods dogged Davis for the rest of his life.

On April 19, when Varina wrote to Jefferson from Abbeville, she mentioned the "fearful news" that "fills me with horror"—but she wasn't referring to Lincoln's death, which she still didn't know about, but instead about the recent Confederate military disasters, which included the disbanding of General Lee's army and the surrender of General Longstreet's corps. Then she turned her attention to her husband and his well-being. "Where are you—how are you—What ought I to do with these helpless little unconscious charges of mine are questions which I am asking myself always. Write to me of your troubles freely for mercy's sake—Do not attempt to put a good face upon them to the friend of your heart, I am so at sea . . ."

On April 19, General Wade Hampton wrote to President Davis from Hillsborough, North Carolina, encouraging him to continue the fight from Texas. "Give me a good force of cavalry and I will take them safely across the Mississippi, and if you desire to go in that direction it will give me great pleasure to escort you. My own mind is made up as to my course. I shall fight as long as my Government remains in existence . . . If you will allow me to do so, I can bring to your support many strong arms and brave hearts—men who will fight to Texas, and who, if forced from that State, will seek refuge in Mexico rather than in the Union."

The crowds standing before the East Front of the Capitol would have to wait outside all night before they could see Lincoln. No visitors would be allowed to enter the rotunda until morning. As at the White House the night before the public viewing there, Lincoln would rest alone. Only his honor guard watched over him. Townsend described the scene: "At night the jets of gas concealed in the spring of the dome were lighted up, so that their bright reflection upon the frescoed walls hurled masses of burning light, like marvelous haloes, upon the little box where so much that we love and honor rested on its way to the grave. And so through the starry night, in the fane of the great Union he had strengthened and recovered, the ashes of Abraham Lincoln, zealously guarded, are now reposing."

If the doors to the rotunda had been thrown open that night, thousands of people would have poured in to see him. Lincoln's private, silent night in the rotunda was an intermission to the great drama, a pause that allowed the tension to build. For those thwarted by the long lines at the White House, the morning of April 20 would be their last chance to see him. Many onlookers who had crowded the East Front to watch Lincoln's coffin ascend the stairs did not leave the grounds once it vanished inside. Determined to keep an all-night vigil to guarantee their entry to the Capitol the next morning, thousands of people lined up on East Capitol Street.

In Charlotte, Davis tried to fathom all the implications of Lincoln's murder. He must have recalled the many wartime rumors of Yankee plots to kidnap or murder Davis in Richmond. Now it had happened to Lincoln. But who had killed him, and why? What did this news mean for his retreat and for his plans to continue the war? Davis had not heard about the assassination until five days after it happened and he did not know about the funeral and procession that had started Lincoln's transformation into an American saint. Lincoln's death had made Davis's cause even more difficult to sustain. Davis could not imagine the intensity of emotions unleashed by the assassination. But had he sensed it in any way, he might have increased the speed and urgency of his journey south.

Close to midnight on the day of Lincoln's White House funeral, Edwin Stanton telegraphed Major General Dix in New York City and told him that the route of the death pageant was now set. "It has been finally concluded," Stanton wrote, "to conform to the original arrangements made yesterday for the conveyance of the remains of the late President Abraham Lincoln from Washington to Springfield, viz, by way of Baltimore, Harrisburg, Philadelphia, New York, Albany, Buffalo, Cleveland, Columbus, Indianapolis, Chicago to Springfield."

The doors to the Capitol were thrown open on the morning of April 20. People passed between two lines of guards on the plaza, entered the rotunda via the East Front, split into two lines that passed on either side of the open coffin, and exited through the West Front. The experience was quick. Visitors were not allowed to linger, and they walked through the rotunda at the rate of more than three thousand an hour. At 10:00 A.M. a heavy rain soaked more than ten thousand people waiting in line. The mourners did not include the Lincoln family but did include Petersen house boarders George and Huldah Francis: "We saw him the last time in the Capitol the day before he was carried away . . ." In the dimly lit rotunda only the sound of rustling dresses and hoopskirts broke the silence. At 6:00 P.M. the doors were closed and the public viewing ended. The people, if permitted, would have kept coming all through the night.

Jefferson Davis awoke on the morning of April 20 with continued resolve. For him, Lincoln's death had changed nothing. Indeed, in Davis's mind, the ascendancy of Andrew Johnson made it more imperative to stave off defeat. If the South surrendered to Johnson, his vengeance would be more terrible than any suffered under Lincoln. Davis had made his decision: The Civil War would go on as long as he lived. But he must have also known that Lincoln's murder placed his life in greater danger. If he came under the hand of Union troops, he might be fated to join Lincoln in death.

While Lincoln lay in state in the Capitol rotunda, Stanton received a telegram from Governor Andrew Curtin of Pennsylvania, regarding Mary Lincoln:

> *HARRISBURG,*
> *April 20, 1865.*
>
> *Hon. E. M. STANTON,*
> *Secretary of War:*
> *I am as yet unadvised as to whether Mrs. Lincoln will accompany the remains. In case she does, will you oblige me by presenting my compliments to her, and say that I will of course expect herself and her family to make my house her home during her melancholy sojourn here. May I beg the favor of an answer?*
> *A. G. CURTIN*

Governor Curtin did not know Mary Lincoln's condition. Overwrought, she had still not left her room, or viewed the president's remains, and had not attended the White House funeral. She had refused almost all visitors, even close associates of the president from Illinois, including Orville Browning, and high officials of her husband's administration. She had already begun her final mental descent into postwar instability without the president to save her from drifting away.

Stanton replied quickly to Curtin's inquiry:

WAR DEPARTMENT,
Washington City,
April 20, 1865.

Governor CURTIN,
Harrisburg:

Your kind and considerate message will be immediately communicated to Mrs. Lincoln. By present arrangements neither she nor her sons will accompany the funeral cortege, she being unable to travel at present.

EDWIN M. STANTON,
Secretary of War.

Just nineteen hours before the train was scheduled to leave Washington, Stanton received a request to add Pittsburgh to the route. He replied to this, and all other last-minute requests, "The arrangements already being made cannot be altered . . ." Stanton did not announce his choice of men to travel on the train with the remains until April 20, the day before they were scheduled to depart. He also released two dramatic and historic documents. Their content could not have been more different. The first was the key order to Assistant Adjutant General Townsend establishing the protocols for the train that would transport the president's remains. Stanton was a stickler for detail, and the elaborate process he laid out left nothing to chance.

WAR DEPARTMENT,
Washington City,
April 20, 1865.

Bvt. Brig. Gen. E.D. Townsend,
Assistant Adjutant-General, U.S. Army:

SIR: You will observe the following instruction in relation to conveying the remains of the late President Lincoln to Springfield,

Ill. Official duties prevent the Secretary of War from gratifying his desire to accompany the remains of the late beloved and distinguished President Abraham Lincoln from Washington to their final resting place at his former home in Springfield, Ill., and therefore Assistant Adjutant-General Townsend is specifically assigned to represent the Secretary of War, and to give all necessary orders in the name of the Secretary as if he were present, and such orders will be obeyed and respected accordingly. The number of general officers designated is nine, in order that at least one general officer may be continually in view of the remains from the time of departure from Washington until their internment.

The following details, in addition to the General Orders, No. 72, will be observed:

1. The State executive will have the general direction of the public honors in each State and furnish additional escort and guards of honor at places where the remains are taken from the hearse car, but subject to the general command of the departmental, division, or district commander.

2. The Adjutant-General will have a discretionary power to change or modify details not conflicting with the general arrangement.

3. The directions of General McCallum in regard to the transportation and whatever may be necessary for safe and appropriate conveyance will be rigorously enforced.

4. The Adjutant-General and the officers in charge are specially enjoined to strict vigilance to see that everything appropriate is done and that the remains of the late illustrious President receive no neglect or indignity.

5. The regulations in respect to the persons to be transported on the funeral train will be rigorously enforced.

6. The Adjutant-General will report by telegraph the arrival and departure at each of the designated cities on the route.

7. The remains, properly escorted, will be removed from the

Capitol to the hearse car on the morning of Friday, the 21st, at 6 a. m., so that the train may be ready to start at the designated hour of 8 o'clock, and at each point designated for public honors care will be taken to have them restored to the hearse car in season for starting the train at the designated hour.

8. A disbursing officer of the proper bureau will accompany the cortege to defray the necessary expenses, keeping an exact and detailed account thereof, and also distinguishing the expenses incurred on account of the Congressional committees, so that they may be reimbursed from the proper appropriations.

EDWIN M. STANTON,
Secretary of War.

Stanton did not order Townsend to draft and telegraph back to Washington detailed reports of the funeral ceremonies in each city. Townsend would have his hands full commanding the funeral train, with little time to spare for preparing detailed narratives of all the ceremonies unfolding along the route. It would be enough if he informed the War Department that the train was running on schedule and reported whenever he arrived in or departed from a city. Stanton would rely on other sources, including the journalists traveling on the train, and stories published in the Washington, Baltimore, Philadelphia, and New York newspapers, for detailed coverage of the obsequies in each city.

Assistant Adjutant General Edward D. Townsend, commander of Lincoln's funeral train.

Stanton's second historic order that day was a public proclamation, signed by him, offering an unprecedented $100,000 reward for the capture of the assassin John Wilkes Booth and of his coconspirators John Surratt and David Herold. Six days after the assassination, the murderer was still at large. And anyone who dared help Booth during his escape from justice would be punished with death. Large broadsides announcing the reward went up on walls all over Washington and New York City.

Jefferson Davis would enjoy his liberty a while longer; he was fortunate that Stanton's fearsome proclamation did not yet implicate him as an accomplice to Lincoln's murder or offer a cash reward for his capture. But if Stanton was not prepared to accuse Davis of Abraham Lincoln's murder, the newspapers were.

On the same day that Stanton issued the reward, Davis, writing from Charlotte to General Braxton Bragg, wondered if Lincoln's murder might help his cause: "Genl. Breckinridge . . . telegraphs to me, that Presdt. Lincoln was assassinated in the Theatre at Washington . . . It is difficult to judge of the effect thus to be produced. His successor is a worse man, but has less influence . . . [I] am not without hope that recent disaster may awake the dormant energy and develop the patriotism which sustained us in the first years of the War."

Davis busied himself with other military correspondence, including dispatches to General Beauregard on April 20 indicating a scarcity of supplies. "General Duke's brigade is here without saddles. There are none here or this side of Augusta. Send on to this point 600, or as many as can be had." In another dispatch Davis asked for cannons and more men, but the replies he received did not indicate that there were any to be sent.

Things were breaking down elsewhere, too. On the evening of April 20, Breckinridge wrote from Salisbury, North Carolina: "We have had great difficulty in reaching this place. The train from Charlotte which was to have met me here has not arrived. No doubt seized by stragglers to convey them to that point. I have telegraphed commanding officer at Charlotte to send a locomotive and one car without delay. The impressed train should be met before reaching the depot and the ringleaders severely dealt with."

Davis replied promptly: "Train will start for you at midnight with guard."

In Richmond, Robert E. Lee was at home as a private citizen. He still wore the Confederate uniform and posed in it when Mathew Brady showed up to take his photograph, but he had no army to command. He knew that Davis was still in the field, trying to prolong the war. Lee disagreed with that plan. Any further hostilities must, he believed, degenerate into bloody, lawless, and ultimately futile guerilla warfare. Better an honorable surrender than that. Lee and Davis had enjoyed a good wartime partnership, and he knew the president valued his judgment. On April 20, General Lee composed a remarkable letter to his commander in chief, urging him to surrender.

> *Mr. President:*
>
> *The apprehension I expressed during the winter, of the moral condition of the Army of Northern Virginia, have been realized. The operations which occurred while the troops were in the entrenchments in front of Richmond and Petersburg were not marked by the boldness and decision which formerly characterized them. Except in particular instances, they were feeble; and a want of confidence seemed to possess officers and men. This condition, I think, was produced by the state of feeling in the country, and the communications received by the men from their homes, urging their return and the abandonment of the field . . . I have given these details that Your Excellency might know the state of feeling which existed in the army, and judge of that in the country.*

From what I have seen and learned, I believe an army cannot be organized or supported in Virginia, and as far as I know the condition of affairs, the country east of the Mississippi is morally and physically unable to maintain the contest unaided with any hope of ultimate success. A partisan war may be continued, and hostilities protracted, causing individual suffering and devastation of the country, but I see no prospect by that means of achieving a separate independence. It is for Your Excellency to decide, should you agree with me in opinion, what is proper to be done. To save useless effusion of blood, I would recommend measures be taken for suspension of hostilities and the restoration of peace.

I am with great respect, yr obdt svt
R. E. Lee
Genl

In the confusion after Appomattox, Davis never received the letter. If he had, its sentiments would have failed to convince him to end the war. Even if Davis had received it, and if he agreed with Lee's view that resistance east of the Mississippi was futile, he still had faith in a western confederacy on the far side of the Mississippi. Yes, he agreed with Lee on the impropriety of fighting a dishonorable guerilla war. He would not scatter his forces to the hills and sanction further resistance by stealth, ambush, and murder. But Davis, unlike Lee, still believed he could prevail with conventional forces.

Davis and Lee did not communicate again until after the war was over. Indeed, the arrival in Charlotte that very day of several cavalry units gave Davis new hope. According to Mallory:

> No other course now seemed open to Mr. Davis but to leave the country, as he had announced his willingness to do, and his immediate advisers urged him to do so with the utmost promptness. Troops began to come into Charlotte, however . . . and there was much talk among them of crossing the

Mississippi and continuing the war. Portions of Hampton's, Duke's, Debrell's, and Fergusson's commands of the cavalry were hourly coming in. They seemed determined to get across the river and fight it out, and whenever they encountered Mr. Davis they cheered and sought to encourage him. It was evident that he was greatly affected by the constancy and spirit of these men, and that he became indifferent to his own safety, thinking only of gathering together a body of troops to make head against the foe and so arouse the people to arms.

On Friday, April 21—one week after the assassination—Edwin Stanton, Ulysses S. Grant, Gideon Welles, Attorney General James Speed, Postmaster General William Dennison, the Reverend Dr. Gurley, several senators, members of the Illinois delegation, and various army officers arrived at the Capitol at 6:00 A.M. to escort Lincoln's coffin to the funeral train. Soldiers from the quartermaster general's department, commanded by General Rucker, the officer who had led the president's escort from the Petersen house to the White House,

[FREE TRANSPORTATION.] War Department,

ADJUTANT GENERAL'S OFFICE,

Washington, April , 1865.

Hon. Joseph Bailey is *invited to accompany the remains of the late President,* ABRAHAM LINCOLN *from the City of Washington to Springfield, Illinois.* with free transportation back.

BY ORDER OF THE SECRETARY OF WAR:

E. D. Townsend

Assistant Adjutant General.

A CONGRESSMAN'S TICKET TO RIDE ABOARD THE LINCOLN FUNERAL TRAIN.

removed the coffin from the catafalque in the rotunda and carried it down the stairs of the East Front. The statue of Freedom atop the Capitol looked down upon the scene from her omniscient perch.

Four companies of the Twelfth Veteran Reserve Corps stood by to escort the hearse to the train. This was not supposed to be a grand or official procession. There were no drummers, no bands, and no cavalcade of thousands of marchers. It was just a short trip from the East Front plaza to the Baltimore and Ohio Railroad station at First Street and New Jersey Avenue, a few blocks north of the Great Dome. But that did not deter the crowds.

Several thousand onlookers lined the route and surrounded the station entrance. Although this last, brief journey in the capital—Lincoln's third death procession in Washington—was not part of the official public funeral events, Stanton supervised it himself to ensure that the movement of Lincoln's body from the Capitol to the funeral train was conducted with simplicity, dignity, and honor.

Earlier that morning, another hearse had arrived at the station before the president's. It had come from Oak Hill Cemetery in Georgetown, where they had unlocked the iron gates of the Carroll vault. When the soldiers carried Abraham Lincoln aboard his private railroad car at 7:30 A.M., Willie was already there, waiting for him. Lincoln had planned to collect the boy himself and take his coffin home. Now two coffins shared the presidential car.

The railroad car used to transport Lincoln's body was not built as a funeral car. Constructed over a period of two years at the U.S. Military Rail Road car shops in Alexandria, Virginia, the car, named the *United States,* was built as a luxurious vehicle intended for use by the living Lincoln. Although the presidential car had been completed in February prior to Lincoln's 1865 inauguration, he never rode in it or even saw it. The elegant interior, finished with walnut and oak, and upholstered with crimson silk, contained three rooms—a stateroom, a drawing room, and a parlor or dining room. A corridor ran the length of the car and gave access to each room. The exterior was painted chocolate brown, hand-rubbed to a high sheen, and on both sides of the car hung identical oval paintings of an eagle and the coat of arms of the United States. As soon as Stanton knew that Lincoln's body would be carried home to Illinois by railroad, he authorized the U.S.M.R.R. to modify the car, decorate it with symbols of mourning, and build two catafalques so that it could accommodate the coffins of the president and his son.

Willie would have enjoyed calculating the railroad timetable for this trip. He used to delight his father by calculating accurate timetables for imaginary railroad journeys across the nation.

Members of the honor guard took their places beside Lincoln's coffin. Under the protocols established by Stanton and Edward Townsend, the president of the United States was never to be left alone. "There was never a moment throughout the whole journey," Townsend recalled, "when at least two of this guard were not by the side of the coffin."

The hearse and horses that had carried Lincoln's body down Pennsylvania Avenue from the White House to the Capitol, and from there to the train station, were not being boarded onto the train. Instead, in every city where the train was stopping for funeral services, local officials were required to provide a suitable horse-drawn hearse to transport the coffin from the train to the site of the obsequies.

At 7:50 A.M. Robert Lincoln boarded the train, but he planned to leave it after just a while and return to Washington to wrap up his father's affairs. Mary Lincoln did not come to the station to see her husband off—nor did she permit Tad to go. He should have gone to the station and then ridden with his father and brother all the way back to Illinois.

After Willie's death, Tad and Abraham were inseparable. Sometimes Tad fell asleep in the president's office, and Lincoln lifted the slumbering boy over his shoulder and carried him off to bed. Tad loved to go on trips with his father and had relished their recent visit to City Point and Richmond. He loved to see the soldiers, and he enjoyed wearing—and posing for photos in—a child-size Union army officer's uniform, complete with a tiny sword, that Lincoln had given him. Tad would have marveled at the sights and sounds along the 1,600-mile journey. And he would have been proud of, and taken comfort from, the tributes paid to his father.

A pilot engine departed the station ten minutes ahead of the funeral train to inspect the track ahead. At 7:55 A.M., with five minutes to spare, Lincoln's secretaries, John G. Nicolay and John Hay, arrived from the White House and boarded the train to ensure that all was in order. In all, about 150 men were on the train that morning. The manifest included the twenty-nine men—twenty-four first sergeants and four officers—from the Veteran Reserve Corps who would serve as the guard of honor; nine army generals, one admiral, and two junior officers also serving in the guard of honor; a number of senators, congressmen, and delegates from Illinois; four governors; seven newspaper reporters; David Davis, an old friend of Lincoln's and a justice on the U.S. Supreme Court; and Captain Charles Penrose, who had accompanied Lincoln to Richmond.

With so many dignitaries present at the station, the crowd failed to recognize two of the most important men on the passenger manifest. In the days to come, the success or failure of this vital mission would turn in large part upon their work. To accomplish it, they alone would have unfettered access to the president's corpse at any time of the day or night. These were the body men, embalmer Dr. Charles Brown and undertaker Frank Sands. For the next thirteen days, it was their job to keep at bay death's relentless companion, the decomposing flesh of Abraham Lincoln.

At exactly 8:00 A.M. the wheels of the engine Edward H. Jones began to revolve and the eight coaches it pulled began to move.

When Lincoln's train left Washington, the special funeral duties of Benjamin Brown French were done. It had been the most incredible seven days he had ever witnessed during his long tenure in Washing-

Lincoln's funeral car.

ton. He wanted to retrieve a souvenir as a tangible link to the historic week in which he had played an important part. He wrote a letter to Quartermaster General Meigs:

> It is my intention to have the mausoleum, intended for the remains of Washington, beneath the crypt of the Capitol, thoroughly cleaned & properly fitted, and to place in it the catafalco on which the body of our late beloved President lay in the rotunda, there to be preserved as a memento.
>
> The cloth which covered it—made and trimmed by the hands of my wife—was taken with the remains. I should be very glad, when it has done all its duty, that it may be returned to me, to be placed upon that sacred memorial. Will you be pleased, if you conveniently can, to have it so ordered.

While Lincoln's funeral train moved north without incident, the Confederate trains continued to experience difficulties. John C. Breckinridge had more railroad problems and communicated them to Davis in Charlotte at 9:00 A.M. on the twenty-first: "Paroled men and stragglers seized my train at Concord. Operator reports that engine and tender escaped, and will be here presently. I have telegraphed General Johnston to guard the bridges and organize these men to receive subsistence and transportation . . ."

But train troubles were the least of his worries. General Joe Johnston had lost all interest in prolonging the war. At this point two things occupied Johnston's mind: how to negotiate the surrender of his army to General Sherman, and how to get his hands on some of the Confederate gold. On April 21 he asked Breckinridge for money: "I have heard from several respectable persons that the Government has a large sum of gold in its possession. I respectfully and earnestly urge the appropriation of a portion of that sum to the payment of the army, as a matter of policy and justice. It is needless to remind you that the troops now in service have earned everything that the Government can give them, and have stood by their colors with a constancy unsurpassed—a constancy which enables us to be now negotiating with a reasonable hope of peace on favorable terms."

Lincoln's train would reach Baltimore in four hours, and in the days ahead, the train was scheduled to stop many times for official honors, processions, ceremonies, and viewings, but, for the most part, those plans were just dry words, miles, and timetables printed on paper. These documents said nothing of other things to come: spontaneous bonfires, torches, floral arches, hand-painted signs, banners, and masses of people assembled along the way at all hours of the day or night. No government official in Washington had ordered these public manifestations. Stanton did not expect the train itself to take on a life of its own and to become a venerated symbol in its own right.

The train's progress fed not just on firewood and water but on human passions to animate its momentum. At each stop it took aboard the tone and temper of each town and its people. The moving train was like a tuning fork, or an amplifier. The more time it spent on the road, and the greater distance it traveled, the more it picked up the sympathetic vibrations of the nation's pride and grief. It intensified and harmonized the emotions of the people. It became more than the funeral train for one dead man. It evolved into something else. What happened was not decreed. Nor could it be resisted.

Somewhere between Washington and Springfield, the train became a universal symbol of the cost of the Civil War. It came to represent a mournful homecoming for all the lost men. In the heartbroken and collective judgment of the American people, an army of the dead—and not just its commander in chief—rode aboard that train.

In every city where the train stopped—or even just passed through—the people knew it was coming and had read newspaper accounts of the events that had occurred in other cities that preceded it up the line. This built excitement into a fever pitch and created a desire to outdo the honors already rendered in other cities. General Townsend felt the change. Parents held out their sleepy-eyed infants and even uncomprehending babes in their arms, so that one day they could tell their children, "*You* were there. You saw Father Abraham pass by."

Mourning ribbon worn by members of the U.S. Military Rail Road.

Baltimore was a strange but necessary destination. Maryland had remained in the Union, but it was anti-Lincoln and pro-Confederate. Four years earlier, when president-elect Lincoln passed through the city, he had to do so secretly in order to avoid assassination. The threats from the Baltimore conspiracy were real. A gang of dozens of men had sworn to kill Lincoln. Disloyal, rioting mobs would soon attack and kill

Union soldiers in the street. Lincoln arrived in the city in the middle of the night and had to change trains, which involved uncoupling his car from one train, using horses to pull it one mile along the tracks to another station, and then coupling it to a second train. This was the moment of maximum danger. His enemies did not know he had gained safe passage through Baltimore until after he was gone and had arrived in Washington.

Lincoln's escape in Baltimore led to public ridicule and false charges: The president had adopted a disguise; he was a coward who had abandoned his wife and children to pass through the city on another train. Cartoonists caricatured Lincoln sneaking through town wearing a plaid Scotch cap and even kilts. Later, he regretted skulking into Washington. It was not an auspicious way to begin a presidency.

Baltimore was home ground for John Wilkes Booth, and he recruited some of his conspirators there. Indeed, a letter found in Booth's trunk on the night of April 14 suggested that he had multiple conspirators in the city and that he might have sought sanctuary there. It might have been considered obscene to stop the train there, to carry Lincoln's murdered body into the city that had wished him so much ill and that might revel in his assassination.

Before leaving Washington, General Townsend had sent a telegram to General Morris, who was in command there, giving him advance warning to prepare for the president's arrival in Baltimore on April 21 and to receive the remains in person.

After a brief stop at Annapolis Station, where Governor A. W. Bradford joined the entourage, the train arrived at Baltimore's Camden Station at 10:00 A.M. Townsend telegraphed Stanton promptly: "Just arrived all safe. Governor Bradford and General E. B. Tyler joined at Annapolis Junction." The once unruly city showed no signs of trouble. No lurking secessionists uttered verbal insults against Lincoln or the Union. Instead, thousands of sincere mourners, undeterred by a heavy rain, surrounded the station and awaited the president. The honor guard carried the coffin from the car, placed it in the hearse parked on Camden Street, and the procession got under way, marching to the rotunda of the Merchants' Exchange. Brigadier General H. H. Lockwood commanded the column, and a number of army officers, including Major General Lew Wallace, who would soon serve as a judge on the military tribunal convened to try Booth's accomplices, brought up the rear.

The hearse, drawn by four black horses, was designed for the ideal display of its precious passenger. According to a contemporary account, "The body of this hearse was almost entirely composed of plate glass, which enabled the vast crowd on the line of procession to have a full view of the coffin. The supports of the top were draped with black cloth and white silk, and the top of the car was handsomely decorated with black plumes."

It took three hours for the head of the procession to reach Calvert Street. The column halted, the hearse drove to the southern entrance of the exchange, and Lincoln's bearers carried him inside. There they laid the coffin beneath the dome, upon a catafalque, around which, Townsend observed, "were tastefully arranged evergreens, wreaths, calla-lilies, and other choice flowers." Flowers, heaps of flowers, a surfeit of striking and fragrant fresh-cut flowers, would become a hallmark of the funeral journey. Soon, the lilac, above all other flowers, would come to represent the death pageant for Lincoln's corpse.

The catafalque was made especially for Lincoln. City officials had studied newspaper accounts of the White House funeral two days earlier, and they paid special attention to the descriptions of the extravagant decorations and grand bier. On April 20, while mourners in Washington viewed the remains at the U.S. Capitol, carpenters and other tradesmen in Baltimore built a catafalque to rival the one in the East Room. A contemporary account recorded every detail:

> It consisted of a raised dais, eleven feet by four feet at the base, the sides sloping slightly to the height of about three feet. From the four corners rose graceful columns, supporting a cornice extending beyond the line of the base. The canopy rose to a point fourteen feet from the ground, and terminated in clusters of black plumes. The whole structure was richly draped. The floor and sides of the dais were covered with black cloth, and the canopy was formed of black crepe, the rich folds drooping from the four corners and bordered with silver fringe. The cornice was adorned with silver stars, while the sides and ends were similarly ornamented. The interior of the canopy was of black cloth, gathered in fluted folds. In the central point was a large star of black velvet, studded with thirty-six stars—one for each State in the Union.

In Baltimore there would be no official ceremonies, sermons, or speeches; there was no time for that. Instead, as soon as Lincoln's coffin was in position, and after the military officers and dignitaries from the procession enjoyed the privilege of viewing him first, guards threw the doors open and the public mourners filed in. Over the next four hours, thousands viewed the remains. The upper part of the coffin was open to reveal Lincoln's face and upper chest. Lincoln's enemies could have masqueraded as mourners and come to gloat over his murder, but the crowd would have torn them to pieces.

In Baltimore, Edward Townsend established two rules that became fixed for every stop during the thirteen-day journey. "No bearers, except the veteran guard, were ever suffered to handle the president's coffin," he declared. Whenever Lincoln's corpse needed to be removed from the train, loaded or unloaded from a hearse, or placed upon or removed from a ceremonial platform or catafalque, his personal military guard would handle the coffin.

Each city would furnish a local honor guard to accompany

the hearse and to keep order while the public viewed the body, but these men did not lay hands upon the coffin. Townsend also forbade mourners from getting too close to the open coffin, touching the president's body, kissing him, or placing anything, including flowers, relics, or other tokens, in the coffin. Any person who violated these standards of decency would be seized at once and removed.

At about 2:30 P.M., with thousands of citizens, black and white, still waiting in line to see the president, local officials terminated the viewing, and Lincoln's bearers closed the coffin and carried it back to the hearse. A second procession delivered the remains to the North Central Railway depot in time for the scheduled 3:00 P.M. departure for Harrisburg, Pennsylvania. The orderly scene in the Monumental City, as Baltimore was called, was a good omen for the long journey ahead. The first stop had gone well. General Townsend dispatched a telegram to Stanton: "Ceremonies very imposing. Dense crowd lined the streets; chiefly laboring classes, white and black. Perfect order throughout. Many men and women in tears. Arrangements admirable. Start for Harrisburg at 3 p.m."

The train stopped at the Pennsylvania state line, and Governor Andrew Curtin; his staff; U.S. Army general Cadwalader, commander of the military department of Pennsylvania; and assorted officers came aboard. Maryland governor A. W. Bradford received them in the first car. En route to Harrisburg the train stopped briefly at York, where the women of the city had asked permission to lay a wreath of flowers upon Lincoln's coffin. Townsend could not allow dozens of emotional mourners to wander around inside the train and hover about the coffin. He offered a compromise: He would permit a delegation of six women to come aboard and deposit the wreath. While a band played a dirge and bells tolled, they approached the funeral car in a ceremonial procession, stepped inside, and laid their large wreath consisting of a circle of roses and, at the center, alternating parallel lines of red and white flowers. The women wept bitterly as they left the train. Soon, at the next stop, their choice flowers would be shoved aside in favor of new ones.

The train arrived at Harrisburg at 8:20 P.M. Friday, April 21, and Townsend reported to his boss: "Arrived here safely. Everything goes on well. At York a committee of ladies brought a superb wreath and laid it on the coffin in the car." Here too, as in Baltimore, no funeral services or orations were on the schedule. To the disappointment of the crowds waiting at the station, a reception for the remains had to be canceled. "A driving rain and the darkness of the evening," General Townsend noted, "prevented the reception which had been arranged. Slowly through the muddy streets, followed by two of the guard of honor and the faithful sergeants, the hearse wended its way to the Capitol."

Undeterred by the severity of the storm, thousands of onlookers joined the military escort of 1,500 men who had been standing in the rain for an hour and followed the hearse to the state house. To the boom of cannon firing once a minute, the coffin was carried inside and laid on a catafalque in the hall of the house of representatives. Lincoln was placed on view until midnight. Viewing resumed at 7:00 A.M. on Saturday. For the next two hours, double lines of mourners streamed through the rotunda. The coffin was closed at 9:00 A.M., and at 10:00 A.M. a procession began to escort the hearse back to the railroad depot. This became the grand march the public had hoped to witness the previous night.

A military formation led the way. Then came the hearse, accompanied by the guard of honor from the train plus sixteen local, honorary pallbearers. There followed a cavalcade of passengers from the funeral train, including the governor of Pennsylvania, various generals and officers, elected officials, fire and hook and ladder companies, and various fraternal groups including Freemasons and Odd Fellows.

On April 22, Jefferson Davis was still in Charlotte. Lincoln's murder had put his life in great danger, but he still considered the cause more important than personal safety. Indeed, the idea of "escape" was anathema to him. In his mind, he was still engaged in a strategic retreat, not a personal flight.

Davis was not alone in his desire to continue the fight. Wade Hampton wrote to him again from Greensborough and encouraged him to make a run for Texas. "If you should propose to cross the Mississippi River I can bring many good men to escort you over. My men are in hand and ready to follow me anywhere . . . I write hurriedly, as the messenger is about to leave. If I can serve you or my country by any further fighting you have only to tell me so. My plan is to collect all the men who will stick to their colors, and to get to Texas."

Varina Davis, safe in Abbeville, South Carolina, wondered where her husband was. On April 22 she wrote to him via courier that she had not received any communication from him since April 6, when he was still in Danville, Virginia. "[I] wait for suggestions or directions . . . Nothing from you since the 6th . . . the anxiety here intense rumors dreadful & the means of ascertaining the truth very small send me something by telegraph . . . the family are terribly anxious. God bless you. Do not expose yourself."

The funeral train was scheduled to leave Harrisburg at noon on April 22, but the hearse arrived at the station almost an hour early. General Townsend telegraphed Stanton from the train depot shortly before pulling out: "We start at 11.15 [A.M.] by agreement of State authorities. It rained in torrents last night, which greatly interfered with the procession, but all is safe now."

On the way from Harrisburg and Philadelphia, the train passed through Middletown, Elizabethtown, Mount Joy, Landisville, and Dillersville. At Lancaster twenty thousand people, including Congressman Thaddeus Stevens and Lincoln's predecessor, former president James Buchanan, paid tribute. The train pushed north through Penningtonville, Parkesburg, Coatesville, Gallagherville, Downington, Oakland, and West Chester. At every depot, and along the railroad tracks between them, people gathered to watch the train pass by. For miles before Philadelphia, unbroken lines of people stood along both sides of the tracks and watched as the train went by them.

When the train arrived at Broad Street Station in Philadelphia at 5:00 P.M. on Saturday, April 22, it was greeted by an immense crowd. The *Philadelphia Inquirer* explained the reason: "No mere love of excitement, no idle curiosity to witness a splendid pageant, but a feeling far deeper, more earnest, and founded in infinitely nobler sentiments, must have inspired that throng which, like the multitudinous waves of the swelling sea, surged along our streets from every quarter of the city, gathering in a dense, impenetrable mass along the route . . . for the procession."

A military escort, including three infantry regiments, two artillery batteries, and a cavalry troop, had arrived at the depot by 4:00 P.M. in preparation. A vast crowd had assembled along the parade route, and as soon as the engine rolled into the depot, a single cannon shot announced to the city that Lincoln had arrived. Minute guns began to fire.

At 5:15 P.M. the hearse, drawn by eight black horses, got under way. With the military escort leading the way, the huge procession took almost three hours to reach the Walnut Street entrance on the southern side of Independence Square. There, members of the Union League Association had assembled to receive the coffin and guide its bearers to the catafalque inside Independence Hall. According to one account, "the Square was brilliantly illuminated with Calcium Lights, about sixty in number, composed of red, white and blue colors, which gave a peculiar and striking effect to the melancholy spectacle."

As minute guns continued firing and the bells of Philadelphia tolled, Lincoln's body was carried into the sacred hall of the Ameri-

President Lincoln's hearse in Philadelphia.

can Revolution and placed on a platform with his feet pointing north. The entire interior of Independence Hall was shrouded with black cloth. It hung everywhere: from the walls, from the chandelier over the coffin, and from most of the historical oil paintings. The white marble statue of George Washington remained uncovered, and it stood out like a ghost in the blackened room.

Honor guards pulled back the American flag that had covered the coffin during the procession and the undertakers removed the lid to reveal Lincoln's face and chest. Looming near the president's head was a monumental metal object, the most renowned and beloved symbol of the American Revolution. They had laid the slain president at the foot of the Liberty Bell. It was a patriotic gesture that stunned the crowd.

On February 22, 1861, ten days before taking the oath of office, president-elect Lincoln told a Philadelphia audience: "I have never had a feeling politically that did not spring from the Declaration of Independence . . . that which gave promise that in due time the weights should be lifted from the shoulders of all men, and that *all* should have an equal chance . . . Now, my friends, can this country be saved upon that basis? . . . If it can't be saved upon that principle . . . if this country cannot be saved without giving up on that principle . . . I would rather be assassinated on this spot than to surrender it."

A contemporary writer described the memorial scene:

> On the old Independence bell, and near the head of the coffin, rested a large and beautifully made floral anchor, composed of the choicest [japonicas and jet-black] exotics . . . Four stands, two at the head and two at the foot of the coffin, were draped in black cloth, and contained rich candelabras with burning tapers; and, again, another row of four stands, containing candelabra also, making in all eighteen candelabras and one hundred and eight burning wax tapers.
>
> Between this flood of light, shelving was erected, on which were placed vases filled with japonicas, heliotropes, and other rare flowers. These vases were twenty-five in number.
>
> A delicious perfume stole through every part of the Hall, which, added to the soft yet brilliant light of the wax tapers, the elegant uniforms of the officers on duty, etc., constituted a scene of solemn magnificence seldom witnessed.

Newspaper accounts failed to describe the practical purpose of the sweet-smelling flowers, but they were there for a reason. Lincoln had been dead a week, and the embalmers were fighting a ticking clock. They had slowed but could not stop the decay of his flesh. Fragrant flowers would mask the odor.

Dignitaries viewed Lincoln first, from 10:00 P.M. until midnight. Then the public surged in, entering via temporary stairs through two windows and exiting, via a second set of stairs, through the windows facing Independence Square. The coffin was closed at 2:00 A.M. on Sunday, April 23. Many of those who failed to glimpse the president stood outside Independence Hall for the rest of the night to be sure they would be admitted when the doors reopened in several hours.

Beginning at 6:00 A.M. Sunday, authorities had announced, the public would be admitted until 1:00 A.M. Monday.

By late morning on Sunday, the line of mourners extended as far west as the Schuylkill River and east to the Delaware River. "After a person was in line," reported the *Philadelphia Inquirer,* "it took from four to five hours before an entrance into the Hall could be effected." After the long wait, mourners were given only a few seconds to view Lincoln: "Spectators were not allowed to stop by the side of the coffin, but were kept moving on, the great demand on the outside not permitting more than a mere glance at the remains."

The vast crowds had become dangerous and the *Inquirer* reported alarming incidents: "Never before in the history of our city was such a dense mass of humanity huddled together. Hundreds of persons were seriously injured from being pressed in the mob, and many fainting females were extricated by the police and military and conveyed to places of security. Many women lost their bonnets, while others had nearly every article of clothing torn from their persons."

On that Sunday, April 23, Jefferson Davis and his entourage attended church in Charlotte. The minister's fire-and-brimstone sermon, which according to Burton Harrison denounced "the folly and wickedness" of Lincoln's murder, seemed to be aimed at Davis. "I think," Davis said with a smile, "the preacher directed his remarks at me; and he really seems to fancy that I had something to do with the assassination." Despite his predicament, the president had not lost his sense of humor.

Later that day, Davis wrote a long, thoughtful letter to Varina that revealed his state of mind twenty-one days since he had left Richmond. Sanguine, less hopeful, more realistic, but not beaten yet, Davis apologized to his beloved companion for taking her on the lifelong journey that had led to this fate.

My Dear Winnie

I have asked Mr. Harrison to go in search of you and to render such assistance as he may . . .

The dispersion of Lee's army and the surrender of the remnant which had remained with him destroyed the hopes I entertained when we parted. Had that army held together I am now confident we could have successfully executed the plan which I sketched to you and would have been to-day on the high road to independence . . . Panic has seized the country . . .

The loss of arms has been so great that should the spirit of the people rise to the occasion it would not be at this time possible adequately to supply them with the weapons of War . . .

The issue is one which is very painful for me to meet. On one hand is the long night of oppression which will follow the return of our people to the "Union"; on the other the suffering of the women and children, and courage among the few brave patriots who would still oppose the invader, and who unless the people would rise en masse to sustain them, would struggle but to die in vain.

I think my judgement is undisturbed by any pride of opinion or of place, I have prayed to our heavenly Father to give me wisdom and fortitude equal to the demands of the position in which Providence has placed me. I have sacrificed so much for the cause of the Confederacy that I can measure my ability to make any further sacrifice required, and am assured there is but one to which I am not equal, my Wife and my Children. How are they to be saved from degradation or want . . . for myself it may be that our Enemy will prefer to banish me, it may be that a devoted band of Cavalry will cling to me and that I can force my way across the Missi. and if nothing can be done there which it will be proper to do, then I can go to Mexico and have the world from which to choose . . . Dear Wife this is not the fate to which I invited [you] when the future was rose-colored to us both; but I know you will bear it even better than myself and that of us two I alone will ever look back reproachfully on my past career . . . Farewell my Dear; there may be better things in store for us than are now in view, but my love is all I have to offer and that has the value of a thing long possessed and sure not to be lost. Once more, and with God's favor for a short time only, farewell—

Your Husband.

Back in Philadelphia, the funeral procession left Independence Hall at 1:00 A.M. on Monday, April 24. This escort—the 187th Pennsylvania infantry regiment, city troops, the honor guard, the Perseverance Hose Company, and the Republican Invincibles—was much smaller than the one that welcomed Lincoln to Philadelphia. Despite the late hour, thousands of citizens from every part of the city joined the march. It took three hours, until almost 4:00 A.M., to reach Kensington Station. Townsend kept Stanton up to date: "We start for New York at 4 o'clock [A.M.]. No accident so far. Nothing can exceed the demonstration of affection for Mr. Lincoln. Arrangements most perfect." The funeral train departed a few minutes later, en route to New York City.

Thousands of people lined the tracks on the journey to New York City. The train encountered large crowds at Bristol, Pennsylvania, and across the New Jersey state line at Morristown. At 5:30 A.M., the train made a brief stop at Trenton before continuing through Princeton, New Brunswick, Rahway, Elizabeth City, and Newark. The train reached Jersey City, New Jersey, at 9:00 A.M. There the presidential car was uncoupled from the train and rolled onto a ferryboat.

As a young man, Lincoln had floated on a flatboat down the Mississippi River to New Orleans; now he was crossing the Hudson River in a flatboat on his way into New York. The ferry landed in Manhattan at the foot of Desbrosses Street. He was back in the city that had

THE EXTRAVAGANT NEW YORK CITY FUNERAL HEARSE. ON THE RIGHT, CITY HALL IS DRAPED IN MOURNING.

helped make him president and that had given him so much trouble during the war.

This was to be the biggest test of the funeral pageant since it had left Washington. New York City had the biggest population, the greatest crowds, and the most volatile citizens in the North. New Yorkers loved a good riot, as they demonstrated on a number of occasions, including the Astor Street Shakespeare riot in the 1840s and, most recently, the Civil War draft riots. Given the strong Copperhead presence in the city, many believed that Manhattan cried crocodile tears for the fallen president. But mourning Unionists outnumbered Lincoln's enemies on the streets of New York in April 1865.

The procession went from Hudson Street to Canal, to Broadway, and then to City Hall.

The hearse, which was photographed as it rolled through Manhattan, beggared description. According to one published account,

> it was fourteen feet long at its longest part, eight feet wide and fifteen feet one inch in height. On the main platform, which was five feet from the ground, was a dais six inches in height, at the corners of which were columns holding a canopy, which, curving inward and upward toward the centre, was surmounted by a miniature temple of liberty. The platform was entirely covered with black cloth, drawn tightly over the body of the car, and reaching to within a few inches of the ground, edged with silver bullion fringe . . . At the base of each column were three American flags, slightly inclined, festooned, covered with crape. The columns were black, covered with vines of myrtle and camellias. The canopy was of black cloth, drawn tightly, and from the base of the temple another draping of black cloth fell in graceful folds over the first; while from the lower edges of the canopy descended festoons, also of black cloth, caught under small shields. The folds and festoons were richly spangled and trimmed with bullion. At each corner of the canopy was a rich plume of black and white feathers.
>
> The Temple of Liberty was represented as being deserted, having no emblems of any kind in or around it save a small flag on top, at half-mast. The inside of the car was lined with white satin, fluted, and from the centre of the roof was suspended a large gilt eagle, with outspread wings, covered with crape, bearing in its talons a laurel wreath, and the platform around the coffin was strewn with laurel wreaths and flowers of various kinds.
>
> The car was drawn by sixteen gray horses, with coverings of black cloth, trimmed with silver bullion, each led by a colored groom, dressed in the usual habiliments of mourning, with streamers of crape on their hats.

The richness, extravagance, and exaggeration of the sight overwhelmed the senses. New York had outdone all other cities on the funeral route. To anyone in the streets of Manhattan that day, it seemed unimaginable that any city following New York could rival the magnificence of this day.

One newspaper noted that every public place within sight of the procession was crammed with people: "The police, by strenuous exertions, kept the streets cleared, but the sidewalks and the Park were filled with men, women and children, while the trees in the Park were loaded with adventurous urchins."

A self-congratulatory *New York Herald* piece swelled with typical Manhattan pride. "The world never witnessed so grand a collection of well-dressed, intelligent, and well-behaved beings, male and female, as thronged the streets of New York yesterday and gathered around the bier of the leader of the nation."

City Hall had been transformed beyond recognition.

"There was no trace of the interior architecture to be seen on the rotunda of the City Hall," recalled the main chronicler of the New York funeral.

> Niche and dome, balustrade and paneling were all veiled . . . The catafalque graced the principal entrance to the Governor's Room. Its form was square, but it was surmounted by a towering gothic arch, from which folds of crape, ornamented by festoons of silver lace and cords and tassels, fell artistically over the curtained pillars which gave form and beauty to the structure.
>
> The arch seemed lost in the vast labyrinths from which it rose. A spread eagle was perched above it. Beneath this aerial guardian was a bust of the dead President in sable drapery. Then came a ubiquitous display of black velvet, studded with beautiful silver stars in filigree lace, which reflected light over the suits of woe and gloom of which they were the national ornaments . . .
>
> Beneath the canopy, near the honored dead, were busts of

> Washington, Jackson, Webster and Clay—all resting on high pedestals. The vicinity of the catafalque was also the scene of elaborate and artistic mourning. All the furniture, the statues and the portraits of the Governor's room were in character with the sad scenes around them [and all] were covered with crape. The statue of George Washington, near which Mr. Lincoln received his friends four previous years, was elaborately draped, and the chandeliers were covered with black cloth.

For the most part New Yorkers behaved well and respected Lincoln's remains. Some, however, could not control their emotions. One report observed: "The deportment of the people was very different from that of the crowds which usually assemble in great cities. No gladsome laugh, no familiar greeting, no passing jests. Grief was denoted on every countenance. Many would have pressed close to the coffin, if but to touch it with their fingers, were they permitted. Frequent attempts were made by ladies to kiss the placid lips of the corpse."

During the viewing at City Hall, some people tried to do more than touch Lincoln. Some actually wanted to place mementos in the coffin. Captain Parker Snow, a commander of polar expeditions, presented to General Dix some relics of Sir John Franklin's ill-fated expedition. They consisted of a tattered leaf of a prayer book, on which the first word legible was "martyr," and a piece of fringe and some portions of uniform. These relics were found in a boat lying under the head of a human skeleton. What possible connection did these bizarre relics have to Abraham Lincoln? They did not belong in the coffin, and Dix refused to place them there. Such practices, if tolerated, would have turned Lincoln's coffin into a traveling gypsy cart overflowing with antiquarian oddities that would have weighed more than Lincoln's corpse.

The coffin was closed at 11:00 A.M. on Tuesday: "With practiced fingers," wrote one eyewitness, "the undertaker, Mr. F. G. Sands,

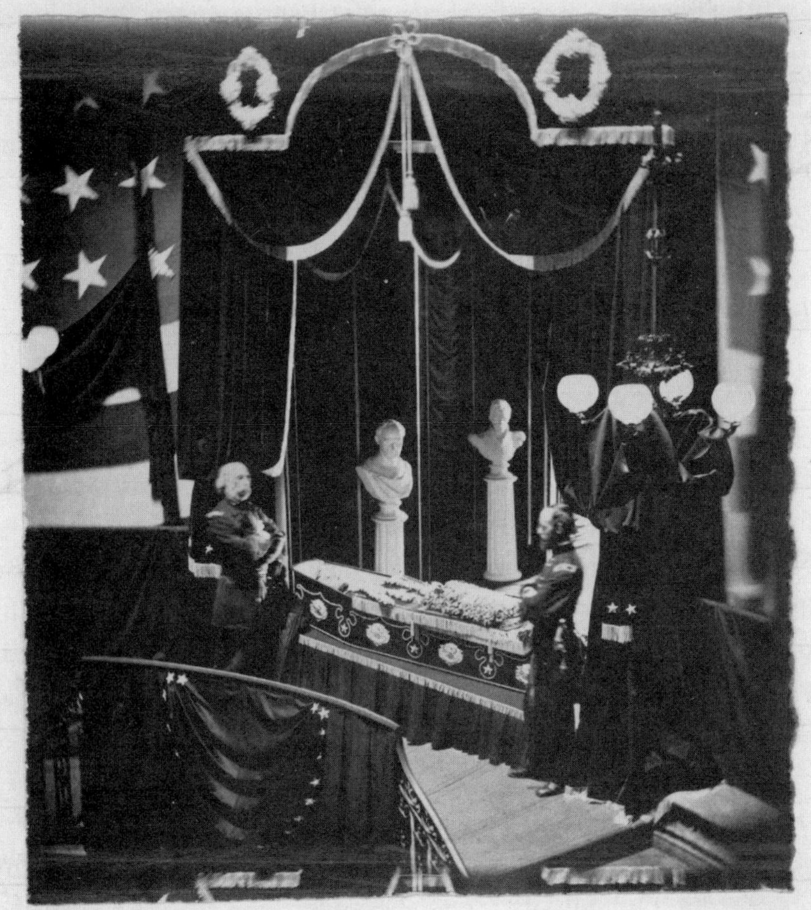

THE NOTORIOUS GURNEY IMAGE, TAKEN INSIDE NEW YORK CITY HALL. EDWARD D. TOWNSEND STANDS AT THE FOOT OF THE COFFIN IN THE ONLY SURVIVING PHOTOGRAPH OF LINCOLN IN DEATH.

and his assistant, Mr. G. W. Hawes, removed the dust from the face and habiliments of the dead . . . and the lid was silently screwed down without form or ceremony, and with none but a few officers and orderlies and a couple of reporters as witnesses. The . . . bearers, eight in number, sergeants of the Veteran Reserve, stationed themselves on each side of the coffin, and remained there motionless as statues awaiting further orders."

At 12:30 P.M. the hearse, drawn by sixteen white horses, began traveling uptown to the depot of the Hudson River Railroad on Twenty-ninth Street. One hundred and twenty-five people had viewed the corpse. Five hundred thousand stood along the procession route.

A TIME FOR WEEPING, BUT VENGEANCE IS NOT SLEEPING, read one of the signs posted along the parade route.

The hearse stopped for an oration at Union Square, and by 3:00 P.M., the head of the procession arrived at the Hudson River Railroad depot. But it took another half hour for the hearse to reach the embarkation point.

The spectacular events in New York City outdid all previous public honors for the late president, even those conducted in the national capital. The *New York Times* congratulated the people of Gotham: "As a mere pageant, the vast outpouring of the people, the superb military display, the solemn grandeur and variety thrown into the procession by the numberless national, friendly, trade and other civic societies; the grand accomplishment of music; and, above all, the subdued demeanor of the countless multitude of onlookers, made the day memorable beyond the experience of the living generation."

CHAPTER EIGHT

"He Is Named for You"

The funeral train departed New York City at 4:15 P.M. on Tuesday, April 25. The engine steamed north along the Hudson River. After a few hours, darkness and torch flames intensified the drama. The train passed Fort Washington, Mount St. Vincent, Yonkers, Hastings, Dobbs Ferry, Irvington, Tarrytown, Sing Sing, Montrose, and Peekskill.

At 6:20 P.M., the train stopped at Garrison's Landing, opposite the U.S. Military Academy at West Point. The corps of cadets assembled to honor their fallen commander in chief. They passed through the funeral car and saluted. At 9:45 P.M. thousands of people gathered at Hudson to see the train.

General Townsend was surprised to see so many mourners when he looked out the window:

> The line of the Hudson River road seemed alive with people. At each of the towns by which it passes, the darkness of night was relieved by torches, which revealed the crowds there assembled.

The memorial arch above the tracks at Sing Sing, New York.

At Hudson . . . elaborate preparations had been made. Beneath an arch hung with black and white drapery and evergreen wreaths, was a tableau representing a coffin resting upon a dais; a female figure in white, mourning over the coffin; a soldier standing at one end and a sailor at the other. While a band of young women dressed in white sang a dirge, two others in black entered the funeral-car, placed a floral device on the President's coffin, then knelt for a moment of silence, and quietly withdrew. This whole scene was one of the most weird ever witnessed, its

> solemnity being intensified by the somber lights of the torches, at that dead hour of night.

At 10:55 P.M., East Albany welcomed Lincoln with a torchlight escort that led the funeral train across the river to Albany. Even in the dark, witnesses could make out the signs posted on two houses: THE HEART OF THE NATION THROBS HEAVILY AT THE PORTALS OF THE TOMB. LET US RESOLVE THAT THE MARTYRED DEAD SHALL NOT HAVE DIED IN VAIN. At 1:30 A.M. Lincoln's coffin was placed in the assembly chamber of the state capitol and the middle-of-the-night viewing began. The lateness of the hour did not deter the citizens. On they came, at a rate of seventy viewers per minute, more than four thousand an hour.

Upon leaving New York, Townsend had sent his usual positive report to Stanton, in which he commented that he had "examined the remains and they are in perfect preservation," but did not mention an episode that had occurred while Lincoln's remains were on view at City Hall. When Stanton learned of the incident by reading the newspapers later that night, he became enraged and dispatched a wrathful telegram that threatened to ruin the reputation and military career of the trusted aide he had personally chosen to command the funeral train.

"I see by the New York papers this evening that a photograph of the corpse of President Lincoln was allowed taken yesterday at New York," Stanton wrote. "I cannot sufficiently express my surprise and disapproval of such an act while the body was in your charge. You will report what officers of the funeral escort were or ought to have been on duty at the time this was done, and immediately relieve them and order them to Washington. You will also direct the provost-marshal to go to the photographer, seize and destroy the plates and any pictures and engravings that may have been made, and consider yourself responsible if the offense is repeated."

At the bottom of the handwritten screed, Stanton scrawled a message to Major Eckert at the War Department telegraph office: "Please order this telegram to be delivered to-night, and if the escort has left New York order it to be forwarded to Albany."

Stanton had assumed, no doubt, that close-up images had been made of Lincoln's face. That was not an unusual custom in nineteenth-century America. It was common for mourners, especially the bereaved parents of deceased infants or children, to commission photographers to preserve for eternity the faces of the loved and lost. Indeed, some enterprising cameramen specialized in corpse photography, often posing dead infants cradled in the arms of their parents, as though in not death but slumber. But Stanton was likely thinking about the condition of Lincoln's body. By the time he was photographed in New York, Lincoln had been dead for nine days. Civil War mortuary science could not preserve his body indefinitely. Yes, the undertakers rode aboard the train, but there were limits to the art. Stanton no doubt feared that horrific images depicting Lincoln's face in a state of gruesome decay would be distributed to the public.

When Townsend arrived in Albany, he had not seen Stanton's telegram. He sent the secretary of war another sunny report: "We have arrived here safely. Words cannot describe the grandeur of the demonstration in New York and all along the Hudson River. The outpouring of popular feeling, quiet and unaffected, is truly sublime."

Stanton's telegram reached Townsend on the morning of April 26 and shattered the general's sense of aesthetic sublimity. He knew his boss well, including his propensity for angry tirades. If Stanton sounded this ill-tempered on paper, Townsend could only imagine how furiously he was raging back in Washington. And once Stanton learned the full story, Townsend feared, Lincoln's god of war would become apoplectic.

It was Townsend, and no one else, who had allowed Lincoln's corpse to be photographed. He decided, before others could report the details of what he had done, to confess and accept the consequences. He immediately telegraphed Stanton: "Your dispatch of this date is received. The photograph was taken while I was present, Admiral Davis being the officer immediately in charge, but it would have been my part to stop the proceedings. I regret your disapproval, but it did not strike me as objectionable under the circumstances as it was done. I have telegraphed General Dix your orders about seizing the plates. To whom shall I turn over the special charge given me in order to execute your instructions to relieve the officers responsible, and shall Admiral Davis be relieved? He was not accountable."

When Stanton learned that Townsend had permitted the photographs to be made, he decided not to relieve him of command. The train was on the move, in the middle of a synchronized, complicated cross-country journey, and no one on the train possessed better organizational skills to command it than Townsend.

The secretary of war sent a more tempered reply: "As Admiral Davis was not responsible there is no occasion to find fault with him. You being in charge, and present at the time, the sole responsibility rests upon you; but having no other officer of the Adjutant General's Department that can relieve you and take your place you will continue in charge of the remains under your instructions until they are finally interred. The taking of photographs was expressly forbidden by Mrs. Lincoln, and I am apprehensive that her feelings and the feelings of her family will be greatly wounded."

Townsend, offended at the insinuation that he had disobeyed an order from the martyred president's widow, could not resist defending his reputation and replied: "Your dispatch just received. I was not aware of Mrs. Lincoln's wishes, or the picture would not have been taken with the knowledge of any officers of the escort. It seemed to me the picture would be gratifying, a grand view of what thousands saw and thousands could not see."

But Townsend had not yet told Stanton everything. It was bad enough that he had allowed the photographs. Even worse, he had posed in the pictures while standing next to President Lincoln's body. Stanton might have considered this perceived pursuit of personal publicity unforgivable. On April 20, before the funeral train departed Washington, Stanton had issued Townsend detailed, written instructions on the mission, including the admonition that "the Adjutant-General and all the officers in charge are specially enjoined to strict vigilance to see that everything appropriate is done and that the remains of the late illustrious President receive no neglect or indignity." Could there be a greater indignity than for a commissioned officer of the U.S. Army, a general no less, the man "specially assigned to represent the Secretary of War, and to give all orders in the name of the Secretary as if he were present," to pose for souvenir photographs with the corpse of the assassinated president? Perhaps only that Admiral Davis, personal representative of Secretary of the Navy Gideon Welles, had also posed for the photos.

Major General Dix, who had come in from New York City to Albany, discussed with Townsend how to handle the still irate Stanton. Dix advised the adjutant general to confess everything. Townsend, invoking the protective umbrella of his superior officer, telegraphed Stanton again.

"General Dix, who is here, suggests that I should explain to you how the photograph was taken," Townsend wrote. "The remains had just been arranged in state in the City Hall, at the head of the stairway, where the people would ascend on one side, and descend on the other. The body lay in an alcove, draped in black, and just at the edge of a rotunda formed of American flags and mourning drapery. The photographer was in a gallery twenty feet higher than the body, and at least forty distant from it. Admiral Davis stood at the head and I at the foot of the coffin. No one else was in view. The effect of the picture would be general, taking in the whole scene, but not giving the features of the corpse."

On April 26 events in two places far from New York dwarfed in importance the dispute between Stanton and Townsend. Jefferson Davis, still in Charlotte, learned that General Joseph Johnston had surrendered his army to General Sherman. It was essential that Davis abandon the state and cross the border into South Carolina. Mallory stressed the point: "His friends . . . saw the urgent expediency of getting further south as rapidly as possible, and after a week's stay in Charlotte they started with an escort of some two or three hundred cavalry."

And on this day, before dawn, at a farm near Port Royal, Virginia, federal cavalry caught up with Lincoln's assassin, John Wilkes Booth, and shot him dead.

Before Davis departed Charlotte, he wrote to Wade Hampton: "If you think it better you can, with the approval of General Johnston, select now, as proposed for a later period, the small body of men and join me at once, leaving General Wheeler to succeed you in command of the cavalry."

Then, in haste, Davis wrote a letter to Varina:

> *Charlotte*
> *April 26. 1865*
> *There is increasing hazard of desertion among the troops. The Cavalry is now the last hope, and how long they will adhere in sufficient numbers to offer resistance is doubtful. I will organize what force of Cavalry can be had. Hampton offers to lead them, and thinks he can force his way across the Mississippi. The route will be too rough and perilous for you and children to go with me. It may be that a safer deposit can be made of your heavy baggage in the neighborhood where you now are than further West—The tide of war will follow me. There will be more quiet out of the track and behind it. I will leave here by or before tomorrow, but will be compelled to move slowly. Will try to see you soon.*
> *Jeffn Davis*

Back in New York City, the man who had photographed Lincoln in his coffin, Thomas Gurney, proprietor of one of Manhattan's most prominent studios, T. Gurney & Son, was getting worried. He had taken unprecedented, newsworthy, and commercially valuable pictures. No other American president had ever been photographed in death. Since Lincoln's assassination no one, not the famous Mathew Brady or Alexander Gardner of Washington, not any other photographers in Baltimore, Harrisburg, or Philadelphia along the route between Washington and New York City, had succeeded in photographing the president in his coffin.

Gurney hoped to gain publicity by distributing photographic prints to the press to reproduce as newspaper woodcuts, and he hoped to profit by selling to the public mass-produced photographic cartes de visite and large-format photographic prints suitable for framing. On April 26, Gurney sent an urgent telegram, not to Stanton, but to a man he thought might be more sympathetic, Charles A. Dana, the assistant secretary of war: "A dispatch to General Dix directs the seizure and destruction of the photographs taken by us of President's remains. We have obtained delay until 10 o'clock in hope of securing a revocation of the order. We shall see Mr. Beecher and Mr. Raymond, and hope the Secretary will see the propriety of waiting until all the facts are in his possession. In the meantime can you not assist us?"

Gurney reached out to Henry Ward Beecher, the widely known clergyman, abolitionist, and author, and Henry J. Raymond, the famous editor of the *New York Times,* and asked them to lobby Stanton and prevent the seizure and destruction of the glass-plate negatives. They agreed and both telegraphed the War Department.

Beecher wrote: "Messrs. Gurney, photographers, wish me to ask you to so far modify your order to General Dix respecting the negatives taken of President Lincoln as to order him to hold them without breaking until Gurney can present to you the facts of the case. They do not intend to have the face represented." He was joined by Raymond who wrote: "I respectfully join in Mr. Beecher's request that General Dix may postpone destroying the negatives of President Lincoln taken by Gurney & Son till they can see you."

A telegram from the War Department arrived at Gurney's studio, saving the negatives from destruction for the time being, but only if Gurney surrendered all the glass plates and agreed to abide by Stanton's decision once he determined whether or not to smash them.

Gurney surrendered the glass-plate negatives, plus all the albumen-paper photographs which he had already printed from them. He had no choice. In the aftermath of Lincoln's murder, emotions were running high. Across the country more than two hundred people had been shot, stabbed, lynched, or beaten to death for making anti-Lincoln statements or for praising his assassin. Stanton had ordered the indiscriminate arrest of more than one hundred people, including the owners of Ford's Theatre, as suspects in the crime. In Baltimore a mob attacked a photography studio based on rumors that the proprietor was selling images of the infamous John Wilkes Booth.

If Gurney had attempted legal action, no court would have recognized his First Amendment right to protect and publish his photographs. If he failed to surrender them voluntarily, the War Department would have raided his studio and seized them. He complied. The next day an army general notified Stanton from New York City that the offending images were in government custody.

Soon the War Department would ban the sale of other photographs it found offensive. In an order dated May 2, 1865, Major General Lew Wallace, future author of *Ben-Hur,* suppressed images of the assassin: "The sale of portraits of any rebel officer or soldier, or of J. Wilkes Booth, the murderer of President Lincoln, is forbidden hereafter in this department. All commanding officers and provost marshals are hereby ordered to take possession of such pictures wherever found exposed for sale, and report the names of the parties so offending, who will be liable to arrest and imprisonment if again guilty of a violation of this order." This unlawful and pointless directive proved impossible to enforce and was soon rescinded.

Stanton's suppression of the corpse photographs did not succeed entirely. He had wanted to prevent Gurney's images from surfacing in any form—including being copied into woodcuts or engravings—but the photographer had managed to get prints into the hands of a few artists. At least two newspapers published front-page interpretations of the scene, and Currier & Ives published a fine engraving based partly on Gurney's work. But Gurney's negatives were never seen again. Perhaps Edwin M. Stanton had them brought to his office in Washington and, after viewing them, smashed them into unrecognizable shards. Perhaps he sequestered the plates in a secret hiding place, where, to this day, they languish in some forgotten, dust-covered War Department file box, possibly alongside the long-lost, never published autopsy photographs Stanton commanded Alexander Gardner to take of John Wilkes Booth's body.

Stanton could not resist preserving for himself at least one image of Lincoln's corpse. Almost a century after the president's death and burial, a sole surviving photographic print made from one of Gurney's negatives was discovered in an old archive, which was traced back to Stanton's personal files. Perhaps Stanton saved it for history. Or perhaps he intended for it never to be seen and to remain his private memento, for his eyes only, a vivid reminder of the spring of '65 and the "coffin that slowly passes."

On Wednesday afternoon, the train left the Albany depot and as it proceeded past the cities, towns, and villages on its way to Buffalo, people turned out in multitudes and the crowds got thicker wherever the train was scheduled to pass. The *New York Tribune* described a mood so solemn that it was as though a funeral had occurred "in each house in central New York." Little Falls was the next stop, and a local band played a dirge while the women of the city presented flowers for the coffin. A written tribute accompanied their gift: "The ladies . . . through their committee, present these flowers and the shield, as an emblem of the protection which our beloved President ever proved to the liberties of the American people. The cross, of his ever faithful trust in God, and the wreath as the token that we mingle our tears with those of the afflicted nation."

Thereafter the train passed through Amsterdam, Fonda, Palatine Bridge, Rome, Green Corners, Verona, Oneida, Canastota, Chittenango, Kirkville, and Manlius. At 11:15 P.M., it made a short stop at Syracuse, where veteran soldiers paid honors, a choir sang hymns, and a little girl handed a small bouquet to a congressman on the train. A note attached to the flowers read: "The last tribute from Mary Virginia Raynor, a little girl of three years of age."

In Rochester at 3:20 A.M. on Thursday, a collection of military units stood in a line on the north side of the station, and on the south side stood the mayor, twenty-five members of the common council of Rochester, and former president of the United States Millard Fillmore, who got on board and rode to the next stop, Buffalo.

Sometime after the sun rose Thursday morning, tolling bells and booming cannon awoke the citizens of Buffalo who had not already assembled at the railroad depot. Abraham Lincoln had arrived. At 8:00 A.M. a modest procession, which included President Fillmore, escorted the hearse to St. James Hall. The marchers included Company D of the Seventy-fourth Regiment, which, four years earlier, had acted as president-elect Lincoln's escort when he passed through the city in February 1861 on his way to Washington. After the assassination, Buffalo officials, unaware that the train would come through their city, had already honored Lincoln with a grand funeral procession on the day of the White House funeral in Washington. They decided against staging a second one today. They did not want to exhaust the emotions of their citizens. At 9:35 A.M., after the hearse reached the Young Men's Association building, Lincoln's bearers removed his coffin from the vehicle and carried it up the steps into St. James Hall.

Under a simple canopy of drooping black crepe, they laid the coffin on a dais while the Buffalo St. Cecelia Society, a musical group, sang "Rest, Spirit, Rest." Women from the Unitarian Church placed an anchor of white camellias at the foot of the coffin. For more than ten hours, from a little past 9:30 A.M. until the coffin was closed at 8:00 P.M., thousands of people, including many from Canada who had crossed the border for the occasion, viewed the remains.

At some point while the crowds passed by the corpse in the coffin, news reached Buffalo by telegraph that electrified Townsend and the mourners standing in line: John Wilkes Booth had been taken. "Here," Townsend recorded, "we first received intelligence of the capture and death of Booth, the assassin." His body was en route by boat to Washington. Some of the same doctors who performed Lincoln's autopsy now waited there to dissect Booth's corpse.

The president's coffin was closed at 8:00 P.M. and forty-five minutes later, the procession left St. James Hall under military escort. Many of the viewers who had seen Lincoln's body waited outside so that they could follow the hearse to the railroad depot and watch the train depart a few minutes past 10:00 P.M.

On the night that Lincoln's train pulled out of Buffalo, Jefferson Davis was staying in Yorkville, South Carolina. And he was still taking his time. His journey south was more like a farewell pageant than a speedy flight. His lack of urgency worried Stephen Mallory: "[T]wo days after . . . [leaving Charlotte we] reached Yorkville, South Carolina, traveling slowly and not at all like men escaping from the country."

Wade Hampton wanted to lead his cavalry to the president's side, but he was a conflicted man. He confessed his dilemma in a letter to General Johnston:

> By your advice I went to consult with President Davis . . . After full conference with him, a plan was agreed on to enable him to

> leave the country. He charged me with the execution of this plan, and he is now moving in accordance with it. On my return here I find myself not only powerless to assist him, but placed myself in a position of great delicacy. I must either leave him to his fate, without an effort to avert it, or subject myself to possible censure by not accepting the terms of the convention you have made. If I do not accompany him I shall never cease to reproach myself, and if I go with him I may go under the ban of outlawry. I choose the latter, because I believe it to be my duty to do so . . . I shall not ask a man to go with me. Should any join me, they will . . . like myself, [be] willing to sacrifice everything for the cause . . .

And Davis definitely had reason to worry because now that Stanton had Booth, he could focus on Jefferson Davis. Calvary units were already looking for Davis and wanted to kill or capture him. On Wednesday, April 27, one day after Booth was shot and killed, and after Confederate major general Joe Johnston surrendered his army in North Carolina, Stanton telegraphed Major General George Thomas about the Confederate president and his rumored treasure:

> The following is an extract from a telegram received this morning from General Halleck, at Richmond: "The bankers have information to-day that Jeff. Davis' specie is moving south from Goldsborough in wagons as fast as possible. I suggest that commanders be telegraphed through General Thomas . . . to take measures to intercept the rebel chiefs, and their plunder. The specie is estimated at $6,000,000 to $13,000,000." [S]pare no exertion to stop Davis and his plunder. Push the enemy as hard as you can in every direction.

Thomas forwarded the telegram the same day to Union cavalry major general George Stoneman: "I want you to carry out these instructions as thoroughly as possible."

Thomas dispatched a second telegram to Stoneman with additional orders:

> If you can possibly get three brigades of cavalry together, send them across the mountains into South Carolina to the westward of Charlotte and toward Anderson. They may possibly catch Jeff. Davis, or some of his treasure. They say he is making off with from $2,000,000 to $5,000,000 in gold. You can send Tillson to take Asheville, and I think the railroad will be safe during his absence. Give orders to your troops to take no orders except those from you, from me, and from General Grant.

When Stoneman received these telegrams, he ordered troops to pursue Jefferson Davis, and on April 27 he telegraphed orders to General Tillson:

> I want the Eighth and Thirteenth Tennessee, Miller's brigade, all sent to Ashevile, and as soon as they are concentrated at that point I wish the following instructions carried out by General Brown, commanding the Second Brigade: Move via Flat Rock or some other adjacent gap to the headwaters of the Saluda River; follow down this river to Belton or Anderson. From that point scout in the direction of Augusta, Ga. The object of sending you to this point is to intercept Jeff. Davis and his party, who are on their way west with $5,000,000 to $6,000,000 of treasure, specie, loaded in wagons . . . If you can hear of Davis, follow him to the ends of the earth, if possible, and never give him up.

As the Union prepared to cast a wide net to snare its prey, Lincoln rode through New York State, into the darkness of the night. Edward Townsend sensed that the train had begun to leave behind waves of emotion that swelled by the hour: "As the President's remains went farther westward, where the people more especially claimed him as their own, the intensity of feeling seemed if possible to grow deeper. The night journey of the 27th and 28th was all through torches, bonfires, mourning drapery, mottoes, and solemn music."

The engine pushed on through New Hamburg, North Evans, Lakeview, Angola, and Silver Creek. At 12:10 A.M., Friday, April 28, the train passed through Dunkirk on the shore of Lake Erie. There, thirty-six young women representing the states of the Union appeared on the railway platform. They were dressed in white, and each wore a broad, black scarf resting across the shoulder and held a national flag in her right hand. This tableau proved so irresistible that when officials in other cities read about it in the newspapers, they copied the idea for their local tributes.

The train passed through Brocton, stopping at 1:00 A.M. in Westfield where, during Lincoln's inaugural journey, he spoke to Grace Bedell, a little girl who had during the campaign of 1860 written him a letter encouraging him to grow a beard to make him more appealing to women, who would then, the child promised, make their husbands and brothers vote for him. Lincoln grew the beard and won the election. Now, four years later, a delegation of five women led by a Mrs. Drake, whose husband, an army colonel, had been killed the previous year in Grant's futile frontal assault at Cold Harbor, came aboard bearing a wreath of flowers and a cross. The cross bore the motto "Ours the Cross; Thine the Crown." Sobbing, they approached Lincoln's closed casket and were allowed, as a special military courtesy to the war widow, to touch it. They "considered it a rare privilege to kiss the coffin."

At North East, Pennsylvania, the funeral train stopped to allow General Dix and his staff to disembark. He had traveled with the remains since Philadelphia. Before Dix began his return to New York City, he sent a telegram to Stanton, telling him, "Everything has been most satisfactory."

The train crossed the Ohio state line and passed through Conneaut, Kingsville, Ashtabula, Geneva, Madison, Perry, Painesville, Mentor, Willoughby, and Wickliffe, where Governor John Brough received the funeral party. Major General Joseph Hooker, now commanding the Northern Department of Ohio, also boarded there.

Abraham Lincoln had once given the command of the Army of the Potomac to the boastful general. "You have confidence in yourself," the president had written to Hooker, "which is a valuable, if not an indispensable quality . . . But . . . I have heard . . . of your recently saying that both the Army and the Government need a Dictator." Lincoln put Hooker in his place: "Of course it was not *for* this, but in spite of it, that I have given you the command. Only those generals who gain successes, can set up dictators. What I now ask of you is military success, and I will risk the dictatorship . . . And now, beware of rashness. Beware of rashness, but with energy, and sleepless vigilance, go forward, and give us victories."

Hooker failed, and after the disaster at Chancellorsville in May 1863, Lincoln fired him. When the funeral train crossed the Ohio line into Indiana, Hooker did not disembark. He rode it all the way to Springfield. What must he have thought as he contemplated the flag-draped coffin of the man who had placed in his hands the power to win the war? The train stopped again at Euclid, Ohio, to pick up some of Cleveland's leading citizens, who had requested the honor of escorting Lincoln's remains into their city.

On April 28, Davis and his entourage stopped at Broad River, South Carolina, to rest and enjoy a lunch they brought with them. The conversation turned to the subject of how the war had ruined them. John Reagan's home in Texas had been wrecked and partly burned, Judah Benjamin's property in Louisiana had been seized by the federals, as had John C. Breckinridge's property in Kentucky. Stephen Mallory's home in Pensacola, Florida, had been burned by Union soldiers. Reagan remembered them using dark humor to lift their spirits. "After we had joked with each other about our fallen fortunes the President took out his pocket-book and showed a few Confederate bills, stating that they constituted his entire wealth." Davis told his cabinet he was pleased that none of them had profited from his service.

Reagan had seen Davis's scrupulous principles in action two years earlier in 1863, when an officer brought word to Davis that his beloved plantation, Brierfield, situated on the Mississippi River near Vicksburg, would fall into the hands of Grant's forces within a few days. Losing Brierfield would be a financial catastrophe. Friends urged Davis to order Confederate forces to rush to his plantation to rescue his slaves and other property and move them to a safe location. Although he hated to lose his valuables, he bristled at the suggestion: "The President of the Confederacy cannot employ men to take care of his property."

Later, when Union forces threatened his hill house in Jackson, Mississippi, the location of his fine and extensive library, Davis again refused to use his official position to protect his private property. "Thus," testified Reagan, "in his unselfish and patriotic devotion to the cause so dear to his heart he permitted his entire property to be swept away."

Lincoln's train arrived at Cleveland's Euclid Street Station on Friday morning, April 28. Edward Townsend sent word to Washington: "The funeral train arrived here safely at 7 o'clock this morning." Ever since Stanton's scathing rebuke for the Lincoln corpse photography episode, Townsend made no more comments in his dispatches to the secretary of war. From that point on, he was all business, stating only what time the train arrived and departed from the remaining cities on the route.

Thirty-six cannons fired a national salute to the president. At that moment, if General Townsend was looking out the window of his car, he witnessed a bizarre display, perhaps the strangest sight of their journey so far. A woman, identified by the press only as "Miss Fields, of Wilson Street," had erected an arch of evergreens near the tracks, on the bank of Lake Erie. As the train passed, Miss Fields, attired in a costume, stood under her arch and struck poses and attitudes of the Goddess of Liberty in mourning.

In the days leading up to the arrival of Lincoln's remains, Cleveland's public officials and leading citizens had engaged in an orgy of bureaucratic busyness. It began simply enough. First, the city council created a committee of five men—the mayor, the city council president, and three others—that met on April 19, the day of the White House funeral, to prepare for the train's arrival. Then the Board of Trade created its own committee to "cooperate" with the city council's General Committee of Arrangements, which responded by increasing the size of its committee from five to twenty-three men. That committee met on April 22 and created nine subcommittees: "On Location of Remains"; "On Reception"; "On Procession"; "On Military"; "On Entertainment"; "On Music"; "On Decoration"; "On Carriages"; "To Meet the Remains."

At its next meeting, the General Committee of Arrangements established a "Civic Guard of Honor," then divided that group of dozens of men into six "squads." Every leading gentleman in town craved the honor of serving on one of these committees. In just a few days, Cleveland had created more levels of bureaucracy to receive the remains in one city than the War Department needed to plan and staff the entire thirteen-day trip of the funeral train halfway across America.

The good citizens were so busy forming committees, subcommittees, and lesser divisions they failed to realize that, until the subcommittee of "Location of Remains" pointed it out, they did not have one public building or hall in all of Cleveland big enough to accommodate the viewing of the president's remains. They would have to construct a new building in little more than a week. How was it possible? Saner heads prevailed, and somebody suggested a temporary outdoor pavilion. They could make it look like a Chinese pagoda. No one would forget *that*.

The committee members were also so distracted that they failed to set aside hotel rooms for the elected officials and members of the U.S. military escort traveling aboard the train. The passengers did not live on the train, which had no sleeping cars. Such cars joined the train from time to time but did not eliminate the need for proper accommodations. The escorts stayed in hotels and dined in restaurants along the route. The Cleveland hotels were so overbooked that even the commanding general of Abraham Lincoln's funeral train could not find a room. Townsend recalled the episode: "To a gentleman, a stranger to me, who kindly lent me his room at a hotel, I was indebted for fifteen hours' unbroken sleep, to bring up arrears."

In South Carolina, Jefferson Davis crossed the Broad River at Scaife's Ferry, and then the Tyger River at Gist's Ferry. That day, in a letter dated April 28, Varina Davis, then in Abbeville, replied to Jefferson's letter of April 23, in which he had chastised himself for bringing her to ruin. She dismissed his apology, reminding him that she had never expected a life of privilege and ease: "It is surely not the fate to which you invited me in brighter days, but you must remember that you did not invite me to a great Hero's home, but to that of a plain farmer. I have shared all your triumphs, been the only beneficiary of them, now I am but claiming the privilege for the first time of being all to you now these pleasures have past for me . . . I know there is a future for you." But not, she thought, in South Carolina, Georgia, or Florida. Varina advised him to give up the cause east of the Mississippi River. "I have seen a great many men who have gone through [Abbeville]—not one has talked fight—A Stand cannot be made in this country. Do not be induced to try it—As to the trans Mississippi, I doubt if at first things will be straight, but the spirit is there, and the daily accretions will be great when the deluded of this side are crushed out between the upper, and nether millstone."

Federal officials may have fantasized that the Confederate president was fleeing with millions of dollars in looted gold, but Davis was down to his last gold coin—and even then he gave it away. John Reagan watched him do it:

> On our way to Abbeville, South Carolina, President Davis and I, traveling in advance of the others, passed a cabin on the roadside, where a lady was standing in the door. He turned aside and requested a drink of water, which she brought. While he was drinking, a little baby hardly old enough to walk crawled down the steps. The lady asked whether this was not president Davis; and on his answering in the affirmative, she pointed to the little boy and said, "He is named for you." Mr. Davis took a gold coin from his pocket and asked her to keep it for his namesake. It was a foreign piece, and from its size I supposed it to be worth three or four dollars. As we rode off he told me that it was the last coin he had, and that he would not have had it but for the fact that he had never seen another like it and that he had kept it as a pocket-piece.

Officially, the president of the Confederacy was now personally penniless, and that might possibly hinder his escape down the road. Davis might need to buy food, pay for lodgings, bribe a Yankee soldier or a Confederate guerilla, pay his way across the Mississippi River, or secure an ocean-bound vessel in Florida. Poverty jeopardized his chances of success. Bestowing his last gold piece to the infant was a symbolic gesture. It was the casting off of all worldly goods. Yes, his caravan traveled with several hundred thousand dollars in gold and silver—not the majority of the Confederacy's funds—but Davis considered that treasure sacrosanct and unavailable for his personal use. That money belonged to the Confederate government, not its president. Now the only riches he possessed were the residual love and goodwill of the people. He hoped that, in the days ahead, as he pushed deeper into the Southern interior, the people there would show him better hospitality than he had received in Greensboro and Charlotte. His aides assured him that it would be so. In South Carolina and Georgia, they promised, the people still loved him and believed in the cause.

In Cleveland, the hearse transported Lincoln's coffin to the public square where the pagoda—the city fathers called it the Pavilion—had been erected. The wood structure, which measured twenty-four by thirty feet, and fourteen feet high, was an amazing confection of canvas, silk, cloth, festoons, rosettes, golden eagles bearing the national shield mounted at each end of the building, and "immense plumes of black crepe." And, as at every other venue along the journey, the interior was stuffed with all manner of flowers. Evergreens covered the walls, and thick matting carpeted the floor to deaden into silence the sound of all footsteps. Over the roof, stretched between two flagpoles, was a streamer that bore a motto from Horace: "Extinctus amabitur idem" (Dead, he will be loved the same). And to set the somber mood, it was raining, "dripping like tears on the remains of the good man in whose honor the crowd had gathered," according to a journalist's account written at the time.

The embalmer opened the coffin and judged the body ready for viewing. According to one sympathetic chronicler of the ceremonies, "the features were but slightly changed from the appearance they bore when exposed in the Capitol at Washington." But the journey had begun to take its toll on the corpse. Lincoln's face turned darker by the day, and the embalmer tried to conceal this with fresh applications of chalk-white potions. All through the day and night the people came, one hundred thousand of them, before the gates to the

IN CLEVELAND, CROWDS WAIT TO VIEW LINCOLN'S CORPSE IN THE CELEBRATED "PAGODA" PAVILION.

square were shut at 10:00 P.M. The coffin was closed at 10:10 P.M., and one hour later it was carried to the hearse.

Just as Lincoln's remains departed the scene the rain, which had been heavy throughout much of the day, turned into a torrential downpour. The water spoiled the decorations, and the mourning crepe cried streaks of black tears. From the railroad station Townsend telegraphed Washington at 11:30 P.M., Friday, April 28: "The funeral train is ready, and will start at midnight." A *New York Times* correspondent confirmed Townsend's earlier observation that something was happening as the train continued west: "Everywhere deep sorrow has been manifested, and the feeling seems, if possible, to deepen, as we move Westward with the remains to their final resting place."

The downpour lasted for most of the night as the train steamed from Cleveland to the Ohio state capital, Columbus. But the foul weather could not deter the people from turning out along the tracks. According to one contemporary account, "Bonfires and torches were lit, the principal buildings draped in mourning, bells tolled, flags floated at half-mast, and the sorrowing inhabitants stood in groups, uncovered and with saddened faces gazing with awe and veneration upon the cortege as it moved slowly by."

Five miles from Columbus, the passengers witnessed a pitiful tribute that stood out in stark contrast to all the elaborate, official processions and ponderous orations that had gone on before. Those who saw it were taken aback by its heartfelt simplicity: "An aged woman bare headed, her gray hairs disheveled, tears coming down her furrowed cheeks, holding in her right hand a sable scarf and in her left a bouquet of wild flowers, which she stretched imploringly toward the funeral car." Her gesture was as eloquent as a cannonade of one hundred minute guns, the tramp of one hundred thousand mourners marching through the great cities of the North, and as richly decorated hearses and death chambers. Abraham Lincoln would have noticed her. She was an eerie reminder of his aged, pioneer stepmother, who had survived him and awaited his return to the prairies. "I knowed when he went away he'd never come back alive," she'd said upon hearing of the assassination.

The train pulled into the Union Depot at Columbus early on Saturday, April 29. It was as it had been in Cleveland: a reception committee of elected officials, military officers, and leading citizens; an escort to the capitol building by a massive military and civic procession; and lying in state in another death chamber bedecked with the now predictable and overflowing quantities of flowers and mourning decorations. Lincoln's bearers removed his coffin and placed it in yet another fabulous hearse, this one topped with a canopy that resembled a Chinese pagoda. The organizers back in Cleveland must have taken that as a tribute to their unforgettable pavilion. The hearse drove off to the state capitol and at 9:30 A.M. Lincoln's coffin was laid upon the catafalque. As usual, the president's honor guard left behind on the train the smaller, second coffin that had accompanied Lincoln's in the presidential car from Washington.

In the press accounts of the funeral pageant, little mention was made of Willie Lincoln. His coffin was never unloaded from the train. He did not ride in the hearse with his father in any of the funeral processions. His closed coffin—he had been dead for three years—did not lie next to the president's at the public viewings. The national obsequies were for the head of state. But in Columbus, Willie Lincoln was not forgotten. General Townsend was the recipient of the gesture: "While at Columbus I received a note from a lady, wife of one of the principal citizens, accompanying a little cross made of wild violets. The note said that the writer's little girls had gone to the woods in the early morning and gathered the flowers with which they had wrought the cross. They desired it might be laid on little Willie's coffin, 'they felt so sorry for him.'"

Of the dozens of mourning songs composed for Abraham Lincoln, only one of them, "The Savior of Our Country," was dedicated to "little Willie." The lyrics described an eerie, father-son reunion in heaven.

> Father! When on earth you fell Father! Was my mother well?
> When I fell your Mother cried! Then unconsciously I died.
> Glory forms our sunlight here! Astral Lamps our Chandelier!
> Rode you here among the stars, In a train of silver cars!
>
> Willie! On the earth look back! Father! Tis a speck of black!
> Robed in Mourning as you see! Mourns the Earth for you and me!
>
> God is Father! God is dear! May I have two Fathers here?
> Father! On our Golden pave Gingles something from your grave!
> Willie! Yes, Four Million Chains, Bring I here where Justice Reigns.

From the Land your Father saves! Chains that bound Four Million Slaves!
Willie! On the earth look back! Father! Tis a speck of black!
Robed in Mourning as you see! Mourns the Earth for you and me!

But the song was wrong. If Willie had looked down from heaven upon the earth, he would have seen, through the night sky, not only a dim "speck of black." He would have seen a ribbon of flame unspooling across the land as torches and bonfires marked his and his father's way home.

On April 29, Davis crossed the Saluda River at Swancey's Ferry, South Carolina. Federal cavalry had a difficult time picking up his trail. Southerners tried to thwart the president's pursuers. One Yankee cavalryman complained: "The white people seemed to be doing all they could to throw us off Davis' trail and impart false information to their slaves, knowing the latter would lose no time in bringing it to us." Later, reports by blacks to Henry Harnden, First Wisconsin Cavalry, and Benjamin D. Pritchard, Fourth Michigan Cavalry, led them directly to the Davises.

General James Wilson outlined his plan for the pursuit. He did not single out one particular unit to capture Davis. Instead, he planned to flood a whole region with manhunters to increase the chance that one unit among many might catch Davis in the net. "Soon after I heard that Johnston had surrendered to General Sherman . . . I received information that Davis, under an escort of a considerable force of cavalry, and with a large amount of treasure in wagons, was marching south from Charlotte, with the intention of going west of the Mississippi River," Wilson reported. He set a number of units in motion with the hope of intercepting Davis at any point he might attempt to pass through Union lines.

Georgia was now the focal point. Wilson knew that after Davis left Charlotte, he would not turn west or east and risk remaining in North Carolina. Those routes would not lead him to the banks of the Mississippi River or to a safe ocean port. The Union had locked down North Carolina's Atlantic coast. Turning west or east would bring Davis into contact with federal troops. There was only one place to go—down through South Carolina and into Georgia—and General Wilson knew it:

> I immediately directed Brevet Brigadier General Winslow, temporarily in command of the Fourth Division, to march to Atlanta, and from that place watch all the roads north of the mouth of the Yellow River, to send detachments to Newman, Carrollton, and Talladega, as well as to Athens and Washington. Brigadier General Croxton, commanding First Division, was directed to picket the Ocmulgee from the mouth of the Yellow River to Macon, to send his best regiment to the east of the Oconee, via Dublin, with orders to find the trail of the fugitives and follow them to the Gulf or the Mississippi River, if necessary. I directed Col. R. H. G. Minty, commanding the Second Division, to picket the Ocmulgee from [Macon] to Hawkinsville, and the 6th to extend his line rapidly down the Ocmulgee and Altamaha as far as the mouth of the Ohoopee. He also sent a force to Oglethorpe to picket the Flint River and crossings from the Muscogee and Macon Railroad to Albany, and 300 men to Cuthbert, to hold themselves in readiness to move in any direction circumstances might render advisable. A small detachment of men was also sent to Columbus, Georgia.

Wilson also alerted troops in Florida, in case Davis was able to slip through Georgia and make a run for the coast and escape the United States on an oceangoing vessel:

> General McCook, with 500 men of his division, had been previously ordered to Tallahassee, Florida, for the purpose of receiving the surrender of rebel troops in that State. A portion of his command at Albany was directed to picket the Flint River thence to its mouth. He was instructed to send out small scouting parties to the north and eastward from Thomasville and Tallahassee. The troops occupied almost a continuous line from the Etowah River to Tallahassee, Florida, and the mouth of the Flint River, with patrols through all the country to the northward and eastward, and small detachments at the railroad stations in the rear of the entire line. It was expected that the patrols and pickets would discover the trail of Davis and his party and communicate the intelligence by courier rapidly enough to secure prompt and effective pursuit.

In Columbus, former Ohio congressman Job E. Stevenson delivered a memorial address at 4:00 P.M. that was unlike any other given in any city since the train had left Washington. It was a unique cry for vengeance. Stevenson accused the South of many crimes and warned that its people must suffer justice. He proclaimed:

> But he was slain—slain by slavery. That fiend incarnate did the deed. Beaten in battle, the leaders sought to save slavery by assassination. Their madness presaged their destruction . . . They have murdered Mercy and Justice rules alone . . . They have appealed to the sword; if they were tried by the laws of war, their barbarous crimes against humanity would doom them to death. The blood of thousands of murdered prisoners cries to heaven. The shades of sixty-two thousand starved soldiers rise up in judgment against them . . . Some wonder why the South killed her best friend. Abraham Lincoln was the true friend

> of the people of the South; for he was their friend as Jesus is the friend of sinners—ready to save when they repent. He was not the friend of rebellion, of treason, of slavery—he was their boldest and strongest foe, and therefore they slew him—but in his death they die; the people have judged them, and they stand convicted, smitten with remorse and dismay—while the cause for which the President perished, sanctified by his blood, grows stronger and brighter . . . Ours is the grief—theirs is the loss, and his is the gain. He died for Liberty and Union, and now he wears the martyr's crown. He is our crowned President . . . Let us beware of the Delilah of the South.

At 6:00 P.M. the doors to the capitol were closed and a procession escorted Lincoln's body back to the Great Central Railway depot. At 8:00 P.M. tolling bells signaled the train's departure from Columbus. It steamed west through Pleasant Valley, where giant bonfires lit up the country for miles; Unionville; Milford; Woodstock; Urbana, where illuminated color transparencies hung from the arms of a large cross; Piqua, where ten thousand people assembled close to midnight; Covington; Greenville; and New Paris, where more giant bonfires lit the sky.

The cortege crossed into Indiana in the middle of the night. At 3:10 A.M. on Sunday, April 30, it rolled into Richmond, the first town across the border. Despite the late hour, city bells rang and twelve thousand people turned out to watch the train pass under an impressive arch twenty-five feet high and thirty feet wide, while a woman costumed as the Genius of Liberty, flanked by a soldier and a sailor, wept over a mock coffin of Lincoln. The train stopped long enough for a committee of ladies to come aboard and place a floral wreath on each coffin. The motto on Willie's tribute read: "Like the early morning flower he was taken from our midst." All through the night, on the long, rural stretches of open country between the towns, farmers kept watch. The *Indiana State Journal* reported: "All along the line

TERRE HAUTE & RICHMOND RAILROAD.

FUNERAL CEREMONIES

OF THE LATE

PRESIDENT LINCOLN!

To be Observed at Indianapolis Sunday, April 30, 1865.

SPECIAL TRAINS will be run at Half the regular Fare on the above date, according to the following schedule, to carry all persons wishing to participate in the above ceremonies.

GOING EAST.					GOING WEST.			
No. 3	Leave.	No. 1.			No. 2.	Arrive	No. 4.	
6.30	A.M.			Terre Haute,	7.20	P.M.		
6.48	"			Wood's Mill,	7.02	"		
7.00	"			Staunton,	6.50	"		
7.15	"			Brazil,	6.38	"		
7.25	"			Harmony,	6.27	"		
7.41	"			Reelsville,	6.10	"		
8.02	"			Junction,	5.52	"		
8.15	"	7.30	A.M.	Greencastle,	5.45	"	6.20	P.M.
8.30	"	7.46	"	Fillmore,	5.23	"	6.05	"
8.41	"	7.57	"	Coatsville,	5.12	"	5.53	"
8.49	"	8.05	"	Amo,	5.05	"	5.45	"
9.01	"	8.18	"	Clayton,	4.52	"	5.32	"
9.11	"	8.30	"	Cartersburg,	4.41	"	5.18	"
9.20	"	8.40	"	Plainfield,	4.33	"	5.08	"
				Summit,				"
9.30	"	8.52	"	Bridgeport,	4.22	"	4.53	"
9.50	A.M.	9.15	A.M.	Indianapolis,	4.00	P.M.	4.30	P.M.
	Arrive.					Leave.		

Tickets will be on sale at all the principal Stations, for one Full Fare to Indianapolis and return. Passengers must purchase Tickets before taking the Trains, or Full Fare will be collected.

Passengers are specially requested to purchase their Tickets, as much as possible, on Saturday, the 29th inst.

Tickets of Passengers from the Evansville and Albany Roads will be taken on the Evening Express of the 29th inst., and will be good to return on the morning Mail Train of Monday, May 1st; all other Tickets will not be good except on the Sunday Trains.

Trains Nos. 2 and 3 will make all the usual stops, on signal, West, but none East of Greencastle.

Nos. 1 and 4 will make all the usual stops, on signal, East of Greencastle.

CHAS. WOOD, Sec'y. **R. E. RICKER, Sup't.**

INDIANAPOLIS, APRIL 26TH, 1865.

RAILROADS RAN SPECIAL TRAINS TO THE CITIES WHERE LINCOLN'S BODY LAY IN STATE.

the farm-houses were decorated, and their inmates had gathered in clusters, and by a light of bonfires caught a glimpse of the train that was bearing from their sight the remains."

At 7:00 A.M. the train arrived at Union Depot in Indianapolis. In the rain, a hearse fourteen feet long, fourteen feet high, covered with black velvet and decorated with white plumes, silver stars, and a striking silver eagle, drew the president to the rotunda of the Indiana State House. At the Washington Street entrance, the hearse passed under another massive arch, which featured pillars surmounted by busts of Washington, Webster, Clay, and, of course, Lincoln. The coffin was placed on a catafalque over which was suspended a pagoda-like, black canopy studded with silver stars. A bust of Lincoln, wearing a laurel wreath, was placed at the head of the coffin. The public viewing began at 9:00 A.M., and the mourners included a contingent of black Masons displaying a copy of the Emancipation Proclamation and banners bearing the mottoes "Colored Men, Always Loyal" and "Slavery Is Dead." By 10:00 P.M., more than a hundred thousand people had viewed the remains. City officials had planned a grand procession to escort Lincoln's body back to the railroad station, but heavy rain forced its cancellation.

"At midnight the route was resumed for Chicago," Townsend reported. "While the darkness prevailed, the approach to every town was made apparent by bonfires, torches, and music, while crowds of people formed an almost unbroken line."

At 8:30 A.M. on May 1, the train stopped briefly at Michigan City, Indiana, where decorations again made a deep impression on Townsend: "A succession of arches, beautifully trimmed with white and black, with evergreens and flowers, and with numerous flags and portraits of the President, was formed over the railway track." He took particular notice of the mottoes painted on two signs: "Abraham Lincoln, the noblest martyr of freedom, sacred thy dust; hallowed thy resting place" and "The purposes of the Almighty are perfect, and must prevail."

Near the arches Townsend spotted, through the darkness and illuminated by fires, sixteen young "maidens" dressed in white and black and singing "Old Hundred," a popular Civil War song set to a mournful tune. Another group of women attired in white, each carrying a small Union flag, stood on a flower-laden platform, encircling a woman who posed in a motionless tableau as the figure of America. Another sixteen ladies entered the funeral car and reenacted a now familiar ritual—in tears, they placed flowers on Lincoln's coffin.

On one of the arches, a motto spelled with flowers read: "Our guiding-star has fallen." In 1861, Lincoln's election won him the nickname "The star of the west, or the comet of 1861." Now, Walt Whitman called him "O powerful, western fallen star."

That same morning, Varina wrote another letter to Jefferson.

> *Washington*
> *9 O'Clock –*
> *Monday morning [May 1, 1865]*
>
> *My Dearest Banny,*
>
> *. . . I shall wait here this evening until I hear from the courier we have sent to Abbeville—I have given up hope of seeing you but it is not for long—Mr Harrison now proposes to go in a line between Macon and Augusta, and to avoid the Yankees by sending some of our paroled escort on before, and to make towards Pensacola—and take a ship or what else I can . . . still think we will make out somehow. May the Lord have you in his holy keeping I constantly, and earnestly pray—I look upon the precious little charge I have, and wonder if I shall it with you soon again—The children are all well Pie [infant daughter Varina Anne] was vaccinated on the road side, as I heard there was small pox on the road—she is well*

so far—the children have been more than good, and talk much of you . . .

Oh my dearest precious Husband, the one absorbing love of my whole life, may God keep you free from harm.

Your devoted wife.

Not knowing when she would see Jefferson again, Varina had to make a number of decisions by herself to protect her children. Before reaching the Savannah River, her party heard rumors of a smallpox epidemic. Burton Harrison described how Varina reacted: "We started the morning of the second day after I arrived in Abbeville, and had not reached the Savannah River when it was reported that small-pox prevailed in the country. All the party had been vaccinated except one of the President's children." Harrison revealed a telling detail absent from Varina's account: "Halting at a house near the road, Mrs. Davis had the operation performed by the planter, who got a fresh scab from the arm of a little negro called up for the purpose."

In Chicago, minute guns fired to announce the 11:00 A.M. arrival of the funeral train. The cortege pulled into a temporary station at Park Row, one mile north of the railroad depot. Tens of thousands of people had been waiting in the streets for hours. Others observed from buildings. According to one account, "every window was filled with faces, and every door-step and piazza filled with human beings, while every tree along the route was eagerly climbed by adventurous juveniles."

Mourners filled Lake Park, the big stretch of land east of Michigan Avenue, from the street to the shore of Lake Michigan. At the train station, ten thousand children massed behind military units and city officials. The *Chicago Tribune* suggested that upon Lincoln's arrival the waters of Lake Michigan, "long ruffled by storm, suddenly

The Chicago funeral arch.

calmed from their angry roar into solemn silence, as if they, too, felt that silence was an imperative necessity of the mournful occasion." A huge, triple-peaked Gothic funeral arch had been erected at the center of Park Place. Lincoln's coffin was laid near the arch, and thirty-six high school girls, each dressed in white and wearing a black crepe sash, placed a flower on the coffin.

Townsend remembered the biggest arch of the entire journey: "A magnificent arch spanned the street where the coffin was taken from the car, and under this the body rested while a dirge was sung by a numerous band of ladies dressed in white, with black scarves." The honor guard placed the coffin in the hearse, and the procession to the courthouse began.

The hearse, eighteen feet long and fifteen feet high, with a white satin sunburst mounted on black velvet at the head of the coffin, and drawn by ten black horses, moved west on Park Row to Michigan Avenue, then north on Michigan past Randolph Street, then west on Lake Street to the Courthouse Square at Clark Street. The *Chicago Tribune* estimated that more than 120,000 people marched in or witnessed the procession. One of them, Daniel Brooks, a guest of the Chicago Board of Trade, had, as a sixteen-year-old, taken part in George Washington's funeral procession in 1799.

Now Townsend observed "nearly every building on Michigan Avenue . . . was dressed in mourning, and many displayed touching mottoes." One man, Townsend recalled, "who had accompanied the train from Washington, telegraphed to have conspicuously laced on the front of his residence—'Mournfully, tenderly bear him to his rest.' He told me these words were suggested by the really tender care with which the Veteran sergeants—always the bearers—lifted and carried their charge." The hearse stopped at the south door of the courthouse, and the coffin was carried inside and laid upon a catafalque at the center of the rotunda.

The chamber was a confection of black, white, and silver crepe, fabric, velvet, metallic fringe, and more. Other cities had done that. Chicago boasted of an extra "new and solemn" decorative effect: "The roof of the catafalque . . . was a plain flat top of heavy cloth, in which were cut thirty-six stars. Over these were placed a layer of white gauze, and over this several brilliant reflectors, which caused the light to shine through the stars, upon the body below, with a softened, mellow radiance." One of the banners in the room read: "He left us sustained by our prayers; He returns embalmed in our tears."

Public viewing began at 5:00 P.M., and by midnight more than forty thousand people had viewed Lincoln's corpse.

Jefferson Davis spent an uneventful night in Cokesbury, South Carolina, at the home of General Martin Witherspoon Gary. He left there on May 2 before daylight, and at about 10:00 that morning he rode into Abbeville. The townspeople were happy to see him. Captain Willaim Parker, the naval officer safeguarding the Confederate treasure wagon train, had arrived before Davis. Parker turned the gold over to John Reagan, and released his young naval cadets from service. Parker called on Davis and found him alone at the home of Colonel Armistead Burt. They conferred in private for an hour.

"I never saw the President appear to better advantage than during these last hours of the Confederacy," remembered Parker. "He showed no signs of despondency. His air was resolute; and he looked, as he is, a born leader of men."

When Parker revealed that he had disbanded his command of naval cadets, Davis said, "Captain, I am very sorry to hear that," and repeated the words several times. Davis regretted the loss of a single soldier. Parker explained that Mallory had given the order. "I have no fault to find with you, but I am very sorry Mr. Mallory gave you the order."

Davis suggested that they remain in Abbeville for four days, but Parker warned him that if he stayed that long he would be captured. Davis replied that he would never desert the Southern people. "He gave me to understand," recalled Parker, "that he would not take any step which might be construed into inglorious flight. The mere idea that he might be looked upon as fleeing seemed to arouse him."

Davis rose from his chair and began pacing the floor, repeating several times that he would "never abandon his people." Davis's attitude emboldened Parker to speak frankly: "Mr. President, if you remain here you will be captured. You have about you only a few demoralized soldiers, and a train of camp followers three miles long. You will be captured, and you know how we will all feel that." Parker delivered almost an ultimatum: "It is your duty to the Southern people not to allow yourself to be made a prisoner." The naval officer told him how to escape: "Leave now with a few followers and cross the Mississippi . . . and there again raise the standard."

Davis refused, even though, Parker recalled, "I used every argument I could think of to induce him to leave Abbeville."

In the streets outside, order was breaking down, and Davis's presence there did nothing to deter people from breaking into government warehouses. "We witnessed," recalled John Reagan, "the raids made on the provisions by the citizens. I was forced to the thought that the line between barbarism and civilization is at times very narrow."

After Davis met with Parker, he conferred with several cavalry officers. "When we reached Abbeville," reported Reagan, "we were there joined by the remnants of five brigades of cavalry. The President had a conference with their commanders, and sought to learn of their condition and spirit."

Their flesh was weak, and their spirit was *not* willing. Davis could not motivate them to fight on. Stephen Mallory recounted the scene: "The escort was here collected, or so much of it as was left, and upon conversing with its officers, Mr. Davis was candidly apprised by some of them that they could not depend upon their men for fighting, that they regarded the struggle as over. The officers themselves, and a few men, were ready to do anything in their power to secure his safety; but he became satisfied that the escort was almost useless. He was again urged by his friends to push on south for Florida or west for the Mississippi to secure escape from the country; but the idea of personal safety, when the country's condition was before his eyes, was an unpleasant one to him, and he was ever ready to defer its consideration."

CHAPTER NINE

"Coffin That Slowly Passes"

While Davis dallied in South Carolina, the new president, Andrew Johnson, signed a proclamation on May 2 that offered a $100,000 reward for his capture. Johnson doubled the amount that Stanton had offered for Booth. Now Davis was the subject of the highest reward in American history. It would take days for news of the reward to reach the Deep South.

> BY THE PRESIDENT OF THE UNITED STATES OF AMERICA
>
> A PROCLAMATION
>
> WHEREAS it appears, from evidence in the Bureau of Military Justice, that the atrocious murder of the late President, ABRAHAM LINCOLN, and the attempted assassination of the Honorable WILLIAM H. SEWARD, Secretary of State, were incited, concerted, and procured by and between Jefferson Davis, late of Richmond, Virginia, and Jacob Thompson, Clement C. Clay, Beverly Tucker, George N. Saunders, William

Cleary, and other rebels and traitors against the Government of the United States, harbored in Canada:

NOW, THEREFORE, to the end that justice may be done, I, ANDREW JOHNSON, President of the United States, do offer and promise for the arrest of said persons, or either of them, within the limits of the United States, so that they can be brought to trial, the following rewards:

One hundred thousand dollars for the arrest of Jefferson Davis.

Twenty-five thousand dollars for the arrest of Clement C. Clay.

Twenty-five thousand dollars for the arrest of Jacob Thompson, late of Mississippi.

Twenty-five thousand dollars for the arrest of George N. Saunders.

Twenty-five thousand dollars for the arrest of Beverly Tucker.

Twenty-five thousand dollars for the arrest of William C. Cleary, late clerk of Clement C. Clay.

The Provost Marshall General of the United States is directed to cause a description of said persons, with notice of the above rewards, to be published.

In testimony whereof, I have hereunto set my hand and caused the seal of the United States to be affixed.

Done at the city of Washington, this second day of May, in the year of our Lord one thousand eight hundred and sixty-five, and of the Independence of the United States of America the eighty-ninth.

By the President:
ANDREW JOHNSON

Varina wrote to her husband on May 2, giving him the same advice that Parker had: "Do not try to meet me, I dread the Yankees getting news of you so much, you are the countrys only hope, and the very best intentioned do not calculate upon a stand this side of the river. Why not cut loose from your escort? Go swiftly and alone with the exception of two or three . . . May God keep you, my old and only love, As ever Devotedly, your own Winnie."

Mallory decided that there was nothing more he could do to help Davis. At Abbeville he resigned his post as secretary of the navy. His family needed him, he said, and he did not want to flee the country and abandon them. Still, he agreed to remain with Davis's party for a few more days.

By noon on May 2, the line of people in Chicago who were waiting to view Lincoln's remains stretched nearly a mile. They came all day, and when the doors to the courthouse were shut at 8:00 P.M., the thousands of people still waiting in line had to be turned away. The cortege, flanked by marchers bearing torches, exited the square and moved north through Washington and Market streets to the Madison Street bridge, and on to the St. Louis and Alton Railway Depot. The *Chicago Tribune* reminded readers of the city's special relationship with Abraham Lincoln: He practiced in the federal courts; he debated with Douglas, and five years earlier in May cannon and jubilations celebrated his nomination at the Wigwam convention center. The newspaper wrote that its city had "first summoned him from . . . obscurity . . . and demanded that the country . . . recognize . . . him [as] one fit to stand in high places." And now, "he comes back to us, his work finished, the Republic vindicated, its enemies overthrown and suing for peace; but alas! He returns with the crown of martyrdom, the victim of the dastard assassin . . . his calm, sad face was ever turned westward" to Chicago.

"Taken in all," boasted the *Tribune,* "Chicago made a deeper impression upon those who had been with the funeral train from the first than any one of the ten cities passed through before had done . . . seeing how other cities had honored the funeral, there seemed to be no room for more; and the Eastern members of the cortege could not repress surprise when they saw how Chicago and the North-west came, with one accord, with tears and offerings, to help bury 'this Duncan.'"

Townsend remembered that the "cortege left Chicago at half past nine o'clock P.M. As usual, night was forgotten by the people in their anxiety to show all possible respect for him whom they expected; and bonfires and torches threw their uncertain light upon mourning emblems which were destined to stand in their places as memorials for weeks to come."

The excitement aboard the train increased. This was the last night. In the morning, the funeral train would complete its journey.

Lincoln was in his home state now, and the emotions of the people huddled around the fires along the tracks reached a fever pitch. The passengers on the train saw more signs that they passed in the night. ILLINOIS CLASPS TO HER BOSOM HER SLAIN BUT GLORIFIED SON. COME HOME, read a sign posted on a house at Lockport. GO TO THY REST, said the one atop a large arch at Bloomington.

The next stop was Lincoln, Illinois, the first town in America named after Abraham Lincoln. This honor had been bestowed in the 1850s before the man became the president. It was a gesture meant to recognize his work as a lawyer. Tonight an arch over the railroad tracks leading into his town displayed his portrait and the motto "With malice toward none, with charity for all."

During the night of May 2 and through the early morning hours of May 3, the residents of Springfield were restless. They had anticipated Lincoln's homecoming since they heard the news of his death. At first, they had not been sure that he would come home at all. Mary Lincoln had tortured them about the disposition of his remains. She had rejected their initial proposal to bury him downtown, and then she had threatened to deny them his remains entirely and instead keep them in Washington, or send them to Chicago. Mary's ultimatums infuriated her former neighbors. How dare she deny them, they complained, their just reward for their long association with him?

SPRINGFIELD WELCOMES LINCOLN HOME. HIS OLD LAW OFFICE DRAPED IN MOURNING.

But once the citizens knew that Springfield would be his final resting place, they began frenzied preparations.

They had finished hanging the decorations and painting the signs. Crepe and bunting blackened the town. Lincoln's two-story frame house at Eighth and Jackson streets was a decorated masterpiece of mourning. Over the front door of his law office, through which he had passed countless times during his circuit-riding days, hung one of the most stunning and beautifully painted signs seen along the entire funeral route: HE LIVES IN THE HEARTS OF HIS PEOPLE. The townspeople had waited twenty days since Lincoln's death and thirteen days since the train had left Washington. Beginning tomorrow, over the next two days of May 3 and 4, Springfield would show the nation that no town loved Abraham Lincoln more. It was up to Springfield to stage the final act of the death pageant.

That night Judah Benjamin came to William Parker's room around 8:00 P.M. and begged him to call on Davis once more and persuade him to leave Abbeville. Parker agreed and proposed that he and three naval officers depart with Davis and escape to the eastern coast of Florida, where they might seize a boat and sail to Cuba or the Bahamas. Parker presented this strategy to Davis, who again refused. But as soon as Parker left, Davis summoned his cabinet. From that meeting, Mallory sent Parker a note saying that Davis had, in part, changed his mind. He agreed to leave Abbeville that night, but he would not break off from his escort and go with the four naval officers to make a run for the coast.

After Davis met with the cabinet he wrote a letter that evening to his private secretary about his plans. He confessed his low morale and disparaged the troops, which Lee had never done.

Abbeville S.C.
[2] May 65 9 P.M.

To Burton Harrison:
My Dear Sir:

The courier has just delivered yours and I hasten to reply—I will leave here in an hour and if my horse can stand it will go on rapidly to Washington [Georgia]—The change of route was I think judicious under the probabilities of the Enemy's movements.

I can however learn nothing reliable and have to speculate—I think all their efforts are directed for my capture and that my family is safest when furthest from me—I have the bitterest disappointment in regard to the feeling of our troops, and would not have any one I loved dependent upon their resistance against an equal force—

Many thanks for your kind attention and hoping as time and circumstances will serve to see you am as ever your friend

J

At 11:00 P.M., Davis left Abbeville. He, the cabinet, and his personal staff rode at the head of the column, in advance of the cavalry escort. The wagons carrying the Confederate treasury and deposits from the Richmond banks followed, accompanied by Secretary of War Benjamin and the troops.

Lincoln's funeral train steamed into Springfield on the morning of Wednesday, May 3. His journey was now almost complete. He had been on the move since April 21. It was as though, while the train stayed in motion, he wasn't quite dead. Edward D. Townsend dispatched his usual, matter-of-fact telegram to Secretary of War Stanton: "The funeral train arrived here without accident at 8.40 this morning. The burial is appointed at 12 p.m. to-morrow, Thursday [May 4]." The brief text spoke in a detached voice empty of emotion. Townsend concealed the feeling of relief that must have been his as the train rolled into the Springfield depot.

He recorded his true feelings years later in his memoirs:

"Thus closed this marvelous exhibition of a great nation's deep grief. It seemed as though for once the spirit of hospitality and of all Christian graces had taken possession of every heart in every place. Not one untoward event can be recalled. Every citizen rivaled his neighbor in making kindly provision for the comfort of the funeral

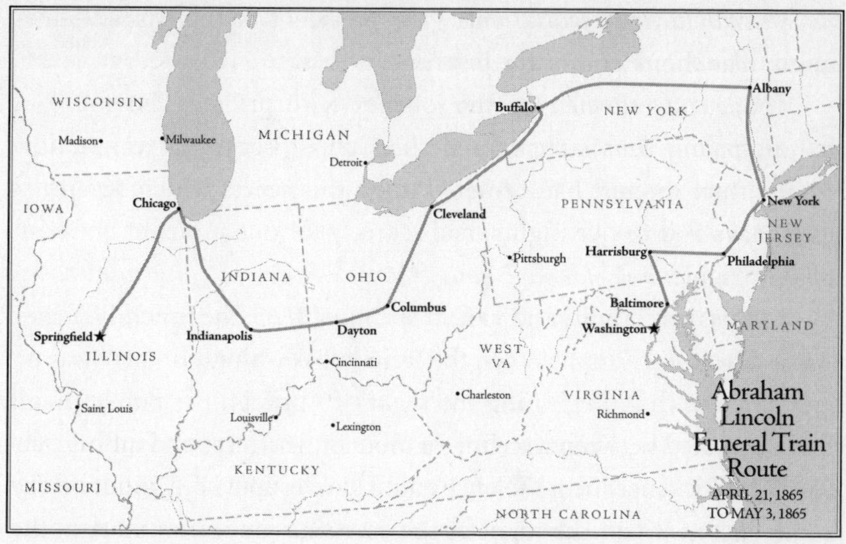

company while in their midst. Unstinted hospitality was not forgotten in the exceeding pains taken with the public displays."

Townsend had done it. Under his command, the funeral train had transported the corpse of Abraham Lincoln 1,645 miles from Washington, D.C., to Springfield, Illinois, and it had arrived on schedule. During its thirteen-day odyssey, the train never broke down, suffered an accident, or deviated from the master timetable. At every stop along the way, the honor guard performed flawlessly. Not once did they falter in their handling of the heavy casket. Carrying the coffin off the train, loading it into the hearse, unloading it from the vehicle, carrying it upon their shoulders to the place of public viewing and laying it down on the catafalque, raising it from the catafalque after the viewing, and then reversing the process to bring Lincoln back to the train required stamina and concentration. Whenever they carried the president's body, whether on level ground, up and down steep winding staircases, or onto a ferryboat, whether in daylight or darkness, in sunshine or a driving rain, the veteran Union army soldiers of the casket team never made a misstep. Now, in Springfield, they would carry the president of the United States upon their shoulders for the last time.

Townsend reflected on the journey with pride: "The guard of honor having thus surrendered their trust, began to realize how closely their interest had centered upon this object which, for [thirteen] days and twelve nights, had scarcely for one moment been out of their sight."

In this tumultuous and violent spring of 1865, the funeral journey was a peaceful triumph. When the train left Washington, there was no guarantee of that. Beginning the night of April 14, the public mood had fluctuated between a feeling of mournful sadness and an urge for bloodthirsty vengeance. Violence could have erupted any point on the route, just as it had the night of the assassination. In an instant, the funeral processions might have degenerated into ugly demonstrations against Jefferson Davis, the Confederacy, and the Southern people.

During those thirteen days in April and May, many Northerners, including Edwin Stanton, continued to believe that Davis and the Confederacy were behind John Wilkes Booth's conspiracy to murder Lincoln, Johnson, and Seward to throw the Union into chaos. Indeed, during Lincoln's death pageant from Washington to Springfield, every member of the cabinet, plus the Chief Justice of the United States, remained under armed guard around the clock to thwart anticipated additional assassinations.

But nothing untoward happened. The millions of Americans who had either viewed Lincoln's corpse, participated in the huge processions, or watched the train pass by remained peaceful. Yes, the petty criminals who always prowl through urban crowds—especially the infamous pickpockets of New York City—preyed on some of the mourners. But the crowds did not beat or murder anyone they judged guilty of insulting the martyred Lincoln. Nor did they cry out for vengeance upon the sight of Lincoln's corpse, or shout anti-Confederate epithets as the cortege rolled by.

Even the signs and banners spoke words of mourning, not vengeance. Only a handful demanded justice—or revenge. Of all the public utterances, from the White House funeral in Washington to the graveside prayer in Springfield, and at all points between, only once did an overwrought orator surrender to an explicit impulse for vengeance. Instead, the bereaved millions adopted Lincoln's second inaugural message of peace and reconciliation as their own.

From Springfield, General D. C. McCallum, the superintendent of the United States Military Railroad, who had ridden the rails all the way from Washington to ensure that everything went as planned, also reported to Stanton that he had accomplished his mission: "The duty assigned me has been completed promptly and safely, and I believe satisfactorily to all parties."

Like Townsend, McCallum understated the meaning of what he had done. The journey of the Lincoln funeral train across America was a tour de force of railroad engineering and military planning. Without the railroads to move troops, rifles, artillery, ammunition, rations, horses, equipment, and other supplies over thousands of miles of standardized track, the Union might not have won the Civil War. Railroad technology had proven to be a key advantage over the Confederacy. Yes, the North might have prevailed in the end, but without railroads, victory would have taken longer, and at a price more dear in blood. Trains helped win the war, and now, at its end, one train, its progress followed by an entire people, helped bring the country together.

Lincoln was home, back at the Great Western Railroad station where his journey began four years earlier, on February 11, 1861, one day before his fifty-second birthday. When he left for Washington that morning, he contemplated that he might never return. He stood on the platform of the last car, looked at the faces of his neighbors, and spoke:

> My friends, no one, not in my situation, can appreciate my feelings of sadness at this parting. To this place, and the kindness of these people, I owe everything. Here I have lived a

> quarter of a century, and have passed from a young man to an old one. Here my children have been born, and one is buried. I now leave, not knowing when, or whether ever, I may return, with a task before me greater than that which rested upon Washington. Without the assistance of the Divine Being who ever attended him, I cannot succeed. With that assistance I cannot fail. Trusting in Him who can go with me, and remain with you, and be everywhere for good, let us confidently hope that all will yet be well. To his care commending you, as I hope in your prayers you will commend me, I bid you an affectionate farewell.

Four years later, he had returned.

Not long after Lincoln's remains reached Springfield, Jefferson Davis arrived in Washington, Georgia. His party had crossed the Savannah River very early in the morning, and while en route to Washington, they were informed that the federal cavalry was at that place, and they were looking for Davis. They stopped at a farmhouse and ate breakfast and fed their horses.

By this time, Davis's escort was war weary and demoralized. They wanted to go home. John Reagan knew what else they wanted: "After they crossed the Savannah River and camped, and before reaching Washington, [Davis's] cavalry, knowing that they were guarding money, demanded a portion of it." If the government on wheels failed to pause here to pay them, they were going to seize the money. "[Breckinridge] told me that after he reached Washington the cavalry demanded that the silver and gold coin, equal to the amount of the silver bullion, should be divided among them, and that he and the officers commanding them found it necessary to yield or to risk their forcibly seizing it."

It was here that Judah Benjamin decided to leave Davis and make his own escape. The president's pace was too leisurely for Benjamin's taste, and he thought he would have a better chance on his own. He had never been comfortable riding a horse and set out in a carriage. Reagan spoke to him before he set off: "I inquired where he was going. 'To the farthest place from the United States,' he announced with emphasis, 'if it takes me to the middle of China.' He had his trunk in the carriage with his initials, J. P. B., plainly marked on it. I inquired whether that might not betray him. 'No,' he replied, 'there is a Frenchman traveling in the Southern States who has the same initials, and I can speak broken English like a French-man.' "

Benjamin's departure deprived Davis's party of the good humor of its court jester in chief. It also suggested how Davis should travel from this point on—alone, or with no more than a couple of aides. Benjamin's strategy served him well in the days ahead. His secretary of state gone, Davis mounted his horse and led the way to Washington, wary of reports of its occupation by the enemy. "We found no Federal cavalry at Washington," recalled Reagan, "where we remained a few days. Before reaching that place, General Breckinridge and myself, recognizing the importance of the capture of the President, proposed to him that he put on soldier's clothes, a wool hat and brogan shoes, and take one man with him and go to the coast of Florida, ship to Cuba, thence by an English vessel to the mouth of the Rio Grande. We proposed [he] take what troops we still had, to go west, crossing the Chattahoochee between Chattanooga and Atlanta, and the Mississippi River, and to meet him in Texas. His reply to our suggestion was: 'I shall not leave Confederate soil while a Confederate regiment is on it.' "

Davis had been willing to abandon Richmond—and its citizens—for the good of the Confederacy, but what he told Reagan was not sound military strategy. If he hoped to avoid capture, his advisers were right. He needed to move fast to the Mississippi River or Florida.

When Davis trotted his horse into Washington, Georgia, late on the morning of May 3, accompanied by an advance party of about forty men, the people welcomed their president as if he rode at the head of a triumphant army. Fate had spared this town during the war, and the citizens, unlike many in neighboring North Carolina, had not turned against the Confederacy. Eliza Andrews described the scene:

> About noon the town was thrown into the wildest excitement by the arrival of President Davis. He is traveling with a large escort of cavalry, a very imprudent thing for a man in his position to do, especially now that Johnston has surrendered, and the fact that they are all going in the same direction as their homes is the only thing that keeps them together. He rode into town at the head of his escort . . . and as he was passing by the bank . . . several . . . gentlemen were sitting on the front porch, and the instant they recognized him they took off their hats and received him with every mark of respect due the president of a brave people. When he reined in his horse, all the staff who were present advanced to hold the reins and assist him to dismount.

A rumor spread that Yankees in pursuit of Davis were advancing on the town from two directions, but there was nothing to it. Another wild rumor spread through town that Davis was there. Once "the president's arrival had become generally known," Eliza wrote with pride, "people began flocking to see him." This was the warm welcome that Greensboro and Charlotte, North Carolina, had withheld from Davis. In Georgia, the people's delight at his presence improved his morale.

Davis received a dispatch from Secretary of War Breckinridge reminding him of their conversation the previous night in Abbeville, urging him to flee, and reporting on the military situation. There were almost no Confederate soldiers in the vicinity to protect Davis: "The troops are on the west side of the Savanah, and guard the bridge," Breckinridge wrote. "A pickett which left Cokesbury after dark last evening reports no enemy at that point. I have directed scouts on the various roads this side of the river. The condition of the troops is represented as a little better, but by no means satisfactory. They cannot be relied upon as a permanent military force. Please let me know where you are."

Varina also sent a letter telling him to not make a stand, to leave his escort, and to flee.

In Springfield, the honor guard removed Lincoln's body from the train, escorted it to the State House, where he had served as a legislator, given his famous "House Divided" speech, and, in another part of the building, set up an office after his election as president. His guards laid the coffin on a catafalque in the Hall of Representatives. Springfield was not a great American city, and its officials knew they could not hope to rival the pageantry displayed in Washington, Philadelphia, New York City, or Chicago. Nor could Springfield match the stupendous crowds or financial resources marshaled by the major cities. Indeed, Lincoln's hometown had to borrow a hearse from St. Louis. However, what the state capital could not offer in splendor, it vowed to lavish in an emotional catharsis that would outdo every other city in the nation.

Few in Springfield were disappointed that Mary Lincoln was not on that train, even if it meant no Lincolns had made the trip from Washington. This morning, for the first time since the funeral train left Washington, the honor guard also removed Willie's coffin from the presidential car.

The embalmer Charles Brown and undertaker Frank Sands opened Lincoln's coffin. He had been dead for eighteen days, and his corpse had not been refrigerated. Only preservative chemicals and makeup had kept him presentable during the journey. At the beginning, at the White House funeral, Lincoln's face looked almost natural. He changed along the way. People had started to notice it as early as New York City. The face continued to darken, making necessary several reapplications of face powder during the trip. Travel dust and dirt had settled on the corpse during each open-coffin viewing, and the body men had to dust his face and black coat faithfully. Lincoln no longer resembled a sleeping man. Now he looked like a ghastly, pale, waxlike effigy.

The doors to the State House opened to the public at 10:00 A.M. on May 3 and stayed that way for twenty-four hours. It was the first round-the-clock viewing of the entire funeral pageant. Mourners ascended the winding staircase to the Representatives' Hall, approached the corpse from Lincoln's left, walked around his head, and then departed down the same stairs. During the night, trains continued to arrive in Springfield, and people without lodgings wandered the streets until dawn.

By 10:00 A.M. on May 4, seventy-five thousand people had passed by the presidential body. The coffin was removed from the capitol and placed in the hearse waiting on Washington Street. The procession began at 11:30 A.M., passing by Lincoln's home at Eighth and Jackson streets, then heading west to Fourth Street, and down Fourth to Oak Ridge Cemetery, about a mile and a half from town. Oak Ridge was not a traditional urban cemetery with tightly spaced headstones lined up in rows. Instead, it was a product of the rural cemetery movement that had swept America, which had transformed old-fashioned graveyards into nature preserves with brooks, sloping valleys, oak trees, and tombs situated in sympathy with the natural landscape.

Lincoln's guards removed his coffin from the hearse, carried it into the limestone tomb, and laid it on a marble slab. Willie's coffin rested near him.

Bishop Matthew Simpson, who had officiated at the White House funeral, delivered the last oration at Oak Ridge Cemetery. "Though three weeks have passed," he reminded his listeners, and "the nation has scarcely breathed easily yet. A mournful silence is abroad upon

The Springfield tomb.

the land." Simpson then set the unprecedented pageant in historical context: "Far more eyes have gazed upon the face of the departed than ever looked upon the face of any other departed man. More eyes have looked upon the procession for sixteen hundred miles or more, by night and day, by sunlight, dawn, twilight and by torchlight, than ever before watched the progress of a procession." It was the end of an era, he said: "The deepest affections of our hearts gather around some human form, in which are incarnated the living thoughts and ideas of the passing age."

Simpson read Lincoln's second inaugural speech at tomb-side. Invoking the president's mantra of "Malice toward none," Simpson proposed forgiveness for the "deluded masses" of the Southern people: "We will take them to our hearts." And we must, said Simpson, continue Lincoln's work: "Standing, as we do today, by his coffin and his sepulcher, let us resolve to carry forward the work which he so nobly begun."

But the bishop scorned Jefferson Davis and other Confederate leaders:

> Let every man who was a Senator . . . in Congress, and who aided in beginning this rebellion, and thus led to the slaughter of our sons and daughters, be brought to speedy and certain punishment. Let every officer . . . who . . . has turned his sword against . . . his country, be doomed to a felon's death. This . . . is the will of the American people. Men may attempt to compromise and restore these traitors and murderers to society again, but the American people will rise in their majesty and sweep all such compromises . . . away, and shall declare that there shall be no peace to rebels.

This shocking, tomb-side lust for revenge echoed Job Stevenson's remarks in Columbus. It would have horrified Lincoln.

Lincoln's pastor, the Reverend Dr. Gurley, who had completed the long journey from the assassination-night deathbed at the Petersen house to the grave in Springfield, again gave the last prayer, which was followed by a funeral hymn he had composed for the occasion. There was nothing more to say. They closed the iron gates and locked Abraham and Willie Lincoln in their tomb. Then everybody went home.

Carl Sandburg evoked the dénouement better than any witness present that day: "Evergreen carpeted the stone floor of the vault. On the coffin set in a receptacle of black walnut they arranged flowers carefully and precisely, they poured flowers as symbols, they lavished heaps of fresh flowers as though there could never be enough to tell either their hearts or his.

And the night came with great quiet.

And there was rest.

The prairie years, the war years, were over."

The coffin was just one of the things that made the Washington, D.C., events the most expensive funeral in American history. In a bound accountant's ledger titled "Funeral Expenses of the late Abraham Lincoln," handwritten in a clerk's neat script on twenty-eight lined, blue-gray pages, is the itemized list of the costs. It is all here—the names of the vendors, the goods or services they provided, and the price. No matter how trivial the purchase or inconsequential the cost, the information did not escape the ledger. These facts and figures, dry and impersonal as they are, and most never published, form a strangely fascinating book of the dead.

The government purchased wagonloads of fabric to hang in mourning—several thousand yards of black cambric, fine white silk, alpaca, cotton velvet, black crepe, black Silesias, and black draping—along with boxes of nails and tacks to attach the textiles to the major public buildings. John Tucker & Co. billed the government $161.00 for "Labor and material at President's House," including "1260 feet of lumber plus 1 gross screws, and 40 pounds of nails plus labor." The wood was used to build the bleachers for the funeral in the East Room. Another firm submitted an invoice of $358.14 for "Preparing East Room for President's Funeral." The largest bill—$4,408.09—was from John Alexander, which covered a variety of goods and services, including more than three thousand yards of fabric, twelve boxes of pins, and thirty packs of tacks, plus "putting front of President's House in mourning ($50.00) [and] East Room ($30.00); upholstering catafalque East Room ($75.00); [upholstering] Funeral Car ($50.00); [upholstering] Rail Road Car ($85.00)." The last expense connected with the White House was not submitted until May 27, five weeks after the funeral: "Removing draping and platform from East Room. $45.20, less for lumber returned."

Fabrics were purchased not only to clothe the public buildings but to dress Mary Lincoln too. On April 19, the day of the funeral, Harper & Mitchell submitted its bill for "1 mourning dress ($60.00); 1 shawl ($25.00); 1 crape veil ($10.00); 5 yards black crape ($20.00); Gloves and handkerchiefs ($7.50); 5 pair hose ($5.00); 1 crape bonnet ($15.00) TOTAL $142.50." Mary Lincoln, who remained in seclusion, had no public use for the black mourning dress and accessories.

Decorating the Capitol and preparing it to receive Lincoln's corpse involved additional costs. Benjamin Brown French submitted a bill for "services superintending the draping of the Rotunda & erection of Catafalque $15.00, plus $2.50 reimbursement for cash paid for ribbon," and John R. Hunt invoiced the government $20.00 for "upholstering catafalque and draping west wing U.S. Capitol." E. H. Litchfield and eight additional carpenters, riggers, gasfitters, and assistants were paid $33.50 to get the rotunda ready for Lincoln to lie there in state on April 19 and 20: "Extra services draping the Dome and lighting the gas and attending the same." And George Whiting submitted bills of $28.50 and $25.05 for "refreshments sent to Capitol for Officers in charge of the President's remains" on the two days that Lincoln lay in state.

Then there were the expenses connected to the president's body. Cooling & Bros. billed the government $6.00 for the hearse used to remove the corpse from the Petersen house, $75.00 for the six horses that pulled the hearse on the funeral day procession, plus one hearse for "removing President and son," on April 21, the day Lincoln's train left Washington. Drs. Brown and Alexander charged $100.00 for "embalming remains of Abraham Lincoln late President of the U.S.," and $160.00 to ride the train: "16 days Services for self & Asst. @ $10.00 per day." The single most expensive item recorded in the ledger was the casket, from the firm of Sands & Harvey: "Coffin covered with fine Broad Cloth, lined with fine White Satin silk trimmed with best mounting, solid silver plate, bullion fringe tassles & etc, heavy lead lining & walnut outside case for the late President, Abraham Lincoln. $1500.00."

The funeral expenses totaled $28,985.31. As an addendum, one last entry was written on an otherwise blank page in the ledger: "Money actually expended in attending the Widow from Washington D.C. to Chicago Illinois. $47.00." That expense closed out the book.

Back in Washington, D.C., the federal government shut down the day of the funeral. Offices were closed, flags flew at half-staff, public buildings remained draped in black, and military officers wore black crepe ribbons around the coat sleeves of their uniforms. John Wilkes Booth lay in a secret, unmarked grave on the grounds of the U.S. Army penitentiary at the Old Arsenal along the river, and at that walled, fortress prison eight of his accused conspirators languished in shackles and hoods awaiting their trial by military tribunal. Those proceedings would begin in a few days. Ford's Theatre remained closed and under guard. Secretary of State William Seward recovered from his wounds. At the White House, Lincoln's office had been sealed like a ship in a bottle, preserved just as the president left it on the afternoon of April 14. His widow continued her refusal to vacate the Executive Mansion, thus denying its proper use by the new president, Andrew Johnson. She had become the subject of much talk. At the Petersen house, Private William Clarke went to sleep each night covered by the same quilt that had warmed the dying president. Soldiers back from the war once again got drunk in saloons, and people dining out at public houses gorged themselves on the delicacy of the day, fresh oysters. George Harrington, back to his routine, went about his usual business at the Treasury Department. And Secretary of War Edwin M. Stanton thought about Jefferson Davis.

All across America, cities marked the hour of Lincoln's entombment. One newspaper stated: "by the open grave of Abraham Lincoln stood this day the American people . . . a nation in . . . mourning looking into the open grave of a President . . . do not forget that open grave, nor the unparalleled crime which caused it to be dug." It was true. On this day, at the precise hour—noon, Springfield time—tens of millions of his fellow citizens "all across this broad land," in the words of his first inaugural address, paused to honor Abraham Lincoln.

In Washington, Georgia, a few hundred of Jefferson Davis's fellow citizens honored him on his second day with them. Eliza Andrews was thrilled to have Davis in her town: "I am in such a state of excitement . . . Father and Cora went to call on the President, and . . . father says his manner was so calm and dignified that he could not help admiring the man. Crowds of people flocked to see him, and nearly all were melted to tears." Not only did the townspeople gather around Davis but they put together an enormous feast. "The village sent so many good things for the President to eat," recalled Eliza, "that an ogre couldn't have devoured them all, and he left many little delicacies, besides giving away a number of his personal effects, to people who had been kind to him."

It was in Washington that Mallory left the president's caravan. Davis understood that it was time for Mallory to return to his family and despite the desperate situation, he took time to compose a warm farewell letter: "It is with deep regret that I contemplate this separation. One of the members of my first cabinet we have passed together through all the trials of war and not the less embarrassing trials to which the Congress has of late subjected the Executive . . . I will ever gratefully remember your uniform kindness and unwavering friendship to myself." Davis did not know it then, but it was the last letter he would write as president of the Confederate States of America.

Next he made what would be his last official appointment. A clerk drafted the document for his signature: "M. H. Clark Esq is hereby appointed Acting Treasurer of the Confederate States, and is authorized to act as such during the absence of the Treasurer."

Davis then finally agreed to take action he should have chosen days ago. "After some delay at Washington," Reagan remembered, "we induced Mr. Davis to start on south with an escort of ten men, his staff officers and secretary, and to leave General Breckinridge to wind up the business of the War Department, and me to close the business of the Post Office Department and the Treasury . . . We were then to go on and overtake him." Davis left Washington in the morning. Reagan stayed behind in the town, planning to catch up with his president sometime that night. After Reagan wasted valuable time burning piles of Confederate currency, he left Washington by midnight.

Eliza Andrews watched Davis ride out the night of May 4: "The President left town about ten o'clock, with a single companion, his unruly cavalry escort having gone on before. He travels sometimes with them, sometimes before, sometimes behind, never permitting his precise location to be known." She had heard rumors that the Union army did not want to capture Davis: "The talk now is . . . that the military authorities are conniving the escape of Mr. Davis . . . The general belief is that Grant and the military men, even Sherman, are not anxious for the ugly job of hanging such a man as our president, and are quite willing to let him give them the slip, and get out of the country if he can. The military men, who do the hard and cruel things in war, seem to be more merciful in peace than the politicians who stay at home and do the talking." Davis's departure from Washington made the feisty and irrepressible Eliza sad: "This, I suppose, is the end of the Confederacy."

But Davis's war years were not over yet. Abraham Lincoln's journey may have ended, but Jefferson Davis was determined to press on. That night Davis's party camped near the Ogeechee River. The next day Union cavalry rode into Washington. Eliza Andrews had been right. Davis's departure *was* the end, at least in her town.

Northern newspaper editorials reflected upon the meaning of the assassination and the funeral train:

> Twenty days after the terrible night on which the assassin's bullet destroyed the most precious life in the American nation,

> the body which that great and good man animated, is deposited in the humble cemetery . . . in ceremonies which are the saddest that may ever be performed on American soil.
>
> What do those twenty days suggest! Twenty days of National mourning; twenty days with flags at half-mast; twenty days with emblems of sorrow on the peoples' dwellings, with sable drapery and solemn mottoes on all public buildings; twenty days of such tokens of love, such tributes as never before were paid to mortal man?

For the previous three weeks, the newspapers had obsessed over the minutiae of the funeral obsequies. No insignificant detail was too obscure to observe and print. They reported it all by naming, in each city, hundreds of people who had participated in the local procession; by identifying every military unit—and its roster of officers—that marched or rode in the cortege; by identifying every public official and dignitary in the parade, down to specifying in which carriages they rode; by identifying every local civic organization in the procession, including descriptions of the costumes worn by their members; by naming every band and every tune played; and by naming every weeping woman who had laid flowers upon the coffin.

The papers took particular interest in the hearses constructed in each city for the president; they described these vehicles with admiring, exacting, and even maniacal detail. The published description of each hearse was so precise it could have served as a verbal blueprint for the construction of an identical replica without the aid of drawings, plans, or photographs. Newspapers in New York and Chicago gave special credit to the local hearse makers who dreamed up these impossible, extravagant vehicles. Lincoln, who ridiculed decorative excess, would likely have laughed at the sight of them, and been embarrassed that his remains were carted about in vehicles so costly in labor and materials. Some of them looked like small houses, not hearses, and they were as big as the log cabin where Lincoln was born, or the one he occupied as a young man.

Reporters had taken note of every sign, banner, floral arch, or bonfire they saw along the railroad tracks, and they copied down mottoes they saw along the procession routes. In New York City, a book published after the obsequies preserved the texts of hundreds of signs and banners, even revealing where each one was observed. A century and a half after Lincoln's funeral procession passed through old Gotham, one can stand at 356 Broadway and know that on April 25, 1865, a sign in the window read "God moves in a mysterious way, His wonders to perform"; or at 555 Broadway, where the sign read "A nation bowed in grief will rise in might to exterminate the leaders of this accursed Rebellion"; or at 759 Broadway, where it read "He was a man, take him for all in all."

Newspapers published every word of every sermon, every prayer, every oration, and every impromptu speech uttered. Moreover, several hundred ministers printed pamphlets that preserved for posterity the sermons they preached on Black Easter, Sunday morning, April 16, and on the next several Sundays.

Whenever the president's corpse was carried off the train for public viewing, newspaper stories swooned with lavish testimonials supplied by awestruck journalists. Their accounts of richly designed catafalques lying amid their exquisitely morbid surroundings sound more like the enthusiasms of florists and interior decorators than observations by seasoned war reporters. These death chambers resembled voluminous Arabian tents erected indoors, fashioned from hundreds of yards of black fabric, accented with silver highlights. Newspapers singled out the visionaries who created these fantastic settings. Those lucky enough to view Lincoln's corpse loved it—they had never witnessed anything as impressive.

Every story mentioned the floral arrangements—their appearance, preciousness, and scent; crosses, anchors, and wreaths of only the choicest japonicas, roses, jet blacks, and other types, either suspended in midair, presented in Greek vases, laid on the coffin, or placed near it on floors carpeted with evergreens.

The combined effect of these black chambers of death, the heaps of beautiful flowers and their overpowering sickly sweet odor, the coffin open to view, and the face of the martyred president, frozen and ghostly pale, must have overwhelmed the senses of the more than one million Americans who experienced it. Jeremiah Gurney's controversial and long-lost New York City photograph, the sole surviving image taken of Lincoln in death, can only hint at the awesome majesty of the scene.

These stories carried every American who read them to Lincoln's side and allowed people to imagine what it must have been like to behold his face, or to watch his coffin pass by. And not just in the cities and states closest to them. Journalists made it possible for the American people to ride aboard that train, and to imagine they had marched in every procession, joined every torchlight parade, heeded every prayer, inhaled the scent of every flower, and wept at the coffin's every opening.

For the one million Americans who had viewed Lincoln in death, the stories reminded them of the wonders they had seen. For the seven million who kept vigil along the route of the passing train, the stories told of places where it had been, and where it had gone.

In the Old Testament, David's child is struck down by an angry God. "Now he is dead, wherefore should I fast?" David asked. "Can I bring him back again? I shall go to him, but he shall not return to me" (2 Samuel 12:23). The Lincoln funeral train fulfilled and simultaneously dethroned the truth of this biblical lament. Yes, he was dead, and the people went to him. But not all of his people could make that journey. So, although dead, Lincoln did return. He returned to the people of Pennsylvania, New Jersey, New York, and Ohio who voted for him in the presidential elections of 1860 and 1864, and to the well-wishers he met during his inaugural journey east in 1861; he returned to Indiana, site of early boyhood memories; he returned to Chicago, city of his political destiny; he returned to the prairies of Illinois and his old clients; and he returned to Springfield, his home for a quarter century.

Lincoln's coffin became a kind of ark of the American Covenant, possessing hidden meanings and mysterious powers. The death pageant was both a civic and a religious event. Through the national funeral obsequies, Americans mourned the death of their president and elevated him to the pantheon of American political sainthood, equaled only by George Washington. They honored his achievements: He had won the war, saved the Union, and set men free. They united behind his principles and vowed to bear the burden of his "unfinished work." And they reaffirmed, by the tributes they paid to him, that his great cause was worth fighting and dying for.

The death rites also had religious significance. Millions of the faithful pondered why God had allowed Abraham Lincoln to be murdered at the height of his accomplishments and glory. Across the country, ministers compared Lincoln's Good Friday assassination to the passion of Christ. Their sermons suggested a divine purpose overshadowing Lincoln's death. God had called him home, some suggested, because his work was done. God took him now, others warned, because he would have been too merciful to rebel traitors. Now was not the time not for Lincoln's mercy, but for justice and vengeance. Herman Melville, author of *Moby-Dick*, sensed this mood. "Beware the people weeping," he wrote, "when they bare the iron hand." Soon, the fates of Jefferson Davis, the Confederate leadership, and John Wilkes Booth's conspirators would be decided by some of the very people now weeping for their martyred hero.

The twenty-day death pageant transfigured Abraham Lincoln from man to myth. On the day he was murdered, he was not universally loved—even in the North. His traveling corpse became a touchstone that offered catharsis for all the pain the American people had suffered and stored up over four bloody years of civil war. For whom did they mourn? For their slain president, of course. But this outpouring of national sorrow could not be for just one man. "Not for you, for one alone; / Blossoms and branches green to coffins all I bring," wrote Walt Whitman. And so they mourned, not for this one man alone but for all of the men; for every son, every brother, every lover, and every husband, and every father lost in that war. It was as though, on that train, in that coffin, they were *all* coming home. Lincoln's death pageant for Abraham Lincoln was a glorious farewell to him and to the three hundred and sixty thousand men of the Union who, like their Father Abraham, had perished for cause and country.

Years later, General Edward Townsend reflected upon what had been, for him, the journey of a lifetime:

> Mr. Lincoln, on his way from Springfield to Washington in 1861, had passed through all the cities where now his mortal remains had rested for a few hours on their way home. At the principal places he had had enthusiastic public receptions. There could not now be wanting many sad contrasts in the memories of those who had participated in the first ovations to the new President, and who now remained to behold the last of him on earth. Can there be imagined one item wanting to perfect this grandest of human dramas? It is entire; it is sublime!

It took a poet, Walt Whitman, to summarize in a few lines the meaning of it all:

When lilacs last in the door-yard bloom'd,
And the great star early droop'd in the western sky in the night,
I mourn'd—and yet shall mourn with ever-returning spring.
O ever-returning spring, trinity sure to me you bring,
Lilac blooming perennial, and drooping star in the west,
And thought of him I love.

O powerful western fallen star!
O shades of night—O moody, tearful night!

Coffin that passes through lanes and streets,
With the tolling bells' perpetual clang,
Here, coffin that slowly passes,
I give you my sprig of lilac.

Not for you, for one alone;
Blossoms and branches green to coffins all I bring.

After Lincoln was in his tomb, the funeral train party broke up. Not all of the travelers would return to Washington together on that train. Yes, the military officers and the escort would ride back as a group and, once they emptied the compartment of armfuls of decaying flowers, they would take the vacant presidential car back to Washington. The train, which departed Springfield on May 5, would not retrace the identical route by which it had come. Torches and bonfires would not light the way home. No crowds bid the funeral train farewell when it left Springfield for its anticlimactic, homeward-bound journey. Curious eyes might have noticed the train as it chugged east, but huge crowds no longer gathered at the depots to watch it pass, and no one fired cannons.

The elected officials, government appointees, and other special passengers who had ridden the train from Washington to Springfield made their own way back. Their special War Department pass entitling them to free, round-trip travel did not restrict them to the presidential train. Many, including members of the Illinois delegation, tarried in Springfield or visited other local points. Some made stops in other states before they returned to Washington.

In late April, Major General James Wilson, the twenty-seven-year-old Union cavalry prodigy, heard rumors that Davis had left Charlotte and was traveling south through the Carolinas, heading for Georgia. The chase for Jefferson Davis was not like the manhunt for John Wilkes Booth. On April 2, the night Davis fled Richmond, Abraham Lincoln still had a war to win. Victory occupied his mind, not the whereabouts of the Confederate president. Once Davis abandoned his capital, Lincoln considered him irrelevant, and certainly not a serious threat to Union military operations. The danger came from the armies of Robert E. Lee and Joseph E. Johnston. To Lincoln, the fleeing Davis was of little tactical or strategic importance. For other reasons, Lincoln did not want to capture Davis at all. To help heal the rift between North and South, Lincoln wanted no treason trials or prison sentences, and certainly no public hangings. He cued his cabinet and several of his generals on his desires. Between April 2 and April 14, Lincoln issued no orders to hunt down Davis. Instead, Lincoln had issued him an unwritten, unofficial free pass to escape.

That changed with the assassination. Now the federal government wanted Davis not only as a traitor but as a suspect in the murder of the president. As Stanton and the War Department began the pursuit of Booth, the Bureau of Military Justice began building a legal case against Davis. But the capture of Booth, not Davis, was the first priority. Within hours of the assassination, the first cavalry units rode out in pursuit of the assassin. His trail was still hot, he could not have ridden far from Washington, and, as far as Stanton knew, he was alone. Small units of soldiers and detectives could be employed effectively to search Maryland and northern Virginia.

By April 14, Davis's trail had gone cold. He had traveled far from Washington, and at some points during his escape, he enjoyed the protection of up to a few thousand armed and mounted men. Davis had traveled too far to enlist Washington detectives, or to send small

The first reward poster for Jefferson Davis.

cavalry units deep into Virginia and the Carolinas in pursuit. And if they located his position, Davis's armed escort could outgun them. No, the chase for Davis would require more men, probing deeper south in a wide screen spread across a few hundred miles of territory. The War Department had to throw out a wide net, and hope that Davis stumbled into it. But because the war was not over, and because Booth remained at large until April 26, the hunt for Davis was postponed and did not begin in earnest until early May.

On May 7, Wilson ordered Colonel Robert Minty, commander of the Second Cavalry Division, to, as Minty stated later, "make immediate arrangements to prevent the escape of Jefferson Davis across the Ocmulgee and Flint Rivers, south of Macon." Minty's old unit, the Fourth Michigan Cavalry, was in Macon. Recruited in Detroit in 1862, the Fourth was an experienced, combat-seasoned, hard-riding regiment. He directed its commander, Lieutenant Colonel Pritchard, to leave that evening in pursuit of Davis. Minty told him where to set up a screen of pickets, but that if he learned that Davis had already crossed the Ocmulgee, to follow and "capture or kill him."

Pritchard and about four hundred men departed Macon at 8:00 P.M., intending to proceed down the south bank of the river for seventy-five to a hundred miles, scouting the country on both sides of the river "as far as the strength of my command would permit for the purpose of capturing Jeff. Davis." After marching seventy-five miles, Pritchard arrived at Abbeville at 3:00 P.M. on May 9. There he encountered Lieutenant Colonel Harnden, commander of the First Wisconsin Cavalry, who told him that a wagon train had crossed the Ocmulgee the previous midnight at Brown's Ferry a mile and a half north of Abbeville.

The Wisconsin men went off down the main road while Pritchard's followed the river route. The two units did not know that they were on a collision course. Before leaving Abbeville, Pritchard divided his command, taking with him 128 of his best riders plus officers. He left Abbeville at 4:00 P.M. and headed toward Irwinville.

On May 5, Jefferson Davis and the few men still traveling with him made a camp near Sandersville, Georgia. The next day, Burton Harrison directed Varina's wagon train to camp off the road near Dublin, Georgia. Around midnight, Davis's party stumbled upon her campsite. More than a month had gone by since Davis had seen his family, but now finally they were reunited. They traveled together on May 7 and that night camped between Dublin and Abbeville, Georgia.

The president took his eight-year-old son, Jefferson Davis Jr., shooting. Unlike the day in Richmond when Jefferson ordered pistol cartridges for Varina and taught her how to shoot a revolver in self-defense, this was for fun. The boy would have no need to defend himself with firearms. Colonel William Preston Johnston observed the target practice. The president "let little Jeff. shoot his Deringers at a mark, and then handed me one of the unloaded pistols, which he asked me to carry." When Davis and Johnston turned their discussion to their escape route, the colonel "distinctly understood that we were going to Texas." Johnston said that he did not think they could get there by going west through the state of Mississippi, suggesting it might be safer to make for the Florida coast and sail through the Gulf of Mexico to the Texas coast. "It is true," Davis replied, "every negro in Mississippi knows me." He guessed that it would be impossible to travel incognito through his home state without being recognized by at least one slave.

On May 8, Davis decided to part from his family and at dawn he rode on with his personal staff and a small military escort. By that night he had made little progress through heavy rains, and Varina's train caught up with him in Abbeville, a speck of a town consisting of just a few buildings. When Harrison finally found Davis, he was sleeping on the floor of an abandoned house. Word of a Yankee cavalry patrol twenty-five miles away in Hawkinsville persuaded Davis that his wife's party should drive on through the bad weather and not stop to rest. He was too tired to leave the house and come outside to see Varina. He would, he assured Harrison, catch up to them after he rested for a while in Abbeville. Later in the night, Jefferson's party followed Varina's, and they reunited before dawn on May 9. The two groups traveled twenty-eight miles together for the day, stopping at 5:00 P.M.

Davis decided to make camp for the night with Varina's wagon train near Irwinville. They pulled off the road, and the pine trees helped conceal their position. President Davis's escort did not set up a defensive camp, circling their wagons in a compact circle, picketing their horses and mules inside the ring, and pitching tents or laying out bedrolls within the perimeter. Davis was not camping on the western

plains of the frontier of his youth, and he expected no attack from Native Americans during the night. Forming a wagon train into a circle made sense on the wide open plains, but not in the Georgia pines.

If Union cavalry discovered his position and charged his camp in force, his small entourage could not outgun them, and if the federals were able to surround a small camp drawn up in a tight circle, it would be difficult for Davis to take advantage of the confusion of battle and escape. So, instead, Davis's party pitched camp with an open plan, scattering the tents and wagons over an area of about one hundred yards. Now, any Yankee who rode into one part of the camp during the night would not be able to see to the other side of it. A small force of eight to twelve enemy cavalrymen could not gain control of the entire camp, and the men guarding the president had a decent chance of outfighting such a small patrol.

If the chosen few of the president's escort had come this far, now that their numbers had dwindled to less than thirty, from a force of several thousand men, they could be trusted to fight to the death to save the president and his family. If there was a fight, then Davis, unless captured at once, could escape into the woods while it was dark.

The arrangement was perfect, but for one oversight. Tonight the camp posted no guards to keep vigil through the dawn. A handful of cavalrymen, led by Captain Givhan Campbell, were out scouting instead of guarding the camp. As the members of the little caravan began to fall asleep, they faced two dangers in the night: attack from ex-Confederate soldiers—ruthless, war-weary bandits bent on plunder—or an attack from the Union cavalry, on the hunt for President Davis. It was no secret that bandits had been shadowing Varina Davis's wagon train for several days, and they could strike anytime without warning. That was the reason Davis had reunited with Varina, instead of pushing on alone.

Davis's aides knew that it was too dangerous for him to continue traveling with his wife's slow-moving wagon train. "Fully realizing that so large a party [of nearly thirty people] would be certain to attract the attention of the enemy's scouts, that we had every reason to believe [that they] were in pursuit of us," recalled Governor Lubbock, "it was decided at noon [May 9] that as soon as we had concluded the midday meal the President and his companions would again bid farewell to Mrs. Davis and her escort."

Davis did not plan to spend the night of May 9 camped with his wife and children near Irwinville. Unless he abandoned the wagon train and moved fast on horseback, accompanied by no more than three or four men, he had little chance of escape. By this time the Union was flooding Georgia with soldiers and canvassing every crossroads, guarding every river crossing, and searching every town.

Furthermore, the federals had recruited local blacks with expert knowledge of back roads and hiding places, to help in the manhunt for the fugitive president of the slave empire. The former slaves relished the task and its irony. Even if Davis did not know it, by May 9 he was at imminent risk of capture, and possible death. Thus, remembered Lubbock, "we halted on a small stream near Irwinville . . . and dinner over, saddled our horses, and made everything ready to mount at a moment's notice."

Burton Harrison spoke for the entire inner circle when he said that Davis needed to separate from the wagon train and entourage:

> We had all now agreed that, if the President was to attempt to reach the Trans-Mississippi at all, by whatever route, he should move on at once, independent of the ladies and the wagons. And when we halted he positively promised me . . . that, as soon as something to eat could be cooked, he would say farewell, for the last time, and ride on with his own party, at least ten miles farther before stopping for the night, consenting to leave me and my party to go on our own way as fast as was possible with the now weary mules.

Harrison proposed that the president take Lubbock, Wood, Johnston, and possibly Reagan with him, and that Harrison remain with Mrs. Davis, the children, and the rest of the wagon train personnel. Davis told his aides that he would leave the camp sometime during the night. "The President notified us to be ready to move that night," affirmed Reagan.

Davis told him that he would eat dinner, stay up late, and leave on horseback after it was dark. He was dressed for the road: dark felt, wide-brimmed hat; signature wool frock coat of Confederate gray; gray trousers; high, black leather riding boots, and spurs. His horse, tied near Varina's tent, was already saddled and ready to ride, its saddle holsters loaded with Davis's pistols.

Harrison felt sick and retired early. "After getting that promise from the President," he remembered, "and arranging the tents and wagons for the night, and without waiting for anything to eat (being still the worse for my dysentery and fever), I lay down upon the ground and fell into a profound sleep."

Harrison was certain that when he awoke, Davis would be gone. Captain Moody stretched a piece of canvas above Harrison's head and lay down beside him. Several of the men, including Reagan, stayed up late talking, waiting for Davis to give the order to depart. It never came.

The delay puzzled Lubbock: "Time wore on, the afternoon was spent, night set in, and we were still in camp. Why the order 'to horse' was not given by the President I do not know."

"For some reason," Reagan said, "the President did not call for us that night, though we sat up until pretty late."

Wood and Lubbock fell asleep under a pine tree no more than one hundred feet from Davis's tent, with Johnston, Harrison, and Reagan sleeping somewhere between them and Davis.

CHAPTER TEN

"By God, You Are the Men We Are Looking For"

Unbeknownst to the inhabitants of Davis's camp, a mounted Union patrol of 128 men and 7 officers—a detachment from the Fourth Michigan Cavalry regiment—led by regimental commander Lieutenant Colonel B. D. Pritchard, was closing in on Irwinville.

Pritchard reached Wilcox's Mills by sunset of May 9, but the horses were spent. He halted for an hour and had the animals unsaddled, fed, and groomed. Then he pushed on in the dark. "From thence we proceeded by a blind woods road through almost an unbroken pine forest for a distance of eighteen miles, but found no traces of the train or party before reaching Irwinville, where we arrived about 1 o'clock in the morning of May 10, and were surprised to find no traces of . . . the rebels."

Pritchard ordered his men to examine the conditions of the roads leading in all directions, but they saw nothing to suggest that a wagon train or mounted force had passed that way. Pritchard left most of his men on one side of town and rode ahead with a few men to the other side where his main body had not been spotted and, posing as

Confederate cavalrymen, they questioned some villagers. "I learned from the inhabitants," Pritchard recounted, "that a train and party meeting the description of the one reported to me at Abbeville had encamped at dark the night previous one mile and a half out on the Abbeville road."

At first Pritchard suspected it was a Union camp—he knew firsthand from an earlier encounter with them that men from the First Wisconsin Cavalry regiment were in the region hunting for Davis too—but then he realized that mounted federals would not be moving with tents and wagons. Whoever was in that camp, they were not Union men. Pritchard left Abbeville and positioned his men about half a mile from the mysterious encampment. "Impressing a negro as a guide," Pritchard recalled, " . . . I halted the command under cover of a small eminence and dismounted twenty-five men and sent them under command of Lieutenant Purington to make a circuit of the camp and gain a position in the rear for the purpose of cutting off all possibility of escape in that direction."

Pritchard told Purington to keep his men "perfectly quiet" until the main body attacked the camp from the front. Tempted to charge the camp at once, Pritchard decided to wait until daylight: "The moon was getting low, and the deep shadows of the forest were falling heavily, rendering it easy for persons to escape undiscovered to the woods and swamps in the darkness." The men of the Fourth Michigan were in place by about 2:00 A.M. For the next hour and a half, they waited in the dark, undetected.

At 3:30 A.M., Pritchard ordered his men into their saddles and to ride forward: "[J]ust as the earliest dawn appeared, I put the column in motion, and we were enabled to approach within four or five rods of the camp undiscovered, when a dash was ordered, and in an instant the whole camp, with its inmates, was ours." Pritchard's men had not fired a shot. Pritchard exaggerated the speed with which his men had captured the camp. "A chain of mounted guards was immediately thrown around the camp," Pritchard claimed, "and dismounted sentries placed at the tents and wagons. The surprise was so complete, and the movement so sudden in its execution, that few of the enemy were enabled to make the slightest defense, or even arouse from their slumbers in time to grasp their weapons, which were lying at their sides, before they were wholly in our power."

But Pritchard was not omniscient. He could only see what events transpired in front of his own eyes. Throughout the camp, individual human dramas unfolded simultaneously as the Fourth Michigan charged the tents and wagons. Before Pritchard's men could gain full control of the camp, and before the colonel even verified that this was Jefferson Davis's camp, gunfire broke out where Pritchard had stationed Lieutenant Purington and twenty-five men. It was a rebel counterattack, Pritchard feared. He spurred his horse past the tents and wagons and rode to the sound of the gunfire.

As Purington faced Davis's camp, awaiting Pritchard's signal, he heard mounted men approaching him from his rear. He stepped out from cover to halt them, and they called out that they were "friends." But they refused to identify themselves and would not ride forward when Purington ordered them to. In response to his repeated command that they identify themselves, one of them shouted, "By God, you are the men we are looking for" and began to ride away. Purington ordered his men to open fire. The First Wisconsin fired back.

Lieutenant Henry Boutell of the Fourth heard the gunfire and rode toward Purington. "Moving directly up the road," said Boutell, "I was met with a heavy volley from an unseen force concealed behind tree . . . and from which I received a severe wound." Another man from the Fourth was shot and killed. As the battle continued, a third man from the Fourth was wounded, and several Wisconsin men were also shot. In the dark they could not see that they wore the same uniform, Union blue cavalry shell jackets decorated with bright yellow piping.

The shooting woke Jim Jones, Davis's coachman, and he gave

the alarm. He roused William Preston Johnston, whom the charging horses had not awakened.

"Colonel, do you hear that firing?" Jim asked.

Johnston sprang up and commanded, "Run and wake the president."

Jones also woke Burton Harrison: "I was awakened by the coachman, Jim Jones, running to me about day-break with the announcement that the enemy was at hand!"

As Harrison sprang to his feet, he heard musket fire on the north side of the creek. He drew his pistol just in time to confront several men from the Fourth Michigan charging up the road from the south. Harrison raised his weapon and took aim.

"As soon as one of them came within range," he remembered, "I covered him with my revolver and was about to fire, but lowered the weapon when I perceived the attacking column was so strong as to make resistance useless, and reflected that, by killing the man, I should certainly not be helping ourselves, and might only provoke a general firing upon the members of our party in sight. We were taken by surprise, and not one of us exchanged a shot with the enemy." William Preston Johnston didn't hear any more gunfire and began to pull on his boots. He walked out to the campfire to ask the cook if Jim had been mistaken.

"At this moment," Johnston reported, "I saw eight or ten men charging down the road towards me. I thought they were guerillas, trying to stampede the stock. I ran for my saddle, where I had slept, and began unfastening the holster to get out my revolver, but they were too quick for me. Three men rode up and demanded my pistol, which . . . I gave to the leader . . . dressed in Confederate gray clothes . . . One of my captors ordered me to the camp fire and stood guard over me. I soon became aware that they were federals."

Lubbock was up too. "We sprang immediately to our feet." Lubbock pulled his boots on, stood up, and secured his horse, which had been saddled all night and was tied near where the governor had laid down his head. It was too late. "By this time the Federal troopers were on us. We were scarce called upon to surrender before they pounced upon us like freebooters."

Lubbock put up a fight, resisting an attempt by two of the cavalrymen to rob him. Reagan saw it all:

> When this firing occurred the troops in our front galloped upon us. The major of the regiment reached the place where I and the members of the President's staff were camped, about one hundred yards from where the President and his family had their tents. When he approached me I was watching a struggle between two federal soldiers and Governor Lubbock. They were trying to get his horse and saddle bags away from him and he was holding on to them and refusing to give them up; they threatened to shoot him if he did not, and he replied . . . that they might shoot and be damned, but that they should not rob him while he was alive and looking on.

A Union officer spotted Reagan and spurred his horse toward the only member of the Confederate cabinet who had volunteered to remain with the president. The postmaster-general readied his pistol.

"I had my revolver cocked in my hand, waiting to see if the shooting was to begin," he remembered. "Just at this juncture the major rode up, the men contending with Lubbock had disappeared, and the major asked if I had any arms. I drew my revolver from under the skirt of my coat and said to him, 'I have this.' He observed that he supposed I had better give it to him. I knew that there were too many for us and surrendered my pistol."

Pritchard rode up to Harrison and demanded to know the source of the shooting. "Pointing across the creek, [he] said, 'What does that mean? Have you any men with you?' Supposing the firing was done by our teamsters, I replied, 'Of course we have—don't you hear the firing?' He seemed to be nettled at the reply, gave the order, 'Charge,' and boldly led the way himself across the creek, nearly every man in his command following."

Still inside Varina's tent, Davis heard the gunfire and the horses in the camp and assumed these were the same Confederate stragglers or deserters who had been planning to rob Mrs. Davis's wagon train for several days.

"Those men have attacked us at last," he warned his wife. "I will go out and see if I cannot stop the firing; surely I still have some authority with the Confederates."

He opened the tent flap, saw the bluecoats, and turned to Varina: "The Federal cavalry are upon us."

Jefferson Davis had not faced a cavalry charge for two decades. The last time he was in battle, he was in command of his beloved regiment of Mississippi Rifles at the Battle of Buena Vista in the Mexican War. There he encountered one of the most frightening sights a man could see on a nineteenth-century battlefield—massed lancers preparing to charge. The lance was not a toy, and at close quarters it could be more deadly than a pistol or saber. Napoleon's lancers had been feared throughout Europe, and Mexican lancers had slaughtered American soldiers with ease in California and Mexico. Indeed, Colonel Samuel Colt had perfected his dragoon revolver for the express purpose of shooting down charging Mexican lancers. At Buena Vista, Davis's regiment was outnumbered and at risk of being overrun in moments. Davis ordered his men form an inverted formation of the letter *V*, allow the Mexicans to charge into the open *V* and, at the last moment, unleash a devastating rifle volley into them. It worked, and the Mississippi Rifles broke the charge. That victory made Davis a hero, and, in a circuitous route, his military fame two decades earlier had led him to this camp in the pinewoods of Georgia. But this morning he had too little advance warning and not enough men to resist the charge of the Fourth Michigan.

Davis had not undressed this night, so he was still wearing his gray frock coat, trousers, riding boots, and spurs. He was ready to leave now, but he was unarmed. His pistols and saddled horse were within sight of the tent. If he could just get to that horse, he could leap into the saddle, draw a revolver, and gallop for the woods, ducking low to avoid any carbine or pistol fire aimed in his direction. He knew he was still a superb equestrian and was sure he could outrace any Yankee cavalryman half his age. Seconds, not minutes, counted now, and if he hoped to escape he had to run for the horse.

On the morning of his capture, Jefferson Davis wore a suit of Confederate gray and not one of Varina's hoopskirts.

John Taylor Wood got free of the cavalry and tried to help Davis: "I went over to the president's tent, and saw Mrs. Davis. [I] told her that the enemy did not know that he was present and during the confusion he might escape into the swamp."

Before Jefferson left, Varina asked him to wear an unadorned raglan overcoat, also known as a "waterproof." Varina hoped the raglan might camouflage his fine suit of clothes, which resembled a Confederate officer's uniform. "Knowing he would be recognized," Varina explained, "I plead with him to let me throw over him a large waterproof which had often served him in sickness during the summer as a dressing gown, and which I hoped might so cover his person that in the grey of the morning he would not be recognized. As he strode off I threw over his head a little black shawl which was round my own shoulders, seeing that he could not find his hat and after he started sent the colored woman after him with a bucket for water, hoping he would pass unobserved."

Jefferson Davis described what happened that morning:

> As I started, my wife thoughtfully threw over my head and shoulders a shawl. I had gone perhaps between fifteen or twenty yards when a trooper galloped up and ordered me to halt and surrender, to which I gave a defiant answer, and, dropping the shawl and the raglan from my shoulders, advanced toward him; he leveled his carbine at me, but I expected, if he fired, he would miss me, and my intention was in that event to put my hand under his foot, tumble him off on the other side, spring into the saddle, and attempt to escape. My wife, who had been watching me, when she saw the soldier aim his carbine at me, ran forward and threw her arms around me. Success depended on instantaneous action, and recognizing that the opportunity had been lost, I turned back, and, the morning being damp and chilly, passed on to a fire beyond the tent . . .

Even before the gun battle ceased, some of the cavalrymen started tearing apart the camp in a mad scramble. They searched the baggage, threw open Varina's trunks, and tossed the children's clothes into the air. "The business of plundering commenced immediately after the capture," observed Harrison. The frenzy on the part of the cavalry suggested that the search was not random. The Yankees were looking for something.

Lubbock said that "in a short time they were in possession of very nearly everything of value that was in the camp. I resisted being robbed, and lost nothing then except some gold coin that was in my holsters. I demanded to see an officer, and called attention to the firing, saying that they were killing their own men across the branch, and that we had no armed men with us . . . While a stop was being put to this I went over to Mr. Davis, who was seated on a log, under guard."

Johnston was not as lucky resisting the plunderers. Several cavalrymen got his horse and his saddle, with the accoutrements and pistols, which his father, General Albert Sidney Johnston, had used at the Battle of Shiloh on the day he was killed in action. Understandably, the son prized his father's personal effects.

Harrison did not want his captors to lay their vulgar hands on the letters from Constance Cary he carried with him all the way from Richmond: "I emptied the contents of my haversack into a fire where some of the enemy were cooking breakfast, and they saw the papers burn. They were chiefly love-letters, with a photograph of my sweetheart."

As the skirmish between the Union regiments died down, Colonel Johnston's guard left him unattended and he walked fifty yards to Varina Davis's tent, where he found the president outside. "This is a bad business, sir," Davis said, "I would have heaved the scoundrel off his horse as he came up, but she caught me around the arms."

"I understood what he meant," Johnston said, "how he had proposed to dismount the trooper and get his horse, for he had taught me the trick." It was an old Indian move that Davis had learned years before when he served out west in the U.S. Army.

Once Davis had been apprehended, John Taylor Wood decided to escape. "Seeing that there was no chance for the President I determined to make the effort." Lubbock and Reagan approved his plan. Wood strolled around the camp, examining the faces of the Union cavalrymen, until "at last I selected one that I thought would answer my purpose." He asked the soldier to go to the swamp with him, where Wood offered him forty dollars. The Yankee grabbed the money and let him go.

Johnston warned another Union officer that they were firing on their own men: "Feeling that the cause was lost, and not wishing useless bloodshed, I said to him: 'Captain, your men are fighting each other over yonder.' He answered very positively: 'You have an armed escort.' I replied, 'You have our whole camp; I know your men are fighting each other. We have nobody on that side of the slough.' He then rode off."

Soon Pritchard and his officers discovered that this was true. There were no Confederate soldiers behind the camp. His men were fighting the First Wisconsin Cavalry, and they were killing each other. Greed for gold and glory may have contributed to the deadly and embarrassing disaster. The troopers of the Fourth Michigan and the First Wisconsin cavalries knew nothing about President Johnson's proclamation of May 2, offering a $100,000 reward for the capture of Jefferson Davis. They were not after that reward money, although once they learned of it, a few days after the Davis capture, they were eager to claim it. No, they wanted a bigger prize—Confederate gold. Every Union soldier had heard the rumors that the "rebel chief" was fleeing with millions of dollars in gold coins in his possession.

The Northern newspapers had reported it, Edwin M. Stanton and a number of Union generals had telegraphed about it, and, no doubt, every last man of the Fourth Michigan and First Wisconsin had heard about it. The lure of the so-called Confederate "treasure train" was irresistible. General James Wilson's broadside proclamation of May 9, which General Palmer had printed and then distributed as handbills in Georgia, intoxicated Union soldiers with dreams of untold riches.

Eliza Andrews had seen the reward posters:

> The hardest to bear of all the humiliations yet put upon us, is the sight of Andy Johnson's proclamation offering rewards for the arrest of Jefferson Davis, Clement C. Clay, and Beverly Tucker, under pretense that they were implicated in the assassination of Abraham Lincoln. It is printed in huge letters on handbills and posted in every public place in town—a flaming insult to every man, woman, and child in the village, as if [the Yankees] believed there was a traitor among us so base as to betray the victims of their malice, even if they knew where they were . . . if they had posted one of their lying accusations on our street gate, I would tear it down with my own hands, even if they sent me to jail for it.

Wilson had promised this: Whoever captured Davis could claim the millions of dollars in gold he was carrying as their reward. But what these man hunters did not know was that Davis was not the one transporting it.

But Davis's pursuers wanted more than gold—they were also after glory. After a patrol of the Sixteenth New York Cavalry tracked down John Wilkes Booth, they, and especially the sergeant who shot the actor, became national heroes. Why shouldn't the men who captured the archcriminal Jefferson Davis be rewarded with the same level of fame?

The fatal skirmish between the two regiments created tensions on both sides. Their failure to capture the rebel treasure exacerbated their anger and humiliation. They blamed each other for the fratricide, accused each other of appropriating leads about Davis's whereabouts during the chase, and fought over the reward money. The Fourth Michigan did not want to share the money with the First Wisconsin. The Wisconsin men claimed that if the Fourth had not fired upon them, they would have been the ones who captured Davis, not the Fourth. The Wisconsin faction accused Colonel Pritchard of conduct unbecoming an officer and a gentleman. The accusation later made it into the press, and Pritchard demanded the right to a public reply. To settle the dispute, one general suggested that the reward money be divided among all the men of both regiments present at the scene that morning.

Davis, who had sacrificed all he owned for the Confederacy, who never sought to profit from his office, and who was captured without a single dollar to his name, must have appreciated the irony. He never commented publicly about the ugly dispute among his captors, but it must have amused him to see Yankees killing one another and then squabbling over money in their greed to claim him as their prize.

It was only after the deadly skirmish that Pritchard realized he had captured the president of the Confederate States of America. Pritchard took an inventory of his prisoners:

> As soon as the firing had ceased I returned to camp and took an inventory of our captives, when I ascertained that we had captured Jeff. Davis and family (a wife and four children), John H. Reagan, his Postmaster-General; Colonels Harrison [Johnston] and Lubbock, aides de camp to Davis; Burton N. Harrison, his private secretary; Major Maurin, Captain Moody, Lieutenant Hathaway, Jeff. D. Howell, midshipman in the rebel navy, and 12 private soldiers; Miss Maggie Howell, sister of Mrs. Davis; 2 waiting-maids, 1 white and one colored, and several servants. We also captured 5 wagons, 3 ambulances, about 15 horses, and from 25 to 30 mules. The train was mostly loaded with commissary stores and private baggage of the party.

Pritchard did not bother to list in his report the names of the handful of common Confederate soldiers who were captured with Davis. They were not important enough. If a junior officer from the Fourth Michigan had not added their names to another tally of the prisoners, the identity of these twelve men who had volunteered to risk their lives to serve their president might have been lost to history.

In the confusion, Davis's aides gathered around him to protect him and his family. The cavalrymen made no attempt to bind or handcuff Davis. Harrison could not believe that Davis had been captured. He believed that Davis had left during the night, hours before: "I had been astonished to discover the President still in camp when the attack was made." The Union soldiers began taunting and insulting Davis, enraging Governor Lubbock: "The man who a few days before was at the head of a government was treated by his captors with uncalled for indignity . . . A private stepped up to him rudely and said: 'Well, Jeffy, how do you feel now?' I was so exasperated that I threatened to kill the fellow, and I called upon the officers to protect their prisoner from insult."

Lubbock praised Varina's demeanor in the presence of her enemies: "[S]he bore up with womanly fortitude . . . her bearing towards [our captors] was such as was to be expected from so elegant, high-souled, and refined a Southern woman." But the governor saw that her family was frightened: "The children were all young, and hovered about her like a covey of young, frightened partridges; while her sister, Miss Maggie Howell, was wonderfully self-possessed and dignified."

Davis's aides had used good judgment on the morning of May 10. No matter how much they might have wanted to open fire on the Union cavalry, they knew they would lose the fight. They might have killed several of the enemy, but the federals, outnumbering them by more than ten to one, could have killed them all and then shot the president. A gunfight at dawn, when visibility was low, might also have had fatal consequences for Mrs. Davis and the children. The president's aides had done everything that their honor as Confederate officials and Southern gentlemen had required. Surrender, however hateful, had been the honorable choice. Davis never suggested his men should have done more for him that morning. Indeed, he expressed affection for his inner circle for the rest of his life.

And now, thirty-eight days after he evacuated Richmond, after an epic journey through four states by railroad, ferry boat, horse, cart, and wagon, Davis was a prisoner. Others, including his aides, would speculate for years why Davis hadn't placed his own welfare first and escaped to Texas, Mexico, Cuba, or Europe. Judah Benjamin and John C. Breckinridge did so and had escaped abroad. Burton Harrison always believed that Davis could have escaped—if he had "ridden on after getting supper with our party the night we halted for the last time; had he gone but five miles beyond Irwinville, passing through that village at night, and so avoiding observation, there is every reason to suppose that he and his party would have escaped either across the Mississippi or through Florida to the sea-coast . . . as others did."

Harrison speculated that the reason Davis did not was "the apprehension he felt for the safety of his wife and children which brought about his capture."

No one really knows why Davis failed to leave the camp that night. Perhaps he was tired of life on the run, or maybe his chronic illnesses had weakened him. Maybe he thought a few more hours of stolen rest would not matter. In Virginia, John Wilkes Booth had paid the price for tarrying too long at a farm where he had found respite from his manhunt. The assassin might have reached Mexico if he had not slowed his flight. Delay had cost him his life. Perhaps Davis thought it was too late to escape to Texas and resuscitate a western Confederacy there. He might have feared once he left this camp, he would never see his wife and children again. Perhaps part of him did not want to flee, run away to a foreign land, and vanish from history. Instead he would remain onstage for the drama's last act, waiting for the curtain to fall upon the lost cause. Any explanation is just speculation.

But he failed in his mission, which was not escape but victory. He had not been able to rally the army or the people to continue the war. He did not make it across the great Mississippi River to create a new Confederate empire in the west. But he had done his best.

The war, and the chase for Jefferson Davis, were over. But he was alive. His story might have ended at the little camp near Irwinville, Georgia. The cavalryman he hoped to unseat from the saddle might have shot him. Or, if he had seized the horse and galloped for the woods, he might have been cut down by carbine fire. And if he had escaped the scene, he might have been, an hour, a day, or a week later, shot and killed, unprotected and alone, somewhere in the wilderness of southwest Georgia or beyond. May 10, 1865, was the end for Jefferson Davis's presidency and his dream of Southern independence. But it was the beginning of a new story too, one he began to live the day he was captured. "God's will," he said.

Now Davis would begin a new, twelve-day journey to imprisonment.

John Taylor Wood, who had been hiding in the swamp for three hours, witnessed what happened next. "I was within hearing of the camp on either side of the stream and . . . when they came down for water or to water their horses I was within a few yards of them. The wagons moved off first, then the bugles sounded and the President started on one of the carriage horses followed by his staff and a squadron of the enemy. I watched him as he rode off. Sad fate." Wood fled, embarking on a fantastic odyssey by land and sea to avoid capture by Union forces.

That day in Washington, people did not rush into the streets to celebrate Davis's capture. No one knew about it. Georgia was too far away for the news to travel to the capital on the same day. Instead, the newspapers were filled with headlines and stories about the Lincoln

$360,000

REWARD!

THE PRESIDENT OF THE UNITED STATES

Has issued his Proclamation, announcing that the Bureau of Military Justice has reported upon indubitable evidence that

JEFFERSON DAVIS, CLEMENT CLAY, JACOB THOMPSON, GEO. N. SAUNDERS, BEVERLY TUCKER, and WM. C. CLEARY,

incited and concerted the assassination of Mr. Lincoln, and the attack upon Mr. Seward.

He therefore, offers for the arrest of Davis, Clay and Thompson $100,000 each; for that of Saunders and Tucker $25,000 each, for that of Cleary $10,000.

JAMES H. WILSON,

May 9, 1865. Major Gen. United States Army, Commanding.

A second reward poster for Davis and other Confederate leaders. Neither Davis nor his pursuers learned of the reward until after he was captured.

assassination. May 10 was the opening day of proceedings in the great conspiracy trial.

In Washington the eight defendants charged as Booth's accomplices went on trial before a military tribunal convened at the Old Arsenal penitentiary. Many people believed that if Davis was captured before the trial ended, he would be rushed to Washington and charged as the ninth conspirator in Lincoln's murder. Indeed, the government's first plan was to transport Davis to Washington and imprison him at the Old Capitol prison, two blocks east of the

Great Dome. Since mid-April the Bureau of Military Justice had been building a case against Jefferson Davis based on mysterious documents and questionable witnesses.

It took four days to travel from the capture site to Macon, where General James Wilson had his headquarters. At an encampment along the way Davis learned about the $100,000 bounty on his head. Burton Harrison recalled the moment: "It was at that cavalry camp we first heard of the proclamation offering a reward of $100,000 for the capture of Mr. Davis, upon the charge, invented by Stanton and Holt, of participation in the plot to murder Mr. Lincoln. Colonel Pritchard had himself just received it, and considerately handed a printed copy of the proclamation to Mr. Davis, who read it with a composure unruffled by any feeling other than scorn."

Outside Macon, John Reagan got into a dispute with Colonel Pritchard:

> On the morning of the day we arrived at Macon, while I and the President's staff were taking an humble breakfast, sitting on the ground, Colonel Pritchard came by where we were, and I said to him that I understood we were to reach Macon that morning, that I had not changed my clothing for some time, and requested some clothes which I had in my saddle bags, taken from me when we were captured.
>
> "We have not got your saddle bags," he answered me.
>
> "I am sorry to hear you say that, Colonel," I retorted, "for I know you have them."
>
> He asked how I knew that.
>
> "Because your officers told me of your examining their contents right after our capture," I answered; "and named correctly what was in them."
>
> With some temper he questioned, "Who told you so?"
>
> "Your officers."
>
> "What officers?"

"Since you question the fact," I said, "I will not put them in your power by giving you their names." Then I added, "It does not look well for a colonel of cavalry in the United States Army to steal clothes."

"Sir," he said, "I will put you in irons."

"You have the power to do so," I replied, "but that will not make you a gentleman or a man of truth."

He walked off as if intending to execute his threat, but I heard no more of it.

In Macon, Davis met with General Wilson, who had flashed news of his capture to Washington.

Macon, Ga.,
May 12, 1865—11 A.M.

Lieut.-Gen. U.S. GRANT and Hon. Secretary of War, Washington, D.C.:

I have the honor to report that at daylight of the 10th inst., Col. Pritchard, commanding the 4th Michigan Cavalry, captured Jeff. Davis and his family, with Reagan, Postmaster-General; Col. Harrison, Private Secretary; Col. Johnson, A.D.C.; Col. Morris, Col. Lubbock, Lieut. Hathaway and others. Col. Pritchard surprised them their camp at Irwinsville, in Irwin County, Ga., 75 miles south-east of this place. They will be here to-morrow night, and will be forwarded under strong guard without delay. I will send further particulars at once.

J. H. Wilson, Brevet Major-General.

Once the Davis party arrived in Macon, they received better treatment than they had on the road. Union troops honored Davis by presenting arms upon his arrival. It was the last time that federal troops would honor him. John Reagan commented on their treatment: "When we reached Macon, we were taken to the headquarters of General Wilson, which was a large building that had been used as a hotel. General Wilson invited President Davis, his staff, and myself to dine with him, treating us with courtesy."

When Reagan learned that he and Davis were to be separated, he asked General Wilson if he might remain with the president. "I thereupon observed that President Davis was much worn down, and that, as I was the only member of his political family with him, I might be of some service to him, and requested to have the order changed as to send me on with him. He asked me if I was aware that this might involve me in danger. I told him I had considered that; that we had entered upon the contest together, and that I was willing to end it with him, whatever that end might be."

On the morning of their capture, Davis, Mrs. Davis, Reagan, Harrison, Johnston, and Lubbock remained unbowed and defiant. They were not meek prisoners. They had resisted the plundering of their persons and the baggage train, took umbrage at the crude language with which the soldiers addressed Davis, and scorned their captors as moral and social inferiors. To the Southern mind, the officers and men of the Fourth Michigan Cavalry, some of them immigrants, represented all that was wrong with the North and were living proof of the superior civilization of the South.

To the Davises and their loyal aides, these rude, uncouth, ungentlemanly, thieving Yankee troops confirmed the depravity of the North. Davis and his men refused to show fear and conducted themselves as members of the Southern elite. Their hauteur infuriated their captors. The cavalrymen would find a way to settle the score, not with violence but by degrading Jefferson Davis's most precious possession—his reputation. Thus, within a few days, began the myth that Jefferson Davis was captured in women's clothing. If Davis would not behave like a beaten man, then his captors could humiliate and emasculate him.

Popular images lampooned Davis for allegedly attempting to escape capture dressed as a woman.

John Reagan remembered: "As one of the means of making the Confederate cause odious, the foolish and wicked charge was made that he was captured in women's clothes; and his portrait, showing him in petticoats, was afterward placarded generally in showcases and public places in the North. He was also pictured as having bags of gold on him when captured . . . I saw him a few minutes after his surrender, wearing his accustomed suit of Confederate gray, with his boots and hat on . . . and he had no money."

Davis was taken from Macon to Atlanta by train on May 14, and then he traveled to Augusta, from where he departed for Savannah.

On Sunday, May 14, the stupendous news of Davis's capture appeared in the morning papers in Washington. Benjamin Brown French left his home on Capitol Hill to buy a copy of the *Daily Morning Chronicle*. "When I came up from breakfast I went out and got the *Chronicle*," he recalled, "and the first thing that met my eyes was '*Capture of Jeff Davis'* in letters two inches long. Thank God we have got the arch traitor at last. I hope he will not be suffered to escape or commit suicide. Hanging will be too good for him, double-dyed Traitor and Murderer that he is." Gideon Welles noted the Confederate president's capture in his diary: "Intelligence was received this morning of the capture of Jefferson Davis in southern Georgia. I met Stanton this Sunday P.M. at Seward's, who says Davis was taken disguised in women's clothes. A tame and ignoble letting-down of the traitor."

At the other end of Pennsylvania Avenue, journalist George Alfred Townsend wandered around the White House. He wanted to see, one month after the assassination, what signs of Lincoln remained there. Mary Lincoln had still not moved out, forcing President Andrew Johnson to continue living in his hotel room at the Kirkwood House. Townsend went to the second floor, up the same staircase that Lincoln's body descended the night of April 18. It was as though Lincoln had never left.

"I am sitting in the President's Office," Townsend reported. "He was here very lately, but he will not return to dispossess me of this high-backed chair he filled so long, nor resume his daily work at the table where I am writing.

"There are here only Major Hay and the friend who accompanies me. A bright-faced boy runs in and out, darkly attired, so that his fob-chain of gold is the only relief to his mourning garb. This is little Tad, the pet of the White House. That great death . . . has made upon him only the light impression which all things make upon childhood. He will live to be a man pointed out everywhere, for his father's sake; and as folks look at him, the tableau of the murder will seem to encircle him."

Townsend's eyes scanned the room. His description of Lincoln's empty office was as eloquent as anything that had been uttered downstairs in the East Room, or during the thirteen-day journey of the funeral train: "The room is long and high, and so thickly hung with maps that the color of the wall cannot be discerned. The President's table at which I am seated, adjoins a window at the farthest corner; and to the left of my chair as I recline in it, there is a large table before an empty grate, around which there are many chairs, where the cabinet used to assemble. The carpet is trodden thin, and the brilliance of its dyes lost. The furniture is of the formal cabinet class, stately and semi-comfortable; there are oak book cases, sprinkled with the sparse library of a country lawyer."

Townsend watched while the staff cleared out the office: "They are taking away Mr. Lincoln's private effects, to deposit them wherever his family may abide, and the emptiness of the place, on this sunny Sunday, revives that feeling of desolation from which the land has scarce recovered. I rise from my seat and examine the maps . . . [they] exhibit all the contested grounds of the war; there are pencil lines upon them where some one has traced the route of armies . . . was it the dead President . . . ?"

Townsend walked over to Lincoln's worktable and saw some books there.

> Perhaps they have lain there undisturbed since the reader's dimming eyes grew nerveless. A parliamentary manual, a Thesaurus, and two books of humor, "Orpheus C. Kerr," and "Artemus Ward." These last were read by Mr. Lincoln in the pauses of his hard day's labor. Their tenure here bears out the popular verdict of his partiality for a good joke; and, through the window, from the seat of Mr. Lincoln, I see across the grassy grounds of the capitol, the broken shaft of the Washington Monument, the long bridge and the fort-tipped Heights of Arlington, to catch some freshness of leaf and water, and often raised the sash to let the world rush in where only the nation abided, and hence on that awful night, he departed early, to forget this room and its close application in the abandon of the theater.
>
> I wonder if that were the least of Booth's crimes—to slay this public servant in the stolen hour of recreation he enjoyed

> but seldom. We worked his life out here, and killed him when he asked for a holiday.
>
> I am glad to sit here in his chair . . .

On May 15, the *New York Tribune* touted "Our Special Dispatch" received from Washington the previous day: "The public here manifest the utmost enthusiasm over the capture of Jeff. Davis. Some timid politicians, however, express a wish that he had been shot as Booth was, for fear his possession may be embarrassing to the Government." The editors suggested that he be rushed to Washington for trial with Booth's conspirators. "If he is placed in the prisoner's dock at the court, by the side of Harrold and Payne he will certainly be convicted of complicity in the assassination of Mr. Lincoln." Or, speculated the paper, Davis could be tried for treason. "It is urged strenuously, however, by some in high position, that the dignity of the nation demands that on his arrival here the assassination charge ought to be waived, and he be arraigned and tried for treason, the highest crime known to our laws, and, on conviction, hanged. Secretary Stanton will order Jeff. Davis to be put on a gunboat and forwarded direct to Washington."

The *Tribune*'s editorial page implied that Davis must be hanged, but it opposed a vigilante-style lynching. Let things be done according to the law, the paper cautioned:

> Jefferson Davis is a prisoner of the Government. He surrendered under no capitulation but his own,—which—he being isolated, disguised in one of his wife's dresses, and directly within range of several troopers' revolvers—was too sudden to be otherwise than unconditional. Being a prisoner, we trust that he will be treated as a prisoner, under the protection of the dignity and honor of a self-respecting people.
>
> As we are officially assured that he is proved to be inculpated in the plot which culminated in the murder of

> President Lincoln, we trust he is to be indicted, arraigned and tried for that horrid crime against our country and every part of it. We hope he may have a fair, open, searching trial, like any other malefactor, and, if convicted we trust he will be treated just like any other. We have no faith in killing men in cold blood, or in hot blood either, unless when (as in battle) they obstinately refuse to get out of the way; but we neither expect nor desire that the execution or non-execution of the laws shall depend on their accordance or disagreement with our convictions of sound policy. But let all things be done decently and in order.

As soon as the "Davis in a dress" story began to spread, the great showman P. T. Barnum knew at once the garment would make a sensational exhibit for his fabled "American Museum" of spectacular treasures and curiosities in downtown New York City. He wanted that hoopskirt and was prepared to pay a formidable sum to get it. Barnum wrote to Edwin Stanton, offering to make a donation to one of two worthy wartime causes, the care of wounded soldiers or the care of freed slaves.

> *Bridgeport,*
> *May 15, 1865*
>
> *Hon. E. M. Stanton,*
> *Secretary of War:*
> *I will give $500 to Sanitary Commission or Freedman's Association for the petticoats in which Jeff. Davis was caught.*
> *P. T. Barnum*

It was a hefty sum—a Union army private's pay was $16 a month—and that $500 could have fed and clothed a lot of soldiers and slaves. Still, Stanton declined the offer. Perhaps Barnum should

TRUTH VS. MYTH. LEFT: THE RAGLAN COAT JEFFERSON DAVIS ACTUALLY WORE THE MORNING OF HIS CAPTURE. RIGHT: THE SHAWL AND SPURS DAVIS WORE THE MORNING OF MAY 10, 1865.

have offered more money. George Templeton Strong, a New Yorker who kept a celebrated diary chronicling life in wartime Gotham, wrote that "Barnum is a shrewd businessman. He could make money out of those petticoats if he paid ten thousand dollars for the privilege of exhibiting them."

But the secretary of war had other plans for these treasures. He earmarked the capture garments for his own collection, and had ordered that they be brought to his office, where he was keeping them in his personal safe along with other historical curiosities from Lincoln's autopsy, Booth's death, and Davis's capture. But the arrival in Washington of the so-called petticoats or dress proved to be a big letdown. When Stanton saw the clothes, he knew instantly that Davis had not disguised himself in a woman's hoopskirt and bonnet. The "dress" was nothing more than a loose-fitting, waterproof "raglan," or overcoat, a garment as suited for a man as a woman. The "bonnet" was a rectangular shawl, a type of wrap President Lincoln had worn on chilly evenings. Stanton dared not allow Barnum to exhibit these relics in his museum. Public viewing would expose the lie that Davis had worn one of his wife's dresses. Instead, Stanton sequestered the disappointing textiles to perpetuate the myth that the cowardly rebel chief had tried to run away in his wife's clothes.

Barnum's failure to obtain the actual clothing did not deter artists from using their imaginations to depict Jefferson Davis in the coveted petticoats. Printmakers published more than twenty different lithographs of merciless caricatures depicting Davis in a frilly bonnet and voluminous skirt, clutching a knife and bags of gold as he fled Union troopers. These cartoons were captioned with mocking captions, many of them delighting in sexual puns and innuendoes, and many putting shameful words in Davis's mouth. Ingenious photographers doctored images of Davis by adding a skirt and bonnet.

On May 16, Davis arrived in Savannah, one of the loveliest cities in the South. General Sherman had announced its capture in a famous telegram to Lincoln on December 22, 1864: "I beg to present you as a Christmas gift the city of Savannah with 150 heavy guns & plenty of ammunition & also about 25000 bales of cotton." Now a captive in a captive city, Davis did not know it when he left Savannah, but he would return there someday, in an unexpected, triumphant, even miraculous reversal of fortune. The citizens of Savannah would see him again. Davis was put aboard a vessel bound for Fort Monroe, Virginia.

Now that Davis and all the Confederate armies east of the Mississippi River had surrendered, the Union was ready to celebrate the end of the war in style. On May 18, Grant issued General Order No. 239, announcing that a "Grand Review" of the Army of the Potomac and Sherman's Army of the West would take place over two days in Washington, on Tuesday, May 23, and Wednesday, May 24. This extravaganza was rumored to be bigger than even the April 19 Lincoln funeral procession.

On May 19, Davis, aboard the *William P. Clyde,* neared Fort Monroe.

The same day, General James Wilson recommended that all of the officers and men of the First Wisconsin and Fourth Michigan cavalry regiments engaged in the pursuit of Davis below Abbeville receive medals of honor and that the reward be divided among all of the men actually engaged in the capture, with "ample provision being made for the families of the men killed and wounded in the unfortunate affair between the two regiments."

Stanton wanted his prisoners transported in secret, and he was alarmed when he intercepted telegrams sent to Gideon Welles by two naval officers. Commander Frailey and Acting-Rear-Admiral Radford sought to inform the navy secretary that the *Tuscarora* had convoyed to Hampton Roads the vessel *William Clyde,* with Davis on board. Welles recalled Stanton's feeling that "the custody of these prisoners devolved on him a great responsibility, and until he had made disposition of them, or determined where they should be sent, he wished their arrival to be kept a secret . . . He wished me to . . . allow no communication with the prisoners except by order of General Halleck of the War Department . . . and again earnestly requested and enjoined that none but we three—himself, General Grant, and myself—should know of the arrival and disposition of these prisoners . . . not a word should pass."

Welles scoffed at Stanton's obsession with secrecy: "I told him the papers would have the arrivals announced in their next issue." But Welles indulged his military counterpart: "I, of course, under his request, shall make no mention of or allusion to the prisoner, for the present."

Stanton, Welles, and General Grant, who was also present, discussed what to do with Varina Davis and the other women in custody. Stanton exclaimed that they must be "sent off" because "we did not want them." "They must go South," Stanton declared, and he drafted an order dictating that course. When Stanton read the dispatch out loud, Welles could not resist toying with him: "The South is very indefinite, and you permit them to select the place. Mrs. Davis may designate Norfolk, or Richmond."

Or anywhere. Grant laughed and agreed—"True."

Stanton could not tolerate the former first lady of the Confederacy showing up wherever she wanted. "Stanton was annoyed," Welles saw, and "I think, altered the telegram." Stanton knew that if Varina returned to Washington, she could be an influential and dangerous political opponent.

On Saturday, May 20, Davis was two days out from Fort Monroe. In Washington, the Lincoln funeral train, its work done and now back from the thirty-three-hundred-mile round-trip to Springfield, sat at the United States Military Railroad car shops in Alexandria, Virginia. The death pageant had been a spectacular success. Now, on the heels of the Lincoln funeral pageant, Stanton, Grant, and the War Department worked on the final details for the unprecedented Grand Review—the gigantic, two-day parade, the biggest in American history—of the victorious Union armies up Pennsylvania Avenue.

In Richmond, the population labored to recover from the twin plagues of occupation and fire, not knowing that another calamity was about to befall the city. By 5:30 P.M. Saturday, "portentous clouds" had covered Richmond. Then, as the *Richmond Times* reported, they "burst forth with vivid lightning and stunning thunder." This "sudden and extraordinary storm" poured rain until 4:00 A.M., Sunday, May 21, and then, after a brief respite, continued until 1:00 P.M. "Never within 'the recollection of the oldest inhabitant,'" the *Times* testified, "has such a destructive rainstorm occurred in this city." The wind and water seemed a plague of almost biblical proportions.

Indeed, "the very floodgates of heaven seemed to open. And so great was its effect that the whole valley of the city was soon submerged in water, overflowing all the streams and washing from their banks a number of small houses, trees, & c." Wagons, furniture, supplies, and all manner of stuff were swept away and destroyed by the flood. Some people believed the Confederate capital cursed: sacked by mobs, then burned, then occupied by Yankees, and now engulfed by a great deluge. What punishment would Richmond suffer next?

CHAPTER ELEVEN

"Living in a Tomb"

On Monday, May 22, the night before the Grand Review, Jefferson Davis was incarcerated at Fort Monroe. He did not know whether he would ever see his wife and children again. When he parted with Varina, he told her not to cry. It would, he said, only make the Yankees gloat.

In his captivity, the jailers refused to address him as "President." They called him "Jeffy," "the rebel chieftain," or "the state prisoner." Soon, through insult, isolation, silence, shackling, constant surveillance, sleep deprivation, and dungeonlike conditions, they would seek to humiliate him and break his spirit. He was, in the words of some newspapers, the archcriminal of the age, a man "buried alive" who must never be set free.

Lincoln had once spoken of this kind of imprisonment:

> They have him in his prison house; they have searched his person, and left no prying instrument with him. One after another they have closed the heavy iron doors upon him, and

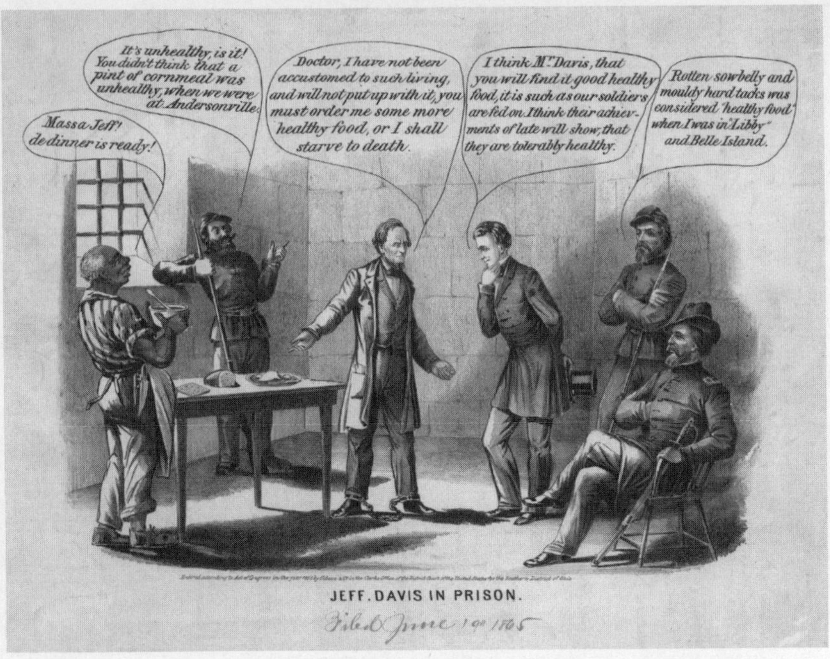

PRINTMAKERS CONTINUED TO RIDICULE DAVIS AFTER HIS IMPRISONMENT.

> now they have him, as it were, bolted in with a lock of a hundred keys, which can never be unlocked without the concurrence of every key; the keys in the hands of a hundred different men, and they scattered to a hundred different and distant places; and they stand musing as to what invention, in all the dominions of mind and matter, can be produced to make the impossibility of his escape more complete than it is.

But Lincoln was not, in these 1857 remarks on the *Dred Scott* decision, speaking of Jefferson Davis. He was, instead, speaking of the "bondage . . . universal and eternal" of the American slave. Now, the leader of the vanquished slave empire found himself locked in a "prison house" as secure as any built during the previous two and a half centuries of American slavery. Millions of his enemies in the North hoped he would never emerge from his dungeon alive.

CONTEMPORARY SKETCH OF DAVIS IN HIS CASEMATE CELL AT FORTRESS MONROE.

The night Davis was placed in his prison cell, Mary Lincoln moved out of the White House. She took with her a suspiciously large number of trunks. Benjamin Brown French was not sorry to see her go. He called on her before she left the presidential mansion: "Mrs. Mary Lincoln left the City on Monday evening at 6 o'clock, with her sons Robert & Tad (Thomas). I went up and bade her good-by, and felt really very sad, although she has given me a world of trouble. I think the sudden and awful death of the President somewhat unhinged her mind, for at times she has exhibited all the symptoms of madness. She is a most singular woman, and it is well for the nation that she is no longer in the White House. It is not proper that I should write down, *even here,* all I know! May God have her in his keeping, and make her a better woman. That is my sincere wish . . ."

No government or military official in Washington regretted that Mary would be absent from the next day's parade.

Early on Tuesday morning, before the Grand Review got under way at 9:00, photographers claimed their positions on the south grounds of the Treasury Building, east of the White House. From there they pointed their cameras up Pennsylvania Avenue to capture the panorama of troops marching toward them, framed by the Great Dome rising in the distance. Other cameramen took photos of the huge crowds gathered in front of Pennsylvania Avenue storefronts. At the Capitol, one photographer aimed his lens at the North Front, where crowds had gathered to watch the troops march up East Capitol Street and swing around the Capitol building on their way down the avenue. When he removed his lens cap, he froze a wondrous, ephemeral moment: blurred figures in motion; a man carrying, on a pole, a sign reading WELCOME BRAVE SOLDIERS, and, strolling through the frame, a young girl wearing a hoopskirt and a straw hat, trailing festive ribbons.

Gideon Welles delayed a trip south to witness the "magnificent and imposing spectacle," and recorded in his diary:

> [T]he great review of the returning armies of the Potomac, the Tennessee, and Georgia took place in Washington . . . It was computed that about 150,000 passed in review, and it seemed as if there were as many spectators. For several days the railroads and all communications were overcrowded with the incoming people who wished to see and welcome the victorious soldiers of the Union. The public offices were closed for two days. On the spacious stand in front of the executive Mansion the President, Cabinet, generals, and high naval officers, with hundreds of our first citizens and statesmen, and ladies, were assembled. But Abraham Lincoln was not there. All felt this.

General Joshua Lawrence Chamberlain, who would receive the Medal of Honor for his defense of Little Round Top at Gettysburg, felt it too when he and his men came opposite the reviewing stand:

"We miss the deep, sad eyes of Lincoln coming to review us after each sore trial. Something is lacking in our hearts now—even in this supreme hour."

From his front porch, Benjamin Brown French watched the Ninth Corps of the Army of the Potomac march west, down East Capitol Street on its way downtown, and to President Johnson's reviewing stand. French, who as commissioner had draped all the public buildings in Washington in mourning for Lincoln, including the Capitol, the Treasury Department, and the White House, now decorated his own house with symbols of joy: "I put a gilded eagle over the front door and festooned a large American flag along the front of the house, the centre being on the eagle, and above the eagle, in a frame placed in the window."

Then he went to the Capitol, climbed the narrow, twisting staircase to the Great Dome, and beheld the magnificent sight: "We went on the dome, from which we could see troops by the thousands in every direction . . . more than 50,000 in sight at one time, as we could see the entire length of Maryland Avenue west, Pennsylvania Avenue east, New Jersey Avenue south, and all of Pennsylvania Avenue west from the Capitol to the Treasury, and they were all literally filled with troops. It was a grand and brave sight."

While Union troops were marching in Washington, others at Fort Monroe entered Davis's cell and told him they had orders to shackle him. Davis saw the blacksmith with his tools and chains. He told them all that he refused to submit to the humiliation and pointed to the officer in charge and said he would have to kill him first. Davis dared his jailers to shoot him. Soldiers lunged forward to grab him, but Davis, exhibiting some of the strength he had displayed decades earlier when wrestling slaves, knocked one man aside and kicked another away with his boot. Then several men ganged up on him, seized him, and held him down while the smith hammered the shackle pins home. This was supposed to have been done in secret, but like many of the events to unfold in the days to come at Fort Monroe, word was leaked to the press. The *Chicago Tribune* reported the scuffle, claiming in one headline that Davis was "On the Rampage." Little did Davis know, his enemies had done him a huge favor.

By May 24, Lincoln's home in Springfield, viewed by thousands of funeral visitors during the first week of May, was no longer a center of attention. The delegations of dignitaries who had lined up in front of the house to pose for dozens of souvenir photographs had all left town. But on this day an anonymous photographer, probably local to Springfield, showed up to make the last known image of the Lincoln home draped in mourning. In the photograph, the black bunting, exposed to the elements for weeks, hangs askew, windswept and weather-beaten. No one poses for the camera, and the big frame house looks abandoned, even haunted. Green leaves—new life—sprout from the tree branches that frame the image. All across the nation, people could not bear to take down their wind-tattered, sun faded, and rain-streaked habiliments of death and mourning. Better, many thought, to allow time and the elements to sweep them away.

The May 29 issue of the *Richmond Times* shocked loyal Confederates who had remained in the city. "Mr. Davis Manacled," announced the headline. The news was several days old, but Davis's humiliation was still raw:

> [He] has manacles on both ankles, with a chain connecting about three feet long. He stoutly resisted the process of manacling. Rather than submit, he wanted the guards to shoot him. It became necessary to throw him on his back and hold him until the irons were clinched by a son of Vulcan. He exhibited intense scorn, but finally caved in and wept. No knives

WEEKS AFTER THE FUNERAL, THE LAST PHOTOGRAPH OF LINCOLN'S SPRINGFIELD HOME DRAPED IN MOURNING, MAY 24, 1865.

> and forks are allowed in his cell; nothing more destructive than a soup spoon. Two guards are in his casements continually. The clanking chains give him intense horror.

The *Times* did not know that by the time it published its report, public outrage, including in the North, had compelled Stanton to telegraph orders to Fort Monroe to have the shackles taken off on May 27 or 28.

The newspaper report described other details of Davis's imprisonment: "The windows are heavily barred, and the doors securely bolted and ironed. Two guards constantly occupy the room with him, while in the other room are constantly stationed a commissioned officer and a guard, all charged with the duty of seeing that the accused does not escape. Davis is not permitted to speak a word to any one, neither is any one permitted to speak a word to him. He is literally living in a tomb."

Feeling in the North was mixed about what should be done with the Confederate president. Some people favored execution. Others suggested mercy, if not for Davis's sake, then the country's. On May 29, the *Richmond Times* reprinted an article from the Springfield, Massachusetts, *Republican* that warned Northerners of the danger of persecuting Jefferson Davis.

> Do we wish to finish the rebellion, to turn out its very ashes? Then make no martyrs. The wounds inflicted in cold blood are what keep animosities alive. At this moment there are a million women in the South who would give all they had to save Jeff. Davis's life, who would conduct and shelter him . . . If his life is taken they are ready to dip their handkerchiefs in his blood, to beg locks of his hair, and to perpetuate for a hundred years the sentiment of vengeance. Unless we present them their grievance, in five years he will be remembered only as the author of innumerable woes.

On June 1, the North held a national day of mourning for Abraham Lincoln. Across the nation on the same day, communities remembered their fallen chief and the funeral pageant. Frederick Douglass praised Lincoln as "the black man's president" in New York City at Cooper Union, and in Providence, Rhode Island, abolitionist William Lloyd Garrison honored the Great Emancipator. The timing was not auspicious for Jefferson Davis. Given the bereaved and vengeful mood of the North, Davis was lucky not to find himself on trial before a military tribunal with the eight assassination conspirators locked up at the Old Arsenal. Although Davis did not sit in the prisoner's dock beside Lewis Powell, David Herold, Mary Surratt, and the others, his reputation did.

But zealous efforts by the government to implicate the Confederate president failed. Indeed, several witnesses who during the trial of the conspirators would give harmful testimony against Davis were exposed as imposters and perjurers. This was the first hint that it might not be so easy to prosecute Jefferson Davis for bloody crimes against the United States. But that did not prevent people from opining on what should be done with him. Through the spring and summer, President Andrew Johnson received many letters advising him to hang Davis or to torture him to death. Few correspondents urged mercy. Hate mail poured in to the Confederate president, taunting him about the terrible doom that must await him.

Two men, one his jailer, the other his doctor, figured prominently in Davis's imprisonment at Fort Monroe. The War Department appointed a Civil War hero, Major General Nelson A. Miles, to take charge of the state prisoner. Miles, only twenty-six, had fought in the battles of Seven Pines, Fredericksburg, and Chancellorsville, had been wounded in each engagement, and would receive the Medal of Honor for his valor at Chancellorsville. He was not awed by or sympathetic to the rebel president, and he became Davis's principal antagonist. It was Miles, acting under the discretionary authority granted him by Stanton, who ordered that Davis be shackled. He would regret that decision.

The chief medical officer of the fort, forty-three-year-old Lieutenant Colonel John J. Craven, was a self-made physician, inventor, and businessman, and he immediately empathized with Davis's plight. He requested an extra mattress and pillow for Davis's iron bed frame, provided tobacco, and objected to the shackles. Craven saw his patient every day, and Davis charmed—even mesmerized—him. Soon they fell into the easy habit of conducting long conversations on a variety of subjects. They became friends, and Craven tried to improve the harsh conditions of Davis's imprisonment in every possible way.

The assassination conspiracy trial continued through June, and its coverage dominated the headlines all month long. Every day, newspapers published transcripts of the previous day's testimony, and the public devoured each new, sensational revelation. The trial of the century stole attention from other events in Washington, including the first anniversary of the funeral for the victims of the Arsenal explosion and fire. On June 17, 1865, a monument to the women was erected over their common grave. The white stone sculpture by artist Lot Flannery depicted a mourning girl with clasped hands and downturned head standing atop a tall pillar inscribed on three sides with the names of the dead. On the front side was a panel carved in deep bas-relief that froze in time the laboratory building at the moment of the explosion, with rays of blinding light, fire, and smoke. Winged hourglasses ringed the monument to remind all viewers that life is fleeting. But no one was present on June 17 to heed that warning. The monument was erected without a dedication ceremony. The dignitaries and crowds who had thronged there one year earlier were absent that day. No reporters wrote stories. Perhaps the national capital was spent, its emotions drained. Perhaps there were no more tears left to shed.

On July 6, the most thrilling news since the capture and death of John Wilkes Booth raced through Washington. All eight defendants in the conspiracy trial had been convicted, three had been sentenced to life in prison, and four would be put to death by hanging the next day. Many Americans, although disappointed that Davis was not the fifth criminal standing on the scaffold at the Old Arsenal on the blazing hot afternoon of July 7, relished the verdicts. By the end of the

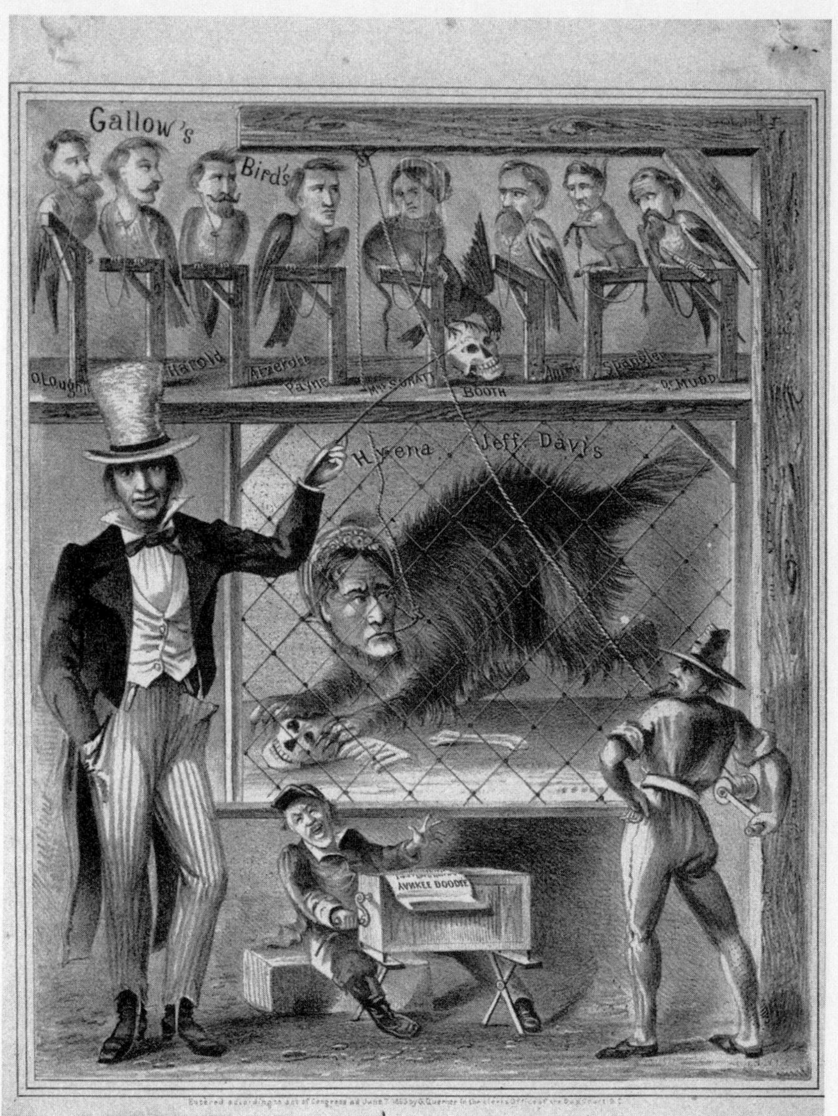

THIS FANCIFUL PRINT DEPICTS DAVIS AS A CAGED HYENA WEARING A LADIES' BONNET. THE LINCOLN ASSASSINATION CONSPIRATORS PERCH ABOVE HIM ON GALLOWS, FORESHADOWING THEIR EXECUTION.

trial, sober government officials had to concede that no credible evidence linked Davis to Lincoln's assassination. The Confederate president felt no empathy for his fallen foe, but would have considered it dishonorable to order his murder, and beneath his dignity to exult in it. If he was to be tried, it would be not for Lincoln's murder but for treason and war crimes.

Nonetheless, the hanging of the conspirators was an ill omen for Jefferson Davis. It showed that the War Department was ready to impose postwar death sentences, even upon Mary Surratt, the first woman executed by the federal government. Soon a Confederate officer, Captain Henry Wirz, would go on trial for crimes committed against Union prisoners of war confined at Andersonville where, allegedly, soldiers had been murdered, starved, and torn apart by vicious dogs. By now Davis had been given access to newspapers, and he must have read accounts of the gruesome hangings—the snap of the rope had not broken the necks of all the condemned and some of them strangled to death slowly—and speculated whether a similar fate awaited him.

On July 13, six days after the execution of the assassination conspirators, New Yorkers meted out symbolic punishment to the rebel chief when they hanged him in effigy. It began with P. T. Barnum. Back in May, when Stanton had refused to sell him the spurious "capture" dress, the brilliant entertainer was inspired to concoct a more exciting exhibit. He created a life-size wax figure of Jefferson Davis, dressed in a bonnet, hoopskirt, and boots, and displayed the mannequin in a tableau surrounded by other life-size figures dressed as Union cavalrymen in the act of apprehending the Confederate president.

This was not the only wax Jefferson Davis that entertained Americans in the summer of 1865. Professor Vignodi, an ambitious talent of the paraffin arts, created a life-size tableau of Lincoln and Booth in the box at Ford's Theatre, a presidential funeral hearse, and, for the benefit of those who missed the opportunity to view the corpse, a replica of Lincoln's coffin with a life-size figure of the dead president resting inside it. He followed up these morbid, self-proclaimed masterpieces with a life-size wax figure of Davis. A third wax figure of the rebel president showed up at a Sanitary Fair in Chicago.

On July 13, a fire broke out at Barnum's huge, four-story American Museum. While flames engulfed the entire building, thousands of New Yorkers rushed to the scene and watched as the live animal exhibits, injured by hideous burns, fled the museum and died agonizing deaths in the street. The *New York Times* described desperate efforts to save the exhibits. "On reaching the main salon, where the wax figures stood, [a performer] found great confusion existing. A man was endeavoring to save a Swiss animated landscape, while others tried to get out various other articles, including the wax figures . . . the crowd rushed to the front windows, and speedily emptied their arms of the gimcrack articles, throwing them indiscriminately into the street." Somebody in the museum tried to rescue President Davis, which amused both the *Times* and the crowd in the streets.

> One man had the JEFF. DAVIS effigy in his arms and fought vigorously to preserve the worthless thing, as though it were a gem of rare value. On reaching the balcony the man, perceiving that either the inanimate Jeff. or himself must go by the board, hurled the scarecrow to the iconoclasts in the street. As Jeff. made his perilous descent, his petticoats again played him false, and as the wind blew them about, the imposture of the figure was exposed. The flight of dummy Jeff. was the cause of great merriment among the multitude, who saluted the queer-looking thing with cheers and uncontrollable laughter. The figure was instantly seized, and bundled off to a lamp-post in Fulton Street, near St. Paul's Church-yard, and there formally hanged, the actors in the mock tragedy shouting the threadbare refrain, commencing the "sour apple" tree.

WITHIN DAYS OF HIS CAPTURE, POPULAR PRINTS RIDICULED THE CONFEDERATE PRESIDENT.

The image of a cowardly Confederate president masquerading as a woman titillated Northerners but outraged Southerners. In Georgia, Eliza Andrews received letters and Northern newspapers from friends in Richmond and Baltimore that outlined the accusation. On August 18 she recorded in her diary her first encounter with the Davis caricatures:

> I hate the Yankees more and more, every time I look at one of their horrid newspapers and read the lies they tell about us . . . The pictures in "Harper's Weekly" and "Frank Leslie's" tell more lies than Satan himself was ever the father of. I get in such a rage when I look at them that I sometimes take off my slipper and beat the senseless paper with it. No words can

> express the wrath of a Southerner on beholding pictures of President Davis in woman's dress; and Lee, that star of light . . . crouching on his knees before a beetle-browed image of "Columbia," suing for pardon! And these in the same sheet with disgusting representations of the execution of the so-called "conspirators" in Lincoln's assassination. Nothing is sacred from their disgusting love of the sensational.

If the first wave of Davis caricatures in newspapers and prints angered Eliza, then the sheet music artwork and satiric lyrics would have infuriated her even more. Davis was pilloried in popular song, many further perpetuating the widespread belief that he had been captured dressed in women's clothing, wearing a bonnet while carrying a large knife and a bag of gold. Lyrics referenced with delight the circumstances of his capture on the run:

> One bright and shining morning, All in the month of May,
> The C.S.A. did "bust" up, and Jeff he ran away;
> He grabb'd up all the specie, And with a chosen band,
> This valiant man skedaddled, To seek some other land . . .
>
> But good old Uncle Sam, Sent his boys from Michigan,
> And in the state of Georgia, They found this mighty man;
> He'd girded on his armor, his SKIRT it was of STEEL,
> But when he saw the soldiers, Quite sick did poor Jeff feel . . .
>
> So when this gallant SHE-ro, Did see the blue coats come,
> He found he had business, A little way from home;
> In frock and petticoat He thought he could retreat,
> But could not fool the Yankees They knew him by his feet.

Another song repeated the accusation that Davis had fled dishonorably, with stolen gold:

> Jeff took with him the people say, a mine of golden coin.
> Which he from banks and other places, managed to purloin:
> But while he ran, like every thief, he had to drop the spoons,
> And may-be that's the reason why he dropped his pantaloons!

For one song, "The Last Ditch Polka," printed sheet music shows a rat with Jefferson Davis's face pictured inside a cage within a prison cell surrounded by chains, guarded by an eagle. Some lyrics, more dark in tone, imagined with delight the punishments awaiting Davis:

> And when we get him up there boys, I'm sure we'll hang him high,
> He will dance around on nothing, in the last ditch he will die.

Another, "Hang Him on the Sour Apple Tree," described as a "sarcastical ballad," has a cover engraving that pictures a noose on a tree and describes the traitor Jefferson Davis getting what he deserves, speaking in Davis's voice: "Now all my friends both great and small, / A warning take from me. / Remember when for 'plunder' you start, / There's a Sour Apple Tree!" In bars and public places all across the North, people gathered and sang the chorus from the most popular song that spring: "We'll hang Jeff Davis from a sour apple tree."

Varina Davis is a prominent character in the story of her husband's capture, rumored to have spoken harshly, defiantly, to the soldiers who captured the president: "His wife now like a woman true, / Said don't provoke the President / Or else he may hurt some of you. / He's got a dagger in his hand."

After several weeks of silence, Varina received her first letter from Jefferson since his capture. It was the beginning of a moving jailhouse correspondence under difficult conditions.

Fortress Monroe, Va.,
21 Aug. '65

My Dear Wife,

I am now permitted to write you, under two conditions viz: that I confine myself to family matters, and that my letter shall be examined by the U.S. Attorney General before it is sent to you.

This will sufficiently explain to you the omission of subjects on which you would desire me to write. I presume it is however permissible for me to relieve your disappointment in regard to my silence on the subject of future action towards me, by stating that of the purpose of the authorities I know nothing.

To morrow it will be three months since we were suddenly and unexpectedly separated . . .

Kiss the Baby for me, may her sunny face never be clouded, though dark the morning of her life has been.

My dear Wife, equally the centre of my love and confidence, remember how good the Lord has always been to me, how often he has wonderfully preserved me, and put thy trust in Him.

Farewell . . . Once more farewell, Ever affectionately your Husband

Jefferson Davis

In October, the conditions of Davis's confinement improved. He was removed from his damp cell in the casemate wall and relocated to private rooms in Carroll Hall in the fort's interior. Better treatment was a sign that he might be staying awhile, and that he would not be leaving soon for trial.

The familiarity between Davis and Craven did not go unnoticed by Miles, and in the fall of 1865, it was rumored that Craven would be replaced. Davis wrote him a letter: "With regret and apprehension

I have heard that you are probably soon to leave this post. To your professional skill and brave humanity I owe it . . . that I have not been murdered by the wanton tortures and privations to which my jailor subjected me. Loaded with fetters when but little able to walk without them, restricted to the coarsest food, furnished in the most loathsome manner . . . and confined in a damp casemate the atmosphere of which was tainted by poisonous exhalations, you came to my relief . . . you have alleviated my sufferings and supplied my wants . . . you have been my protection."

In November, Henry Wirz, the commander of the Andersonville prisoner of war camp, was found guilty of war crimes and hanged at the Old Capitol Prison, across the street from the U.S. Capitol. Photos taken of the execution show Wirz standing on the scaffold, the rope around his neck, with the Great Dome as the backdrop. Davis may have seen woodcuts of the hanging in *Harper's Weekly*. This was the first postwar execution by the federal government for crimes unrelated to the Lincoln assassination. It set an ominous precedent for Davis.

The next month, Davis was allowed his first visit from Rev. Minnegerode. Both men had traveled far since that beautiful April Sunday morning in church nine months ago. The War Department warned the minister to limit his conversation to spiritual matters. The department was possessed by a paranoia that rebel daredevils might break Davis out of prison, and that any visitor might be transmitting secret messages of the plot.

In December, after Craven had one of Davis's tailors send a fine and warm winter coat, Miles and his superiors became incensed. Who was this rebel chief to enjoy such luxuries, and who was this doctor so eager to supply them? Craven was removed as Davis's physician on Christmas Day, and a month later he was mustered out of service and returned to private life. But he had the last laugh. Unbeknownst to Miles, Craven had kept a diary about the conditions of his patient's imprisonment.

Christmas was a hard time for Davis. A year ago, his family had sat around their dining room table in Richmond, feasting on turkey and a barrel of apples that an admirer had sent them as a gift. Now all Varina could offer was a sad letter: "Last Christmas we had a home—a country—and our children—and yet we would not be comforted for our 'little man' [Joseph Evan Davis, who had fallen to his death several months before] was not—This Christmas we have a new child, who has seen but one before." Overcome, she thought of dead sons and cemeteries: "That little grave in Richmond, the other in Georgetown [for Samuel Emory Davis, 1852–1854] is ever fresh to me." Perhaps realizing that her letter had turned too morbid, Varina told Jefferson that her love for him was stronger than sad memories: "But fresher—more enduring still is the love which at this season nearly twenty two years ago filled my heart, and has kept it warm and beating ever since."

As the new year came, the U.S. government still had not decided what it wanted to do with Jefferson Davis. Would 1866 bring him life? Or death?

On January 29, 1866, a young girl in Richmond, Emily Jessie Morton, wrote to Davis to cheer him up:

> *I hope that you will not think me a rude little girl to takeing the liberty of writing to you, but I want to tell you how much I love you, and how sorry I feel for you to be kept so long in Prison away from your dear little children . . . I go to school to Mrs. Mumford where there are upwards of thirty scholars all of which love you very much and are taught to do so. When we go to Hollywood [cemetery] to decorate our dear soldiers graves on the 31st of May your little Joes grave will not be forgotten.*

She told him she loved him so much that her teasing schoolmates called her little "Jefferson Davis."

From the time of Davis's capture in the spring of 1865 to the winter of 1866, Varina Davis waged a relentless one-woman campaign to obtain better treatment for her husband, to visit him in prison, and, ultimately, to gain his freedom. She had been a popular and well-liked figure in antebellum Washington, including among important Northern politicians, and now she used every social and political skill she had learned since her Mississippi girlhood to save her husband. She wrote letters, secured personal meetings, and influenced newspaper coverage.

Six months later, on April 25, 1866, Varina Davis sent a letter to President Andrew Johnson: "I hear my husband is failing rapidly. Can I come to him? Can you refuse me? Answer." Her note alarmed Johnson, who asked Stanton to advise him immediately. The secretary of war, who had kept Jefferson and Varina apart for one year, relented. Miles warned his superiors what a dangerous foe she could be, and a few weeks after she arrived at Fort Monroe, several newspaper articles accused him of punishing his prisoner with inhuman treatment. Enraged, on May 26 Miles forwarded the articles to General E. D. Townsend at the War Department: "It is true I have not made [Jefferson Davis] my associate and confidant or toadied to his fancy . . . [but] the gross misrepresentations made by the press infringes upon my honor and humanity and I am unwilling to let such statements to go unnoticed." The newspaper stories were nothing compared with what was coming. Dr. Craven had written a book.

Varina received permission to visit Jefferson. She arrived at Fort Monroe on May 3, 1866, and brought her little girl with her. She had left the rest of her children in the care of others, deciding her first duty was to save her husband's life. But before she was permitted to see him, the War Department had demanded that she promise in writing that she would not help him escape, or smuggle "deadly" weapons—including pistols, knives, or explosives—into his cell: "I,

Varina Davis, wife of Jefferson Davis," she agreed, "for the privilege of being permitted to see my husband, do hereby give my parole of honor that I will engage in or assent to no measures which shall lead to any attempt to escape from confinement on the part of my husband or to his being rescued or released from imprisonment without the sanction and order of the President of the United States, nor will I be the means of conveying to my husband any deadly weapons of any kind."

The former first lady of the Confederacy might have bristled at the language—she considered it her right, not a "privilege," to see her husband, and she viewed their one-year separation an outrage, but now was not the time to argue. President Johnson had yielded to her will. She signed the document and gave her parole. Now nothing would stop her from reuniting with her beloved "Banny."

This visit was followed a few weeks later when Davis signed a parole that gave him liberty to wander the fort with Varina during the day.

> FORT MONROE, May 25, 1866
>
> For the privilege of being allowed the liberty of the grounds inside the walls of Fort Monroe between the hours of sunrise and sunset I, Jefferson Davis, do hereby give my parole of honor that I will make no attempt to nor take any advantage of any opportunity that may be offered to effect my escape therefrom.
>
> JEFFERSON DAVIS

Varina had won the first round. She had been reunited with her husband. Soon she won another victory—the right to move into the prison and share Jefferson's quarters. If she could not take him home to live, then she and their daughter would live with him at Fort Monroe. Now she prepared for the next stage of her battle with federal authorities—her effort to win his freedom.

In June 1866, a New York publisher released Craven's book under the long-winded title *Prison Life of Jefferson Davis, Embracing Details and Incidents in His Captivity, Particulars Concerning His Health and Habits, Together with Many Conversations on Topics of Great Public Interest.* It caused a sensation and created nationwide sympathy for the imprisoned fallen president, just as Craven hoped. Even most of Davis's enemies did not want to see him languish and die in captivity. As a literary effort, *Prison Life* was riddled with exaggerations and errors. Indeed, when Davis obtained a copy he penciled corrections in the margins of almost two hundred pages. Some critics said that the book was a fraud. It did not matter. As a piece of political propaganda, the book was a work of genius. The month after its publication Joseph E. Davis wrote to his brother: "The prison life by Dr. Craven is I think exerting an influence even greater than expected." In Europe, public opinion favored Davis. The Pope sent him an inscribed photograph and a crown of thorns.

Three months later, Miles would be relieved of his command. The fifteen-month assignment had not been to his liking, but he resented leaving his post under a cloud. His career survived the embarrassment, and he enjoyed future success in the west fighting Indians, and rose in rank until he commanded the entire U.S. Army.

In the summer of 1866, two ghosts from the Lincoln assassination visited Fort Monroe. On June 5, Surgeon General Joseph A. Barnes called upon Davis. Barnes had watched Lincoln die. On August 12, Assistant Surgeon General Charles H. Crane visited Davis. He and Barnes had witnessed the autopsy and watched Curtis and Woodward cut open Lincoln's head and remove his brain. They had come to evaluate Davis's health and to ensure that he did not die while in Union captivity. Stanton wanted no martyrs. Did Davis know what they had seen? No records survive to indicate whether the doctors discussed the assassination with him.

By the fall of 1866, the government had still taken no action to prosecute Davis for treason. He welcomed his trial, whatever its result. If he was acquitted, then the South was not wrong—it did have the constitutional right to leave the Union, and secession was not treason. If he was found guilty, he was happy to suffer on behalf of his people. His death, he believed, would win mercy for the South. The U.S. government wanted neither result. A federal court verdict declaring secession not treasonable would overturn the whole purpose and result of the war. Some of the ablest attorneys in America had offered to defend Davis, and a guilty verdict was by no means certain. And a guilty verdict, followed by Davis's execution, would create a martyr and might inspire the South to rise up again. John Reagan said it would be best for all concerned to release Davis: "I urged that the welfare of the whole country would be subserved by setting him free without a trial; for the South it would be a signal that harsh and vindictive measures were to be relaxed; and for the North it would indicate that they were willing to let the decision of the right of secession rest where it was and not try to secure a judicial verdict . . . the war had passed judgment and that hereafter secession would mean rebellion."

While the government dithered, Davis lingered in legal limbo through the winter of 1867. But by the spring, the federal government finally decided that it wanted Davis off its hands. He would be released on bail, preserving the right to try him at some future time. By prearrangement, his attorneys would initiate proceedings to free him and the government would not oppose them. On May 8, 1867, former president Franklin Pierce visited Davis in prison and congratulated him on his pending release.

On May 10, the second anniversary of Davis's capture, a writ of habeas corpus was served on the commander of Fort Monroe. At 7:00 A.M. on May 11, Burton Harrison, Joseph E. Davis, Jefferson's brother, and several others escorted the former Confederate president, not quite a free man yet, to the landing at Fort Monroe, where he boarded the steamer *John Sylvester* for Richmond. At 6:00 P.M. Davis reached

Rocketts, the same place where, two years earlier, Abraham Lincoln had landed in Richmond to a tumultuous welcome from the city's slaves. Now the white citizens welcomed Davis back to his old capital. As he passed, men uncovered their heads and women waved handkerchiefs. "I feel like an unhappy ghost visiting this much beloved city," Jefferson told Varina. A carriage drove them to the Spotswood Hotel, where they were taken to the same rooms they occupied in 1861.

On Monday morning Davis and his counsel appeared in federal court, in the same building once occupied by his presidential office. The $100,000 bond was signed by an unexpected list of names: Davis, Horace Greeley, Cornelius Vanderbilt, and Gerrit Smith, the famous abolitionist, and one of the "Secret Six" who had backed John Brown. Smith proclaimed that the war had been the fault of North and South: "The North did quite as much as the South to uphold slavery . . . Slavery was an evil inheritance of the South, but the wicked choice, the adopted policy, of the North." After Davis was freed on bail, he left the courtroom and was surrounded by a crowd of supporters; according to his principal legal counsel, Charles O'Connor, "poor Davis . . . wasted and careworn, was almost killed with caresses." Davis returned to the Spotswood, where he and Varina received friends.

On May 13 Davis posted bail, the court released him, and he walked out a free man. In 1867, the Lincoln assassination conspirators were either dead or in prison, Captain Wirz had been executed for war crimes and Jefferson Davis was allowed to leave custody as a free man, never having been tried, let alone found guilty of any crime. The man who led the campaign to divide the nation, the man who gave orders to fight and kill Union soldiers, was never tried. The importance of this cannot be overstated. To Davis's partisans, this meant that no federal court had ever ruled that secession was unconstitutional or treasonous. Thus, they believed, Davis had done no wrong. To Northerners, whatever happened or did not happen to Davis in a court of law, he remained a traitor. If he was freed, it was for prudential reasons, to heal the wounds of war, not to achieve legal justice. His first act, one that would set the tone for the remainder of his life, was one of remembrance. He took flowers to the grave of his son Joseph Evan Davis at Hollywood Cemetery, and while there he also decorated graves of Confederate soldiers. A friend wrote to Varina Davis to assure her that in the joy over the president's release, the people of Richmond had not forgotten their dead son: "Last Friday [June 1], Hollywood was glorified with flowers. The little one who sleeps here was not forgotten. Garland upon garland covered every inch of turf and festooned the marble that bore the beloved name, some with the touching words, 'for his Father's sake.'"

On June 1 a Confederate officer who had served under President Davis sent him a heartfelt letter that described the "misery which your friends have suffered from your long imprisonment," adding that "to none has this been more painful than to me." The letter rejoiced in Davis's freedom: "Your release has lifted a load from my heart which I have not words to tell, and my daily prayer to the great Ruler of the World, is that he may shield you from all future harm, guard you from all evil, and give you the peace which the world can not take away. That the rest of your days may be triumphantly happy, is the sincere and earnest wish of your most obedient faithful friend and servant." The letter was signed by Robert E. Lee.

After his release, Davis was forced to ask himself questions for which he had no immediate answers. What did the future hold? Where would he go? What would he do? How would he live? How would he earn money? Like much of the South, his life was in ruins. He had lost everything. He had no cache of secret gold. His plantation was in ruins, no crops grew there, and he owned no slaves to work the fields. They had all been emancipated. Union soldiers had looted his Mississippi home of all valuables. They even stole his old love letters from Sarah Knox Taylor.

He also had to decide what he must *not* do. His behavior would be scrutinized by Northerners and Southerners alike. He vowed to do nothing to bring dishonor upon himself, his people, or the Confederacy. Because so many Southerners were poor, he decided that he would not shame himself by accepting charity from his supporters while others were in need. He would not speak publicly against the Union or Reconstruction, he decided, out of fear that his words might cause his people to be punished. Nor would he run for public office. He knew without doubt that he could be elected to any political position in the South. But to seek office, he would have to take a loyalty oath to the Union, something he would never do. To swear that oath, to recant his views, to say the South was wrong, would betray every soldier who laid down his life for the cause. He would rather suffer death. And, last, he decided he would never return to Washington, D.C., the national capital he once loved and the scene of many of his greatest achievements and happiest days.

CHAPTER TWELVE

"The Shadow of the Confederacy"

For the first time in his life, Davis needed a job. It was a shocking predicament for a member of the elite, planter class. But he had no choice. He needed money and stability, and the quest for it preoccupied him. For the next two years, he wandered and pursued opportunities that led nowhere. In November 1869, he was offered the presidency of the Carolina Life Insurance Company at the impressive annual salary of $12,000—nearly half the pay of the president of the United States. He took the job. But Davis's days as a "business man" were numbered. An epidemic killed too many customers, and the economic downturn put the company out of business. Davis pursued other moneymaking opportunities, and he considered various schemes that others proposed to him, but he never achieved the financial success that he craved. Failure embarrassed him.

In 1870, the whole South mourned the death of its great general, Robert E. Lee. Davis spoke at the memorial service with sadness and great eloquence, and it was there that he found the true purpose of his remaining days—remembering and honoring the dead. Soon,

Oil portrait of Jefferson Davis as he appeared in the 1870s.

the theme of "The Confederate Dead"—the idea of a vast army of the departed who haunted the Southern landscape and memory—swept the popular imagination. Soon, veterans' groups, historical societies, and women's associations labored to recover the dead from anonymous wartime graves, to build cemeteries for them and to mark the land where they shed their blood with monuments of stone, marble, and bronze. Later, Davis became the symbol of this movement. He was the link between the Confederate living and the dead. For now, Lee's unexpected death was a warning to Davis that he should not wait too long to tell his story. As early as March 30, 1870, Davis told

Burton Harrison that he wanted to write a book: "It has been with me a cherished hope that it would be in my power before I go down to the grave to make some contribution to the history of our struggle." Lee had hoped to do the same. The general had begun to gather documents. He examined his official papers. But before he could write his memoirs, he died.

In the 1870s, Davis hit his stride as a keeper of Confederate memory. He wrote articles. He read histories of the war written by generals and political leaders. He kept up an active correspondence and answered countless inquiries about the conduct of the war. He supported the creation of the Southern Historical Society. The North may have won the war on the battlefield, but the South would not lose it a second time in the books. Davis became the titular head of a shadow government, no longer leading a country, but leading a patriotic cause devoted to preserving the past.

Davis became a fixed symbol in a changing age. He witnessed the passing of an era and the rise of a different, modern America. The U.S. Army fought new wars on the western frontier, and in 1876, when America was set to celebrate its national centennial, the flamboyant Civil War general George Armstrong Custer found death at the Little Bighorn, eleven years after Appomattox. To take advantage of the patriotic fervor during the centennial, grave robbers plotted to kidnap Lincoln's corpse and ransom it for a huge cash payment. They broke into the tomb but were arrested. Davis witnessed the industrialization of the nation, the invention of electric lights, and the first hints of America's future role as a global power. He also witnessed the plight of blacks during Reconstruction in the postwar South, and what happened to them after 1877, when the last of the federal occupying troops returned to the North. In a few years he would read of the assassination of another president, James Garfield, who survived the Civil War only to be shot in the back at a Washington, D.C., train station.

Throughout this era, Davis experienced financial insecurity and

On the front porch at Beauvoir.

domestic instability. He lacked a proper income or a real home. In 1877, he found his sanctuary at last. A longtime friend and widow, Sarah Dorsey, invited Davis to visit her Mississippi Gulf Coast estate, Beauvoir, near Biloxi. The visit became permanent, and Dorsey willed the property to Davis upon her death. Beauvoir became his haven. It relieved him of significant financial distress, gave him a place to live, and allowed him to finish his book. As Davis labored to complete his memoirs, he also found here, fourteen years after the end of the war, and twelve years since his freedom, a kind of peace. As the years passed at Beauvoir, he became more handsome. His face softened.

Photographs from the Gulf coast years capture a gentle smile absent from photos taken earlier in his life, in the 1850s and 1860s.

In 1881, Jefferson Davis published his magnum opus, his two-volume memoir *Rise and Fall of the Confederate Government.* This was not a conventional memoir that tells the story of the subject's life. Instead, *Rise and Fall* was in large part a massive, legalistic, dense, and impersonal defense of state's rights, secession, and Southern independence. It was a dry work of history and politics, not an emotional telling of a riveting life. Yes, partisan publications, including the *Southern Historical Society Papers*, reviewed it favorably. Loyal Confederates purchased more than twenty thousand copies, but the work did not become a national sensation. It was no more than a moderate, regional success. The heavy volumes may have revealed the contents of Davis's analytical mind, but they did not unlock the secrets of his heart.

His memoir done, Davis seemed destined for a quiet life at Beauvoir: receiving guests, dining with friends, writing letters, and sitting on the veranda sporting a jaunty straw hat and enjoying the sea breezes. Visiting journalists seemed surprised when they found not an embittered old man, but a genial host and superb conversationalist.

In June 1882, Davis received the most unusual guest to ever visit Beauvoir—the Irish author Oscar Wilde. During his lecture tour to Memphis, an interviewer found him in his hotel room with a set of Davis's *Rise and Fall* on the table. Wilde said, "Jefferson Davis is the man I would most like to see in the United States." Wilde dispatched a letter to Davis asking if, after Wilde lectured in New Orleans, he might visit Beauvoir. The press learned of the fascination, and on June 23, the *Mobile Register* wrote this about the forthcoming meeting: "We understand that ex-President Davis has invited Mr. Wilde to pay him a visit at Beauvoir . . . and that the aesthete has accepted." Almost half a century—and other experiences—separated the seventy-four-year-old senior statesman from the twenty-six-year-old literary voluptuary. The *Register* could not resist commenting on the bizarre appoint-

THE IRISH AESTHETE OSCAR WILDE VISITS BEAUVOIR.

ment: "It is scarcely conceivable that two persons can be more different than the ex-President of the Confederacy and the 'Apostle of aestheticism,' as known to report; and we confess to sufficient curiosity to desire to know the bent of their coming, protracted interview."

Interviewed in New Orleans in advance of the meeting by the *Picayune*, Wilde spoke well of the Confederacy and Davis: "His fall after such an able and gallant pleading in his own cause, must necessarily arouse sympathy."

At dinner, Varina Davis, her daughter Winnie, her cousin Mary Davis, and Oscar Wilde did most of the talking. Jefferson did not say much as he scrutinized the man who must have appeared to him an odd, even alien, creature. The former president retired early, and afterward, Wilde delighted the literature-loving women until late in the night. The next day, after Wilde departed, Davis rendered his laconic verdict to Varina: "I did not like the man."

Wilde came to the opposite conclusion about his taciturn host. He left a special gift on Davis's desk—an oversized, presentation photograph inscribed: "To Jefferson Davis in all loyal admiration from Oscar Wilde, June '82—Beauvoir." In Davis's world, it was a cheeky gesture. He did not solicit the memento, and its presentation suggested that Wilde considered himself a peer of the Confederacy's first man. It was Davis's introduction to America's burgeoning popular culture of celebrity.

In the 1880s, Davis did not seek out feuds, but when insulted or provoked, he was not one to seethe in silence. He planned to speak at the dedication of the Robert E. Lee mausoleum in Lexington, Virginia, on June 28, 1883. But when he learned that his archenemy, General Joseph E. Johnston, would preside over the event, Davis backed out. He refused to share the stage with the man who had surrendered his army to Sherman in North Carolina, dooming, in Davis's opinion, the Confederacy's last hope east of the Mississippi in April 1865. And it was Johnston who, in an outrageous defamation, had accused Davis of stealing a fortune in Confederate gold. In his youth, Davis would have challenged Johnston to a duel. Instead, his surrogates advised him to remain above the fray while they unleashed a ferocious verbal assault.

In November 1884, Davis counterattacked against charges by General Sherman that Davis had been at the center of a conspiracy to "enslave" the North and not merely win Southern independence. As proof, the general claimed that he had seen secret letters and overheard private conversations implicating Davis. Davis savaged his antagonist. "I have been compelled to prove General Sherman to be a falsifier and a slanderer in order to protect my character against his willful and unscrupulous mendacity . . . He stands pilloried before the public and all future history as an imbecile scold, or an infamous slanderer—As either he is harmless."

By early 1886, Jefferson and Varina were living a quieter life at Beauvoir. Jefferson received inquiries from strangers, or from people he once knew. Many asked if he remembered them. He maintained his passionate interest in Confederate history and how the war was remembered. He hoped to live to celebrate his eightieth birthday in two years. But given his lifelong health problems, he must have felt he that he was lucky to have lived that long. Many times over the years, death had tried to pull him into the grave, but his stubborn body had fought back and willed itself to live. He had lived long enough to write *Rise and Fall,* to correct errors in the record, to survive many of his foes, and to make his own peace with the past. The once-stern expression on his face had softened, and, strangely, the older he got the happier he looked. Yes, he had suffered. He could tally his losses: his first love, Knox Taylor; his first country, the old Union of the United States; his cause of Southern independence; his war; and his presidency; for a time, his liberty; and his sons, now dead, all of them. But in 1886, Davis was serene and at peace at his Gulf Coast sanctuary. His days as a public man seemed at an end. He planned to make no more public appearances, deliver no more speeches before tumultuous crowds, and undertake no more tours of his vanished empire.

Then the invitation came. Would he, the letter from the mayor of Montgomery asked, come to lay a cornerstone for the monument to Alabama's Civil War dead? And perhaps their former president would agree to say a few words in memory of them? Davis could not say no.

When, accompanied by his daughter, Winnie, he boarded the train at the depot half a mile from Beauvoir, he could not have known that he would return from this trip a different man. He did not know his journey would bring to the surface emotions long buried in Southern hearts, and his own. He did not know that by its end, almost a quarter century after the end of the war, the South would love him more than it ever did. Reporters went on the trip, including Frank A. Burr of the *New York World*. Burr had pursued other ghosts from the war. He had written about the Lincoln assassination, and, years after the war, he traveled to Garrett's farm in search of the legend of John Wilkes Booth. Lincoln and Booth were long dead. Now, on this journey, Burr would travel with a living ghost from the past.

On a stop on the way to Montgomery, Davis had two encounters with well-wishers. "At one station," reported the *Atlanta Constitution*, "a soldier with a wooden leg got on board and bidding goodbye to Mr. Davis slapped his wooden leg and said: 'That's what I got from the war but I'm proud of it.' To this Mr. Davis responded with a hearty 'God bless you.' At another station an old colored woman, a former slave of Mr. Davis, was loud in her blessings of her old master." These were the first hints of what lay ahead.

Davis arrived in Montgomery, the first capital of the Confederacy, on April 28. Just before the train rolled into the station, General John B. Gordon, one of Davis's favorite Confederate officers and who now sought a political career, spoke to the crowd: "Let us, my countrymen, in the few remaining years which are left to our great captain, seek to smooth and soften with the flowers of affection the thorny path he has been made to tread for our sake." Then train pulled in, cannon fire erupted, and thousands cheered. A drenching rain thwarted a major public reception at the railroad depot. When Davis disembarked from his car, he got into a carriage that drove him to the Exchange Hotel, where he had spent the night before his inauguration as provisional president. Bonfires and electric lights illuminated the route.

Davis prepared to exit the carriage and walk into the hotel. He stood erect, but stayed in the carriage. He paused and looked over the immense crowd. Then he spoke: "My countrymen, my countrymen, with feelings of the greatest gratitude I tender you my most sincere thanks for your kind reception."

The next day the *New York Times* filed its report. "Dixie Reigns Supreme," the headline blared. "This city has simply gone wild over Jefferson Davis." Davis was welcomed by a "tumultuous crowd that shouts itself hoarse." The *Times* could not understand the symbolism: "The explanation of all this is not easy. There are other incidents as strange and perplexing as those already outlined. The leaders of the throng tell you that it means nothing; that it is but a passing show to please their old chieftan, a day of sound and fury, signifying nothing. It is all a conquered people can offer him . . . it is useless to wonder how much more there is stored up in their hearts."

From the moment he arrived, women flocked to his side, praising him, teasing him, and flirting with him. Some soothed him with fans. They filled his room at the Exchange Hotel with roses. One woman who shook his hand exclaimed, "I am more of a rebel right here than ever before."

Davis's old friend Virginia Clay witnessed some of these encounters: "I saw women shrouded in black fall at [Davis's] feet, to be uplifted and comforted by kind words. Old men and young men shook with emotion beyond the power of words on taking [his] hand."

That night, when Davis spoke at the old capitol, the mayor introduced him as the representative Southern man: "It is with emotions of the most profound reverence that I have to introduce you to that most illustrious type of southern manhood and statesmanship—our honored ex-president, Jefferson Davis."

Davis rose and began to speak. His lips moved, but no one could hear him. The sound of thousands of cheering voices drowned him out. "A cheer long pent up since 1861 rent the air, was taken up by the crowds on the streets, and echoed and re-echoed all over the city," reported one paper. Davis bowed to his right and left and tried to make himself heard above the roar.

"Brethren," he cried, in that same clear, pleasing, distinctive voice that had once charmed Varina the day they met, and that had thrilled listeners when he spoke on the floor of the U.S. Senate. That single word—"brethren"—incited the throng to shout even louder. Women stood on their chairs and, weeping and laughing from joy, flapped their handkerchiefs in the air like the wings of birds, or like little signal flags.

In his seventy-eight years, Davis had never seen anything like it, not even on the day when, twenty-five years ago, he was inaugurated the Confederacy's president. The audience silenced itself and allowed him to speak.

"My friends," he began, "it would be vain if I should attempt to express to you the deep gratification which I feel at this demonstration; but I know that it is not personal, and therefore I feel more deeply grateful, because it is a sentiment far dearer to me than myself." With those words, Davis reminded his listeners that he did not return to the first capital of the Confederacy to claim honors for himself, but to tender honors to others—the Confederate dead. "You have passed through the terrible ordeal of a war which Alabama did not seek . . . a holy war . . . Well do I remember seeing your gentle boys, so small, to use a farmer's phrase, they might have been called seed corn, moving on with eager step and fearless brow to the carnival of death; and I have also looked upon them when their knapsacks and muskets seemed heavier than the boys, and my eyes, partaking of a mother's weakness, filled with tears. Those days have passed. Many of them have found nameless graves . . ." The poetic image of the "seed corn" that never grew to maturity broke the hearts of the parents of those boys. Davis, the old master orator of the U.S. Senate, still knew how to read a crowd. His audience was on the verge of a frenzy.

Then Davis uttered the words that forever more united past and present through the dream of the Lost Cause. "But they are not dead—they live in memory and their spirits stand out a grand reserve of that column which is marching on with unfaltering steps towards the goal of constitutional liberty." Davis transported his listeners back to the beginning of the Civil War:

> I am standing now very nearly on the spot where I stood when I took the oath of office in 1861. Your demonstration now exceeds that which welcomed me then . . . the spirit of Southern liberty is not dead. Then you were full of joyous hope, with a full prospect of achieving all you desired, and now you are wrapped in the mantle of regret . . . I have been promised, my friends, that I should not be called upon to make a speech, and therefore I will only extend my heartfelt thanks. God bless you, one and all, men and boys, and the ladies above all others, who never faltered in our direst need.

His remarks done, Jefferson Davis sat down.

What happened next stunned the reporter from the *Atlanta Constitution*. "Such a cheer as followed the speaker to his seat cannot be described," he noted. "It was from the heart. It was an outburst of nature. It was long continued. Mr. Davis got up again and bowed his acknowledgments. Men went wild for him; women were in ecstasy for him; children caught the spirit and waved their hands in the air." Then a lone Confederate veteran shouted, "Hurrah for Jeff Davis!" as loud as the old rebel could yell and within moments, thousands of voices repeated the salute.

The *Atlanta Constitution* reporter followed Davis back to the Exchange Hotel, and later wrote: "Your correspondent, presuming upon a previous acquaintance formed in Brierfield, called upon the noble old Mississippian in his room. He was greeted by a grasp of the hand which proved that Mr. Davis still had a good grip . . . he expressed himself as in the best of health for one of his years, and judging from his face, in the best of spirits." A procession of well-wishers, starting with the mayor, interrupted the interview. Soon admiring and swooning women filled the former Confederate leader's room.

Davis left Montgomery on April 29, but he did not go home to Beauvoir. He had accepted an invitation to go to Atlanta and speak there at the dedication of a sculpture of Ben Hill. Frank Burr wrote, "At every station along the route from Montgomery Mr. Davis was met by tremendous delegations, who shouted and cheered from the moment the train came in sight until it was out of hearing." Burr penned an extravagant speculation: "All the South is aflame . . . and where this triumphant march is to stop I cannot predict." When Davis arrived in Atlanta, he found fifty thousand people waiting for him at the railroad depot. Eight thousand children lined the route—more than a mile—from the station to the home of his host, his old friend Ben Hill. Two thousand Confederate veterans followed Davis's carriage. Every business closed its doors except for the U.S. Post Office. The authorities in Washington had refused to permit it. At the dedication of the statue on May 1, newspaper editor Henry Grady gave Davis a stirring introduction: "This outcast . . . is the uncrowned king of our people . . . the resurrection of these memories that for twenty years have been buried in our hearts, have given us the best Easter we have seen since Christ was risen from the dead."

Next, Davis traveled by train to Savannah, where he would attend the centennial of a local militia unit, the Chatham Artillery, and participate in the unveiling of two bronze tablets on the monument to the Revolutionary War hero General Nathaniel Greene. This simple forty-foot marble shaft, "the first monument ever erected in the South to a Northern man," said the *New York Times,* had stood in the public square forty years barren of inscriptions. To celebrate the Chatham Artillery, the Georgia Historical Society offered to attach the tablets. The train sped three hundred miles from Atlanta to Savannah, making, by one hour, the fastest time ever recorded between those two cities. The five-car train was decorated with rich bunting, and nailed to each side of the train in large letters was the motto: "He was manacled for us." At the end of Davis's car, the last one, his portrait, captioned "Our President" in flowers formed into the shape of letters, delighted onlookers. Along the tracks, people gathered to watch the train fly by. At every platform, crowds thronged. When the rain stopped at Macon for twenty minutes, Davis spoke of being brought here as a prisoner twenty-one years earlier.

Davis arrived in Savannah that evening for a four-day visit. The streets were impassible. "As in Atlanta," the *New York Times* reported, "flowers were rained upon him from the multitudes." It was a triumphant return to the city from which Davis had been sent by sea to his captivity in May 1865.

By his second day in Savannah, Davis began to lose his strength. The tour had exhilarated him, but he was also exhausted by the travel. The *New York Times* reported on his weakened condition:

> He is beginning to feel the effects of the demands which his people have made upon his waning strength, as well as those which he himself has imposed. The kindly expressions and demonstrations . . . have warmed his heart, so that he wants to meet whatever exactions in the way of speechmaking or handshaking are asked of him. When he started, a week ago, he was reluctant to speak more than a few words of thanks . . . Since then his disposition has changed completely, and he submits himself willingly to the calls from the shouting multitudes for speeches. "Do they want me?" he asked several times yesterday as he lay in his couch in the railway car and heard the rousing cheers at several stations. "All right," he would continue; "tell them I'm coming." And then, taking his stout cane in hand to aid his frail steps, he would walk slowly out to the platform and talk.

By that night, Davis had regained his strength and he spoke to more than ten thousand people from the steps of the Comer residence.

That night was another triumph, but it was also the end of the tour. Davis had not planned it, but he had enjoyed it. When it began, he was not sure that anyone would want to see him again. When it ended, he knew he held a place in Southern hearts.

It was no surprise that the next year, in October 1887, he agreed to return to Macon, which had been but a brief stop during the first tour to attend a Confederate reunion and the Georgia State Fair. This time Varina accompanied him. She marveled at the reception, but worried that it would kill him. A newspaper article warned citizens that this was Davis's "last journey," and that they must do nothing to tax his feeble strength, not even shake his hand. He was seventy-nine and weaker than he had been the previous year. When Varina cautioned him that he might die during the visit, Davis retorted: "If I am to die it would be a pleasure to die surrounded by Confederate soldiers."

Several thousand of them did surround him when they charged the house where he was staying. It was like Pickett's Charge all over again. When they produced a battle flag, David said: "I am like that old flag, riven and torn by storms and trials. I love it as a memento; I love it for what you and your fathers did. God bless you! I am glad to be able to see you again." Then Davis took the flag and waved it through the air. The veterans went wild.

At the climax of the visit, Jefferson and Varina presided over Children's Day at the fair. Beginning with the orphans, the Davises blessed and laid hands upon several thousand children. It was as though, by his touch, he sought to pass on the values of the Lost Cause to a new generation. He returned to Beauvoir, confident that he had made his last public appearance.

In March 1888, Davis accepted an invitation to speak to an audience of young men in Mississippi City, Mississippi. He was prepared to decline, but accepted because the venue was not far from Beauvoir, and the composition of the audience appealed to him. It turned out to be one of the shortest—and one of the most remarkable—speeches of his life:

> Mr. Chairman and Fellow Citizens: Ah, pardon me, the laws of the United States no longer permit me to designate you as fellow citizens, but I am thankful that I may address you as my friends. I feel no regret that I stand before you this afternoon a man without a country, for my ambition lies buried in the grave of the Confederacy. There has been consigned not only my ambition, but the dogmas upon which that Government was based.

Davis sounded like he was about to indulge in bitter sectionalism. But then he changed his tone:

> The faces I see before me are those of young men; had I not known this I would not have appeared before you. Men in whose hands the destinies of our Southland lie, for love of her I break my silence, to speak to you a few words of respectful admonition. The past is dead; let it bury its dead, its hopes and its aspirations; before you lies the future—a future full of golden promise; a future of expanding national glory, before which all the world shall stand amazed. Let me beseech you to lay aside all rancor, all bitter sectional feeling, and to make your places in the ranks of those who will bring about a consummation devoutly to be wished—a reunited country.

Davis reached a milestone on June 3, 1888—his eightieth birthday. He had been inaugurated president of the confederacy twenty-seven years ago; the Civil War had ended twenty-three years ago, and he had survived Abraham Lincoln by the same measure of time; and he had been a free man for the last twenty-one years. Lincoln was only fifty-six when he was assassinated, and he had no time to savor his victory. Davis had almost a quarter century to reflect upon his defeat. From all across the South, from people high and low, congratulations and gifts poured in to Beauvoir. From the state of Mississippi came the gift of a crown, this one fashioned not from thorns, but silver.

One letter, from a former Confederate soldier of no prominence and unknown to Davis, spoke for all the anonymous, faithful veterans who had survived the war.

Lewisville, Arkansas
June 3d. 1888
Hon. Jefferson Davis
Beauvoir, Miss.

My dear Sir:

Permit me to cordially congratulate you upon becoming an octogenarian. As a native of Ponotoc, Miss., and as an ex-confederate, who entered the army at 17 years of age and remained till the last gun had fired, may I not claim a few moments of your time by tendering to you my congratulations on this your eightieth birthday? May Divine Providence bless you with good health and unalloyed happiness.

Many thousands of the old soldiers yet live to congratulate you . . . but many thousands, in the past two decades, have passed over the river . . . and they, and the grandest of all armies, our fallen heroes, with those grand commanders, Lee, Johnston and Jackson, are awaiting the arrival of their Commander in chief. All of us are indeed proud that you have been permitted to remain with us until the ripe age of eighty and we pray earnestly that you may be permitted to enjoy many more years of health, happiness and prosperity . . . My heart goes out to you in your declining years as warmly as when it was beating to the martial tread upon the fields of battle.

. . . I beg you to pardon me for imposing this long letter upon you. I only intended to express to you my joy upon your attaining your 80th birthday—and wishing you a longer life with us, and greater happiness in the beautiful life "over there". . .

With an earnest prayer for your welfare, I am, with great respect,

Yours truly—

W. P. Parks

Dr. Harker's patent medicine company, manufacturers of "the only true iron tonic" that "beautifies the complexion" and "purifies the blood," published a handsome advertising card bearing Davis's image, implying that the venerated former president owed his longevity to its dubious concoction.

In December 1888, former confederate general Jubal Early, a favorite of Davis, visited Beauvoir. He noticed Davis's threadbare clothes and offered to buy him a new suit. No, Davis demurred, he could not accept such generous charity, but he would be willing to accept just the fabric. He would pay his own New Orleans tailor, Mercier, who still kept his measurements on file, to fashion the garment. Early ignored him. He bought the wool and instructed Mercier to cut a simple, elegant suit. It was a fine garment of Confederate gray.

In 1889, Davis received an unexpected communication that jolted him back a half century in time to his youth. A woman from the North, Miss Lee H. Willis, wrote to him from the town of Richview, in Washington County, Illinois. She had discovered an old, lost love letter from him to Sarah Knox Taylor. Would he like it returned to him? The inquiry resurrected long-buried memories from his remote past. He remembered the letters that Knox had written to him, a romantic correspondence lost for decades. Oh yes, "you rightly suppose it has much value to me," he replied in a letter to Willis dated April 13, 1889: "The package containing all our correspondence was in a writing desk, among the books and papers I left in Mississippi when called to Alabama [in 1861 as president], and it would be to me a great solace to recover the letters Miss Taylor wrote to me, and which were with the one you graciously offer to restore."

Memories fifty years old rushed back to him, of the girl he once wrote in December 1834: "Dreams my dear Sarah we will agree are our weakest thoughts, and yet by dreams have I been lately almost crazed, for they were of you . . . I have read . . . your letter to night. Kind, dear letter, I have kissed it often . . . Shall we not soon meet Sarah to part no more? Oh! How I long to lay my head upon that breast which beats in unison with my own, to turn from the sickening sights of worldly duplicity and look in those eyes so eloquent of purity and love." Davis never found the package of lost letters he'd hoped to recover.

Knox Taylor was not the only memory stirring inside Davis. He brooded over many things. "The shadow of the Confederacy," Varina confided to her friend Constance Cary Harrison, "grows heavier over him as the years weigh his heart down." Now, she wrote, "he dwells in the past."

In November 1889, Davis departed Beauvoir alone for one of his periodic trips to inspect his lands at Brierfield, and he became ill during the trip. It was his lungs, with complications from his lifelong foe, malaria. He was rushed to New Orleans, where Varina, summoned by an emergency telegraph, joined him. Through newspaper reports, the whole South kept a vigil at his bedside. For a time Davis rallied, then declined. Near the end, when Varina

The twilight years at Beauvoir.

raised more medicine to his lips, he declined: "I pray I cannot take it." He fell asleep, and shortly after midnight, on December 6, 1889, he died peacefully in his bed, his wife's hands folded into his.

It was nothing like the death of Abraham Lincoln, which was unexpected, sudden, violent, bloody, and when he was in his prime. Neither Lincoln, nor his family, nor the people had time to prepare for it. Lincoln had no time to say good-bye. He enjoyed no final, long look back to recall his life and tally his deeds. Jefferson Davis was granted this privilege. He enjoyed the gift of years to live, to write, and to reflect. He had fallen and yet lived long enough to rise again. And the people of the South had nearly a quarter of a century to prepare themselves for his death. Davis had the chance to review his life in full, and to retrace his journey to the beginning.

For Lincoln, the end was only darkness. But in one way their deaths were the same. Just as April 15, 1865, symbolized to the North more than the death of just one man, so too the death pageant that followed December 6, 1889, was not for Davis alone.

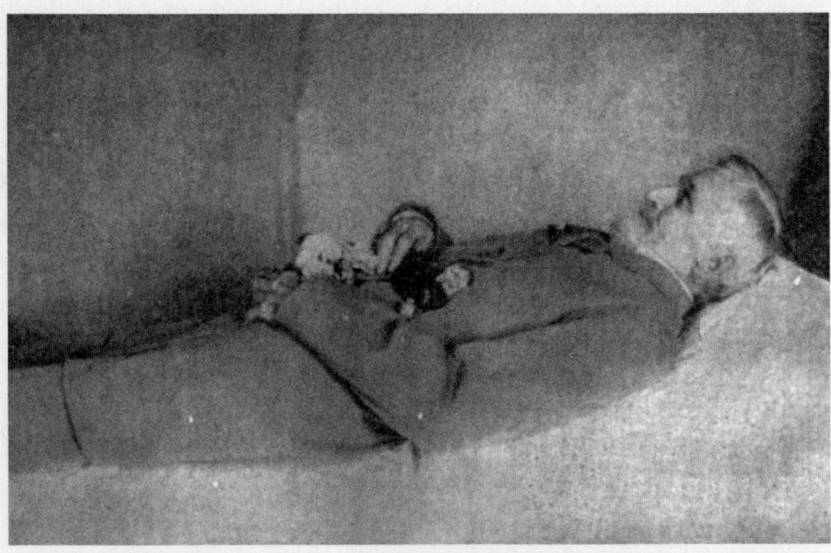

JEFFERSON DAVIS IN DEATH, NEW ORLEANS, DECEMBER 1889.

Jefferson Davis lies in state in New Orleans.

After Davis was embalmed, while he lay in state, a photographer set up his camera and lights to take pictures of the corpse. But there would be no repeat of the scandal that had occurred after Lincoln's remains had been photographed in New York City in April 1865. Varina had given permission for one photographer to make several dignified images of her husband's body, and she allowed at least one print to appear in a memorial tribute volume published by a trusted friend and longtime supporter, the Reverend J. William Jones. These photos of Davis's corpse, taken much closer to the body, and offering far greater detail than Gurney's images of Lincoln, captured Davis in elegant repose, dressed in the coat of Confederate gray that Jubal Early had given him, and clasping in his folded hands sheaves of Southern wheat. His lips formed a faint, subtle smile. Varina also consented to a death mask, an application of wet plaster to the deceased's face that would result in a perfect, life-size, three-dimensional likeness of Davis's visage. Edwin M. Stanton had not allowed a death mask of

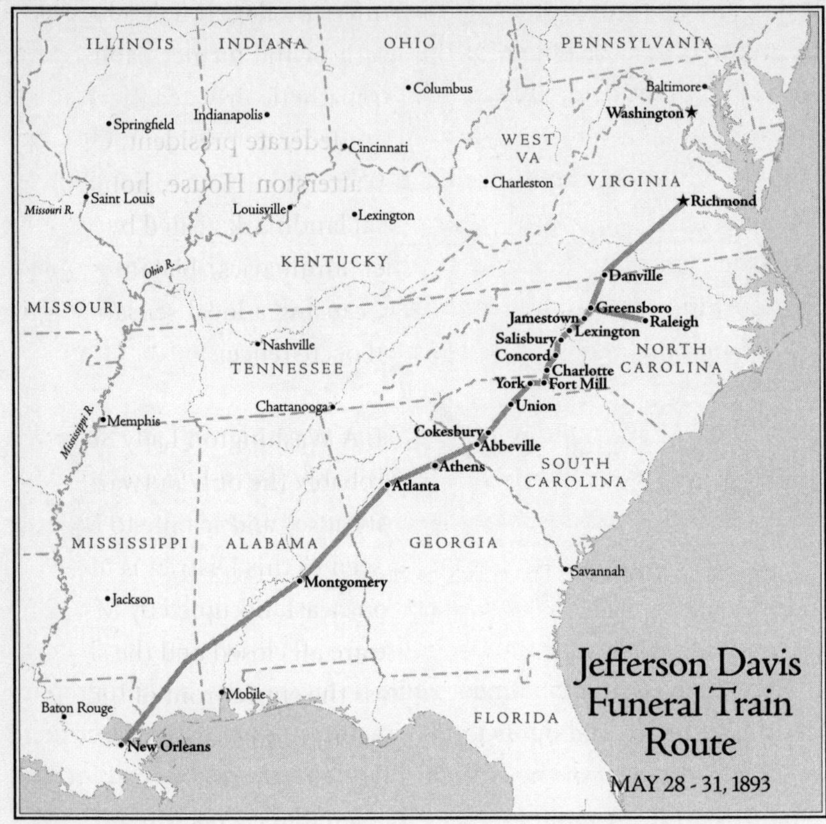

Abraham Lincoln—it was too morbid and disrespectful, he might have reasoned. For Davis, it would be a necessary artifact for the many monuments his supporters intended to build in his honor. Lincoln's two life masks, made in 1860 and 1865, had proven priceless for this purpose. So too would Davis's death mask.

In Washington, the government of the United States took no official notice of the death of Jefferson Davis. The mayor of New Orleans had sent a clever telegram notifying the War Department of the death of, not the former president of the Confederate States of America, but of a former secretary of war of the United States. Protocol dictated that the department fly its flag at half-staff and close for business. But Secretary of War Proctor declined to recognize the passing of

Davis, and he refused to lower the American flag. If federal authorities refused to take notice of the death of the former cabinet officer, that did not deter at least one sympathetic private citizen in the nation's capital from mourning the Confederate president. On Capitol Hill, across the street from the Watterston House, home of the third librarian of Congress, and a local landmark visited by Thomas Jefferson, James Madison, and other luminaries, blatant symbols of mourning adorned a town house, causing a local sensation. The *Washington Post* reported the unusual occurrence:

> A STRIP OF BLACK CLOTH. A Washington Lady Shows Her Grief at Jeff. Davis' Death. Probably the only outward evidence of sympathy for "the lost cause" and tribute to the memory of Jefferson Davis to be seen in this District is to be found at No. 235 Second Street southeast, occupied by Mrs. Fairfax. The shutters of the house are all closed and the slats tightly drawn while running across the entire front of the house, between the windows of the first and second floor, is a broad strip of black cloth looped up with three rosettes—red, white, and red. Mrs. Fairfax is known to have been an ardent supporter of the Confederate cause, and married State Senator Fairfax, of Fairfax County, Va. During the war of the rebellion she gave all the assistance to the South possible, and repeatedly crossed the lines, carrying to the Confederates both information and substantial assistance. It was through knowledge obtained by such expeditions as this that leaders of the Confederate side were enabled to anticipate movements of the Union armies and in a number of instances checkmate them. The house on Second street yesterday attracted more attention, probably, than ever before, and the emblems of mourning were the cause of considerable comment among the neighbors and travelers on that street.

In Alexandria, Virginia, where George Washington had kept a town house, where slave pens and markets once flourished, where Colonel Elmer Ellsworth was killed twenty-eight years ago, and where Lincoln's funeral car was built in the car shops of the United States Military Railroad, the sounds of mourning rose above the city and echoed over the national capital. Old Washingtonians remembered when the bells had tolled for Lincoln. Now they heard them toll for Jefferson Davis.

There was a magnificent funeral in New Orleans, a huge procession just like the one in Washington for Lincoln, and Davis was interred at Metairie Cemetery. But on the day Davis was buried, everyone knew this was only a temporary resting place. It was understood that Varina Davis would in due course select a permanent gravesite elsewhere. Several states, including Mississippi, vied for the honor. In 1891, Varina stunned Davis's home state. She vetoed its proposal and chose Virginia. Davis would return to his capital city, Richmond.

On May 27, 1893, almost three and a half years since Davis's death, his remains were removed from the mausoleum at Metairie Cemetery. A hearse carried his coffin to Camp Street, to Confederate Memorial Hall, a museum crammed with oak and glass display cases filled with Civil War guns, swords, flags and relics. From 5:00 P.M. on the twenty-seventh until the next afternoon, Davis lie there in state. Then a procession escorted his body to the railroad station, where it was placed aboard the train that would take him to Richmond. Like Abraham Lincoln's funeral train, the one for Jefferson Davis that left New Orleans on May 28 did not rush nonstop to its destination. First it paused at Beauvoir, where waiting children heaped flowers upon the casket. The train stopped briefly at Mobile shortly after midnight on May 29 to change locomotives, then stopped at Greenville, Alabama, at 6:00 A.M. People lined the tracks, and wherever the train halted, mourners filled Davis's railroad car with flowers.

Arriving in Montgomery during heavy rains early on the morning of May 29, the train was met by fifteen thousand people. For the first time since Davis's remains left New Orleans, they were removed from the train. A procession escorted the hearse to the old capitol, where in 1886 he experienced his great resurrection. While Davis lie in state in the chamber of the Alabama supreme court, bells tolled and artillery boomed. A large banner reminded mourners that "He Suffered for Us." To anyone who remembered the Lincoln funeral cortege, the sight of another presidential train winding its way through the nation, its passage marked by crowds and banners, and by the scent gunpowder and flowers, the experience was familiar and eerie. Floral arches erected in towns along the route of Davis's final journey proclaimed "He lives in the Hearts of his People." Once, identical words greeted Lincoln's train.

Davis's train reached Atlanta at 3:00 P.M. on May 29, and forty thousand people passed by his coffin. The train pushed on through Gainesville, Georgia; Greenville, South Carolina; and then into North Carolina, where it passed through Charlotte, Salisbury, Greensboro, and Durham. All along the way, people built bonfires, carried torches, rang bells, and fired guns to honor the esteemed corpse. At Raleigh, the coffin was removed again, placed in a hearse, and paraded through town. After lying in state there, Davis was taken to Danville, Virginia, his first refuge after he left Richmond the night of April 2, 1865.

The funeral train arrived in Richmond at 3:00 A.M. on May 31. In the middle of the night, a silent procession escorted the coffin from the station to the state capitol. Artillery fire woke the city later that morning, and twenty-five thousand people, including six thousand children, passed by his casket. At 3:30 P.M. a grand procession escorted his remain to Hollywood Cemetery. There, a simple graveside ceremony and a bugle playing taps ended his last journey.

In preparation for this day, word was sent to Washington, D.C., to Oak Hill Cemetery, where workers opened a small grave that had gone undisturbed since 1854, almost forty years before. The body of Samuel Davis was brought to Richmond to rest by his father. And

The New Orleans funeral procession.

Floral tribute at the Raleigh ceremonies.

The Raleigh, North Carolina, funeral procession, on the way to Richmond.

in Hollywood Cemetery, another grave was opened so that Joseph Davis, buried in 1864, could rest beside his father too.

Jefferson Davis had survived Abraham Lincoln by twenty-four years. Now, his journey also done, he joined him in the grave.

For several years after Davis's reburial, efforts to raise a memorial to him in Richmond floundered. The Jefferson Davis Monument Association, organized in 1890, had, in collaboration with the United Confederate Veterans, been confounded by disagreements about the location, artist, cost, and design of the monument. Finally, in 1893, 150,000 people attended the laying of the cornerstone. Fifteen thousand Confederate veterans marched in the parade. Then the project stalled again, and in 1899 the men of the UCV admitted defeat and handed the task over to the women of the South, who had founded the United Daughters of the Confederacy.

Adopting the motto "Lest we forgot," the UDC rolled out a strategic and passionate campaign to raise the necessary funds. Not only had Davis been the chief executive and chosen leader of the Confederacy, exhorted the UDC in a fund-raising letter, "he was our martyr, he suffered in his own person the ignominy and shame our enemies would have made us suffer. This was thirty-five years ago, and his monument is yet to be built. The women of the South have solemnly sworn to wipe out this disgrace at once."

The campaign worked. As the money rolled in, the women beseeched the famous Southern sculptor Edward Valentine, renowned for his bronze bust of Robert E. Lee and also his Lee sculpture in the chapel at Washington and Lee University, to design the Davis monument and its companion sculpture. In 1907, fourteen years after Davis's reburial in Richmond, his monument was ready. On April 16, Confederate veterans, aided by three thousand children, pulled on two seven-hundred-foot ropes and dragged the eight-foot-tall bronze of Davis along Monument Avenue. Dedication day was scheduled for June 3, Davis's ninety-ninth birthday, and the climax of a major, weeklong Confederate reunion in the city.

Two hundred thousand people stood along the parade route from the capitol to the monument, more people than had lined Pennsylvania Avenue for Abraham Lincoln's funeral procession. An additional 125,000 people gathered near the monument. After an address by Virginia governor Claude A. Swanson, Margaret Hayes, Davis's sole surviving child, unveiled her father's sculpture. Cannon fired a twenty-one-gun salute, an honor traditionally reserved for the president of the United States. Behind the bronze figure of Davis, his right arm outstretched and pointing to the old capitol of the Confederacy, there arose a sixty-foot-tall column topped by a female figure, Vindicatrix. Three mottoes adorn the base: "*Deo Vindici*" (the motto on the wartime Great Seal of the Confederacy); "*Jure Civitatum*" (for the rights of the states); and "*Pro Aris Et Focis*" (for hearth and home). Below the bronze figure a tablet bears the dedication: "Jefferson Davis, President of the Confederate Sates of America, 1861–1865."

After the ceremonies were over, the greatest crowd that had ever assembled in his name went home.

EPILOGUE

The chase for Jefferson Davis and the death pageant for Abraham Lincoln are among the great American journeys. Like the explorations of Lewis and Clark, the settling of the West, the building of the transcontinental railroad, and the landing on the moon, the rise and fall of the two Civil War presidents, each a martyr to his cause, altered our history and added to our myths. The history is well known—620,000 dead, the overthrow of old ways of life, and the end of a great but flawed antebellum empire built upon slavery. When Lincoln and Davis fell from power, they also set in motion two myths—the legend of America's emancipating, secular saint, and the legend of the Lost Cause. The assassination, nationwide mourning, and funeral train for Lincoln; the chase, imprisonment, and long Civil War afterlife of Davis—they haunt American history down to the latest generation.

In the years following Lincoln's funeral, the melancholy curse that afflicted his family would not lift. During the months after Mary left Washington, there were rumors that she had plundered the White House of valuables; and in 1867, a scheme she hatched with Elizabeth Keckly to exhibit her dresses for money—the "old clothes scandal," the press dubbed it—made her a national laughingstock. Tad, the president's constant companion after Willie's death, died of tuberculosis in 1871, when he was eighteen, having survived his father by just six years. The body of another Lincoln was put aboard a train. The tomb in Springfield was opened, and Tad joined Abraham, Willie, and Eddie.

Mary continued to live as an unsettled wanderer, spending much of her time in Europe. Irrationally, she believed herself destitute. She made mad, vicious accusations of dishonesty and theft against her son Robert, which led him to have her committed to a sanitarium for four months in 1875. She posed for the notorious spirit photographer Mumler, who supplied her with the expected image of the ghosts of Abraham and Willie hovering above her. She finally returned to Springfield and moved into the home of her sister, Elizabeth Todd Edwards. It was the house she and Abraham had been married in. As she had after the assassination, Mary spent much time in seclusion in her room, longing for death. She died on July 16, 1882,

THE LONELY YEARS: ABRAHAM AND WILLIE LINCOLN HAUNT MARY IN A FAKE "SPIRIT" PHOTOGRAPH.

surviving her husband by seventeen unhappy years. She joined her family in the tomb. The nation did not mourn her passing.

Robert Todd Lincoln became a prominent attorney, businessman, and government official, but after his death in 1926, the Lincoln line died out within two generations. Today there are no direct descendants of Abraham Lincoln.

Varina Davis rose from Jefferson's deathbed to live a fulfilling life. She helped plan his funeral and took pride in how the South mourned him. Four years later, she oversaw his reburial in Richmond. In 1890, Varina published her book, *Jefferson Davis, Ex-President of the Confederate States of America, a Memoir, by His Wife*. Jefferson had dedicated his memoirs to the women of the Confederacy, and in hers Varina remembered the men: "To the soldiers of the Confederacy, who cheered and sustained Jefferson Davis in the darkest hour by their splendid gallantry, and never withdrew their confidence from him when defeat settled on our cause, this volume is affectionately dedicated . . ." Like her husband, she needed two fat volumes to tell her story. Unlike him, she unburdened her heart. In 1891, in a decision that perplexed many Southerners—and outraged others—she moved to New York City, leaving behind the Southern landscape of her past, but not abandoning her memories of it. She wrote articles, made new friends, maintained a literary salon, and created a Manhattan Confederate circle that included Burton Harrison, who had become a prominent lawyer, and his wife, Constance Cary Harrison. They too wrote about their lives and times in the old Confederacy. Varina Howell Davis died in 1906. All of her sons had preceded Jefferson in death, but through her daughter, Margaret, the line lived on, and today the direct descendants of Jefferson and Varina Davis work to preserve the memory of their ancestors.

Little physical evidence of the Lincoln funeral train survives. The steam engines that performed flawlessly during the sixteen-hundred-mile journey were scrapped long ago. The presidential car performed one more act of service in the summer of 1865. When the wife of Secretary of State Seward died in June—her weak constitution was broken by the bloody assassination attempt in her home—Lincoln's car carried her body back to Auburn, New York. After that journey, it was retired from service—perhaps its use as a funeral car jinxed it as an unlucky conveyance for future presidents—and in 1866 it was auctioned and purchased by the Union Pacific Railroad. Afterward, the president's funeral car enjoyed a few decades of celebrity and then, stripped of its decorations and furniture, and suffering from neglect and decay, it perished in a fire. Souvenir photo postcards from the day depict a pile of collapsed, wooden ribs charred black. The other coach cars vanished, and the funeral train survives only in the dozens of photographs taken of it along the route from Washington to Springfield.

None of the majestic horse-drawn hearses, which had once caused the public to marvel at their size and extravagance, exists today. All of the catafalques save one—the one on which Lincoln's coffin rested when he lay in state at the U.S. Capitol, and upon which dead presidents still repose—are gone. The fabric that covers it today is of a more recent vintage: Benjamin Brown French never got back the black shroud his wife had sewn to drape the catafalque. From all the hearses and catafalques, only a few relics survive, scattered among historical societies and private collections: framed slivers of wood, swatches of black cloth, frayed bits of flag, strips of silver fringe, bullion tassels, dried flowers, and the like. From the New York City funeral, one of the twelve, halberd-topped flag poles mounted to Lincoln's hearse survives, preserved by the assistant undertaker and identified for future generations by his sworn affidavit.

One museum has tried to re-create what it must have been like to experience the Lincoln death pageant and view his corpse. In Springfield, the Abraham Lincoln Presidential Library and Museum built a replica of the city's Hall of Representatives, where the president lay in state. The original room, which still exists, is one block away in the Old State House, where it is fitted out as the legislative chamber Lincoln would recognize from his day. But in the museum's facsimile chamber, it is forever May 3–4, 1865. Black crepe and bunting smother the space, and upon an inclined catafalque rests a replica of Lincoln's coffin. The overall effect is somber and impressive, until one takes a closer look. The corpse is missing; the coffin is closed. How could a *Lincoln* museum, of all places, commit such a spectacular historical error? In Springfield, as in each city where a public viewing occurred, the coffin was open. The American people were desperate to see Lincoln's remains.

At the museum's grand opening, a visitor pointed out the error to an employee. "We know," the official replied. "We did it on purpose. We can't show what it was really like. We can't have an open coffin with a wax figure inside. It would upset the children."

Perhaps the children of 1865 were hardier than today's generation—during the national obsequies, tens of thousands of children viewed Lincoln's remains. Nonetheless, the museum chose to go to great effort and expense to create an exhibit that is not authentic. Strangely, the museum was not reluctant to construct a replica of Lincoln's box at Ford's Theatre, and to place in it a life-size figure of John Wilkes Booth assassinating the president. Nor was the gift shop reluctant to sell to children a plastic, toy replica of the Deringer pistol Booth used to murder Abraham Lincoln, or to place a second life-size figure of Booth not far from the main entrance, within sight of other figures of Lincoln and his family, and Fredrick Douglass. It is bizarre that in the city where Lincoln lies buried, multiple effigies of his assassin stand erect. Thus, the Lincoln museum there enjoys the singular distinction of being the only presidential library and museum in America to boast a waxworks devoted to an assassin.

In a strange twist, another museum in Springfield, the Museum of Funeral Customs, has fabricated not one but several replicas of Lincoln's coffin, which are loaned out for exhibition, and for educational purposes. The lids will open, but the coffins are empty. Wax figures are not included.

Not long after the funeral train left Springfield and returned to Washington, a legend spread of a Lincoln ghost train that rolled down the tracks each spring. An undated, fugitive newspaper clipping, found pasted in an old scrapbook from the late 1860s, is the only surviving evidence of the tale:

A PHANTOM TRAIN
THE DEAD LINCOLN'S YEARLY TRIP OVER THE NEW YORK CENTRAL RAILROAD

A correspondent in the Albany (N.Y.) *Evening Times* relates a conversation with a superstitious night watchman on the New York Central Railroad. Said the watchman: "I believe in spirits and ghosts. I know such things exist. If you will come up in April I will convince you." He then told of the phantom train that every year comes up the road with the body of Abraham Lincoln. Regularly in the month of April, about midnight, the air on the track becomes very keen and cutting. On either side it is warm and still. Every watchman when he feels this air steps off the track and sits down to watch. Soon after the pilot engine, with long black streamers, and a band of black instruments, playing dirges, grinning skeletons sitting all about, will pass up noiselessly, and the very air grows black. If it is moonlight clouds always come over the moon, and the music seems to linger, as if frozen with horror. A few moments after and the phantom train glides by. Flags and streamers hang about. The

> track ahead seems covered with black carpet, and the wheels are draped with the same. The coffin of the murdered Lincoln is seen lying on the centre of the car, and all about it in the air and the train behind are vast numbers of blue-coated men, some with coffins on their backs, others leaning on them.

Many spirits, claimed the storyteller, accompanied Lincoln:

> It seems then that all the vast armies that died during the war are escorting the phantom train of the President. The wind, if blowing, dies away at once, and all over the earth a solemn hush, almost stifling, prevails. If a train were passing, its noise would be drowned in silence, and the phantom train would ride over it. Clocks and watches would always stop, and when looked at are found to be from five to eight minutes behind. Everywhere on the road, about the 27th of April, the time of watches and trains is found suddenly behind. This, said the leading watchman, was from the passage of the phantom train.

More tangible than any ghost trains are the railroad tracks. For the most part, the route followed by the Lincoln funeral train still exists, marked by ancient railroad beds and the villages, towns, and cities on the map when the train passed by. Yes, the aging iron rails forged by the Civil War were replaced long ago, but track locations rarely change, and many of the same railroad beds over which Lincoln's coffin rode are still in the same place, a hundred and fifty years later. Few of the residents who live along the route today know about the torches, bonfires, arches, cannon fire, and huge crowds, or that Lincoln's corpse once passed that way, and perhaps even stopped in their town.

New Yorkers who commute daily from their bedroom communities along the Hudson River, north of Manhattan, have no idea when they return home each night that they are traveling over the

AN EXCEPTIONAL SOUVENIR: A PORCELAIN MEMORIAL OBELISK.

same route taken by the Lincoln funeral train. The stops, Hastings-on-Hudson, Tarrytown, Ossining, Croton, and beyond, are the same ones called out by the conductor on Lincoln's train. Every day in America, thousands of railroad passengers, unbeknownst to them, follow the route of the funeral train.

Other vivid and more venerated evidence of the death pageant survives: the blood relics—locks of Lincoln's hair, tiny pieces of his skull, the probe and other medical instruments, bloodstained pillows and towels, the physicians' bloody shirt cuffs; the fatal bullet, of course; and still more death relics, lurid and macabre ones, best not spoken of. Many of them repose in the Army Medical Museum, or in private collections, handed down from generation to generation, or sold off by the descendants of the ancestors who had once cherished them.

More common than blood relics are the ribbons, timetables, badges, song sheets, broadsides, prints, and photographs produced and sold commercially to millions of mourners of April and May 1865. Even today, it is not unheard of for a silk mourning ribbon, a printed railroad timetable, or an original carte de visite of one of the hearses to turn up at an out-of-print bookstore, antique shop, or estate sale located along the old route of the funeral train.

George Harrington could not have foreseen it, but when he planned the state funeral for Abraham Lincoln, he was planning the funeral for a future president too, one destined to be elevated to that office a century after Lincoln's election, and who, like Father Abraham, would die by an assassin's hand. On November 22, 1963, when the president's body was flown home from Dallas, Texas, to Washington, D.C., a man awaited the landing of Air Force One at Andrews Air Force Base that evening. It was Angier Biddle Duke, U.S. chief of protocol. When Jacqueline Kennedy, still wearing the bright pink suit stained with her husband's blood, stepped off the presidential jet, Duke approached her and spoke one sentence.

"Madam, how may I serve you?" he asked.

"Make it like Lincoln's," she said.

A few nights later, after the funeral at Arlington National Cemetery, as the motorcade headed back to the White House, Jacqueline Kennedy's car broke away from the others. After her vehicle crossed Memorial Bridge, it turned left. Ahead, the thirty-six huge, snowy, marble columns glowed like a classical Greek Temple. Mrs. Kennedy's car braked to a stop on the plaza, and she gazed up at the sculpture of Abraham Lincoln enshrined in his memorial.

Like Abraham Lincoln, Jefferson Davis became a greater legend in death than he had been in life. After he fell from power, his stock rose in the South—"He suffered for us"—and he became not only the defeated Confederacy's "representative man," but also the living catalyst for a new movement, the Lost Cause. He symbolized a collective dream: The South may have lost the war, but it was not wrong, and even in defeat it shone with honor and remained the superior civilization. During Davis's 1886–87 speaking tour, he soared to new heights of glory, surpassing the prestige and fame he once possessed as president of the Confederate States of America. In his old age, it seemed, the South could not have loved him more. Until he died.

The death of Jefferson Davis caused a convulsion of emotion and memory. His funeral, like Lincoln's, represented not just the passing of one man but of an era. Four years after Davis died, the funeral train that carried him from New Orleans to Richmond roused the South and stunned the North. Once more, Americans stood beside railroad tracks, holding signs, bearing torches, and igniting bonfires, waiting for a train to pass by. A tumultuous response welcomed him back to the old capital, where he would reign forever over the dreams of a lost cause. In the 1890s, the White House of the Confederacy was transformed into the Museum of the Confederacy, a shrinelike repository for treasured battle flags, war artifacts, and memories. In 1907, when three hundred and twenty-five thousand people turned out for the dedication of his monument in Richmond, Davis was at the apex of his fame. On that day, his partisans were sure that his name would endure forever and that history would honor him, no less than Lincoln, as a great American.

They were wrong. The twentieth century came to belong to Abraham Lincoln, not Jefferson Davis. His eclipse began as early as 1922, with the completion of the Lincoln Memorial. A grandiose, overwrought monument had been proposed in 1865, not long after the assassination, but, fortunately for the nation, it was never built. The model vanished long ago, but survives in a rare photograph buried in the files of the Library of Congress. Indecision and political squabbling delayed Lincoln's national memorial for fifty-seven years. In the meantime, while Lincoln waited, two monuments had been erected for Davis in Richmond, one at his gravesite in Hollywood Cemetery and the other on Monument Avenue.

The Lincoln Memorial overshadowed these Richmond monuments in physical scale and symbolism. It represented the growing power of the Lincoln legend and the Northern interpretation of the War of the Rebellion. It would not have surprised Davis to know that on the day former president William Howard Taft presented the memorial to President Warren G. Harding, with Abraham Lincoln's son Robert looking on, blacks in attendance were forced to sit in segregated seating. Davis had been dubious of how blacks would fare in postwar America. He believed that once the Union freed the slaves, the North would not welcome them as neighbors or equal citizens. Instead, Davis suspected, Northerners viewed blacks as an abstraction, as a convenient cause they would abandon after the war. Racism and hatred, Davis suggested, were not exclusively Southern phenomena. It took a different kind of Southern senator and president—Lyndon B. Johnson—to redeem Lincoln's promise that had been denied during dedication day, on the steps of his own memorial.

Southerners continued to memorialize Jefferson Davis. His capture site languished in obscurity for years and was, in time, overgrown by pines and brush. It was a quiet, forgotten place. This was no landmark of Confederate glory, and few Southerners cared to visit the spot where Davis's presidency and their last hope for independence had died. In 1894, a local photographer named J. H. Harris, from the town of Tifton in Berrien County, Georgia, went there twenty-nine years later to take, he boasted in his caption, the only photo in the world of the "exact spot" of Davis's last camp. At some point, Davis loyalists marked the place when they hammered into the ground a wooden

JEFFERSON DAVIS CAPTURE SITE.

stake nailed to a crude, handmade sign: SITE OF JEFFERSON DAVIS' CAMP AT THE TIME OF CAPTURE, MAY 10, 1865.

On June 3, 1936, on a spring day seventy-one years after the end of the Civil War, and the 128th anniversary of Jefferson Davis's birthday, the ladies of the United Daughters of the Confederacy, Ocilla, Georgia Chapter, dedicated a handsome monument at the site. Consisting of a large concrete slab bearing a concrete plaque sculpted in bas relief, with a bronze bust of Davis, the main text of the memorial reads: "Jefferson Davis—President of the Confederate States of America. 1861–1865." This monument was meant to celebrate not capture, defeat, or imprisonment, but the "unconquerable heart" of the man who, in enduring those trials, became a beloved symbol to his people. In a dedication-day photograph taken by the U.S. Department of Agriculture Farm Security Administration (a government agency the very existence of which would have roused Davis's skepticism of broad federal authority), the new monument dominates the image, but if you look several feet to its left, low in the frame, the old, handmade wooden sign still holds its ground.

Other monuments to Davis mark the landscape near his birthplace in Kentucky, and in his home state of Mississippi. At the U.S. Capitol, a larger than life bronze sculpture of Davis stands in National Statuary Hall, its presence a tribute to two things: his service as a U.S. senator, and his significant influence on the architecture and modern-day appearance of the Capitol building.

In 2009, America celebrated Abraham Lincoln's two hundredth birthday with great fanfare. President and Mrs. George W. Bush hosted several pre-bicentennial events, including the first black-tie White House dinner ever held in Lincoln's honor. The Library of Congress and the Smithsonian National Museum of American History mounted major exhibitions. The Ford's Theatre Society raised fifty million dollars to renovate the theater and its museum in time for Lincoln's birthday on February 12. The Newseum, located on a stretch of Pennsylvania Avenue overlooking the route of the April 19, 1865, funeral procession, offered an exhibition on the assassination, mourning pageant, and manhunt for Lincoln's killer. Museums in several other cities put on exhibitions. Filmmakers produced several documentaries, and in 2008 and 2009, authors published nearly one hundred books on the sixteenth president. The U.S. Mint and Post Office produced commemorative coins and stamps.

On June 3, 2008, another bicentennial passed almost without notice. Not many Americans were aware of, let alone chose to celebrate, the two hundredth birthday of Jefferson Davis. There were no White House dinners, major exhibitions, shelves of new books, or coins and stamps. Few people know his story. Most have never read a book about him, and no one reads his memoirs anymore. Many people would not recognize his face, and some would not even remember his name. Indeed, when he does make the news, it is more likely in connection with a fevered effort to change the name of some high school named after him a long time ago. Jefferson Davis is the Lost Man of American history.

The Lincoln bibliography, more than fifteen thousand titles, dwarfs the literature on Davis. Lincoln is served by a cottage industry that produces dozens of books, articles, conferences, and lectures every year. There is no Davis cottage industry. The one exception is the *Papers of Jefferson Davis* project, a labor of love and of exquisite scholarly merit. But these volumes are purchased by historians and libraries, not the general public.

What explains the rise and fall of Davis in American popular memory? He lost, and history tends to reward winners, not losers. But there must be more to it than that. Perhaps it comes down to the slaves, the song, and the flag. The Confederate past is controversial. In the spring of 2010, on the eve of the 150th anniversary of the Civil War, the governor of Virginia created a furor by proclaiming Confederate History Month, a celebration condemned by some as, at best, insensitive and, at worst, racist. A historical figure who owned slaves, wished he "was in the land of cotton," and waved the Stars and Bars must today be rebuked and erased from popular memory, not studied. Better to forget. Perhaps, someday, someone will demand that his statue be banished from the U.S. Capitol.

In Richmond, the Confederate White House and the Museum of the Confederacy, two of the finest Civil War sites in the country, are in trouble. Once central to that city's identity, they languish now in semi-obscurity, overshadowed physically by an ugly complex of medical office buildings and challenged symbolically by a competing, sleek new Civil War museum at the Tredegar Iron Works, the former cannon manufactory. The Museum of the Confederacy has fallen on hard times and into local disfavor, dismissed by some as an antiquarian dinosaur, by others as an embarrassing reminder of the racial politics of the Lost Cause. Its very name angers some who insist that perpetuating these places of Confederate history is tantamount to a modern-day endorsement of secession, slavery, and racism. According to numerous newspaper stories, the Museum and the White House are barely hanging on, and have considered closing, or dividing the priceless collection among several institutions. Their failure would be a loss to American history. Unless a benefactor comes forward to save them, their long-term future remains uncertain.

There was one place where the legacy of Jefferson Davis was safe, at his beloved postwar sanctuary, Beauvoir. There, on the Mississippi Gulf, he had found the peace that had eluded him during his presidency and during his unsettled postwar wanderings. In an outbuilding, a three-room cottage he set up as his study, he shelved hundreds of books and piled more on tables. A photograph preserves the interior of this time capsule: books everywhere, his desk and chair where he sat and composed his letters and articles, and where he wrote *The Rise and Fall of the Confederate Government.*

After Davis's death, Beauvoir lived on as a monument, and it became a retirement home for aged Confederate veterans who came to live there. When the last of them died off, Beauvoir became a Davis museum and library. The institution flourished for decades until one day in late August 2005, when Hurricane Katrina hit the Mississippi Gulf hard. The main house, a lovely, nine-room Gothic cottage set upon pillars, was gutted down to the walls. All seven of the outbuildings were destroyed. Countless artifacts were lost, including Davis's Mexican War saddle, as well as the notorious raglan and shawl he wore on the morning he was captured.

DAVIS'S PRIVATE LIBRARY AT BEAUVOIR, PHOTOGRAPHED AFTER HIS DEATH.

His library did not escape the hurricane. On that day, the sanctuary where Jefferson Davis labored to preserve for all time the memory of the Confederacy, its honored dead, and the Lost Cause was, by wind and water, all swept away.

ACKNOWLEDGMENTS

A number of people helped in my pursuit of Abraham Lincoln and Jefferson Davis:

My splendid editor, Henry Ferris, recognized from the start how pairing the final journeys of the two presidents could enhance the power of each story. Our countless conversations and late night editing sessions in his New York office improved the book in immeasurable ways. Henry's assistants, Peter Hubbard, now an editor in his own right, and Danny Goldstein, brought diligence and enthusiasm to the project.

I am also grateful to the rest of my HarperCollins team: Michael Morrison and Liate Stehlik for supporting this book with energy and personal interest, Lynn Grady and Jean Marie Kelly for bringing it to its audience, and miracle-worker Sharyn Rosenblum, the best publicist in the business.

At the Museum of the Confederacy, President S. Waite Rawls III and historian John Coski provided valuable information about Jefferson Davis. Thanks also to Waite for a moving, late afternoon private tour of the Confederate White House.

At the Jefferson Davis Home and Presidential Library at Beauvoir, Mississippi, Chairman Richard V. Forte Sr. and curator Richard R. Flowers answered questions about Davis's last sanctuary and provided the surprising photo of Oscar Wilde. Lynda Lasswell Crist, editor of the *Papers of Jefferson Davis* at Rice University, was a superb guide to the writings of the lost man of American history. Lynda answered questions with good cheer, and supplied numerous documents and transcripts. With the impressive papers project, a model for future historians, she has made a major contribution to the study of American history.

At the Library of Congress, John Sellers is a living treasure who shared his vast expertise on the Lincoln and Civil War manuscript collection. His retirement is a loss to all those who pursue the Lincoln story. In the rare book division, Clark Evans, with his usual effusive charm, made available a number of treasures documenting the final days of Lincoln and Davis. W. Ralph Eubanks, director of publishing and a fine author in his own right, helped me obtain a number of superb photographs and illustrations, as did Helena Zinkham and Barbara Orbach Natanson in the prints and photographs division. I must also thank John Y. Cole, director of the Center for the Book and author impresario of the National Book Festival, for his support and efforts to spread Lincoln scholarship to a wide popular audience at the best book event in America.

At the Surratt Society, Joan Chaconas, Laurie Verge, and Sandra Walia rendered the same generous assistance that they gave to *Manhunt*. Their expertise and good humor make Mary Surratt's country tavern and the James O. Hall Research Center two of the most interesting and informative sites on the Lincoln assassination trail.

Many thanks to my "first readers" Michael Burlingame, Ronald K. L. Collins, and Edward Steers Jr. for reading the manuscript with keen eyes, and making valuable suggestions.

At Ford's Theatre, my friend Paul Tetreault offered good counsel, a public venue to share my research, and the opportunity to participate in the preservation of an American landmark. Paul is a remarkable catalyst and visionary who understands the potential of Ford's as both a working playhouse and a museum that tells the story of Lincoln's life and death. At the National Park Service, Kym Elder, Rae Emerson, and Gloria Swift were always ready to provide assistance, advice, and encouragement.

At the Heritage Foundation, Attorney General Edwin Meese III and Todd Gaziano, director of the Center for Legal and Judicial Studies, provided me with a collegial home during the time I wrote this book. Jessica Kline gave valuable assistance with all computer mysteries. Interns Laura Clauser and Andrew O'Dell helped track down a number of hard-to-find documents and articles.

My friend and literary agent, Richard Abate, grasped the dramatic possibilities of this story about the end days of the two Civil War presidents and made a number of invaluable suggestions on how to think about and tell this tale. He critiqued the manuscript, provided his usual telling insights, and in a number of ways above and beyond the call of an agent, gave this book his "last measure of devotion."

I also thank my television agent at WME Entertainment, Julie Weitz, for her tireless efforts in translating my work into another medium.

My wife, Andrea E. Mays, occupied with her own book on the hunt for Shakespeare's First Folio, read the manuscript several times, made countless editorial improvements, helped sift through the abundance of art works to select the images, and, whenever I got bogged down in the trees, cut me a path to clarity. Andrea lived with this book for more than two years and helped me bring alive the saga of Lincoln and Davis. Our boys, Cameron and Harrison, ages thirteen and eleven, were our companions on visits to historic sites, my assistants at book signings, and coaches on storytelling. "Readers want blood," said Cameron. "And knives," added Harrison.

Finally, my father, Lennart J. Swanson, traveled with me for much of this journey. In a way, he began this book by taking me on an unforgettable trip to Gettysburg when I was ten years old. We have been traveling on that path ever since.

James L. Swanson
Washington, D.C.
July 7, 2010

BIBLIOGRAPHY

A NOTE ON SOURCES

The Lincoln literature is vast. The bibliography of the Civil War is bigger. Thus, the bibliography that follows is selective, not comprehensive. It represents little more than the several hundred books on my own library shelves that I used while researching and writing *Bloody Crimes.*

The cornerstone of any project touching upon Jefferson Davis must be the scholarly and brilliant multivolume set, *The Papers of Jefferson Davis,* edited by Lynda Lasswell Crist. Anyone interested in the life of the Confederate president must begin here, and no book can be written without it. At the time *Bloody Crimes* went to press, volume twelve of the *Papers* covered Davis through December 1870. Future volumes will cover Davis's works through his death in December 1889. Also essential, because the *Papers* refer to it, because it covers Davis's entire life, and because many books cite it, is Dunbar Rowland's ten-volume work published in 1923, *Jefferson Davis, Constitu-*

tionalist. Rowland's work is crammed with invaluable information, including letters to and from Davis.

The best modern biography is William J. Copper Jr.'s *Jefferson Davis, American.* Cooper rescued the Davis story from myth and neglect and is the superior work on its subject. Anyone looking to read just one book about Davis should read Cooper. A valuable companion is his short book *Jefferson Davis: The Essential Writings.* The granddaddy of vintage biographies is Hudson Strode's *Jefferson Davis*, published in three volumes: *American Patriot 1808–1861, Confederate President*, and *Tragic Hero.* While impaired by certain errors, and marked by an anti-Reconstruction point of view, Strode contains valuable material, influenced Davis studies and a number of books, and must be contended with.

Worthy books on the Davis shelf include Felicity Allen, *Jefferson Davis: Unconquerable Heart;* Michael B. Ballard, *A Long Shadow: Jefferson Davis and the Final Days of the Confederacy;* Joan E. Cashin, *First Lady of the Confederacy: Varina Davis's Civil War* (though marked by a postmodern point of view suggesting that Varina suffered from a kind of "false consciousness"); Donald E. Collins, *The Death and Resurrection of Jefferson Davis;* William C. Davis, *Jefferson Davis: The Man and His Hour;* A. J. Hanna, *Flight into Oblivion;* Hermann Hattaway and Richard E. Beringer, *Jefferson Davis, Confederate President;* Robert McElroy, *Jefferson Davis: The Real and the Unreal*; Eron Rowland, *Varina Howell, Wife of Jefferson Davis;* and Robert Penn Warren, *Jefferson Davis Gets His Citizenship Back.*

Difficult to read but impossible to ignore are Jefferson Davis's memoirs, *The Rise and Fall of the Confederate Government*, and Varina's more pleasing *Jefferson Davis . . . A Memoir, by His Wife.*

The ultimate Lincoln book is Michael Burlingame's recent, all-comprehensive, and magisterial two-volume biography, *Abraham Lincoln: A Life.* No future book on the Civil War president can be written without it, and from no other work can a general reader learn so much about Abraham Lincoln. The other vital book of the modern era is Doris Kearns Goodwin's *Team of Rivals: The Political Genius of Abraham Lincoln*. The essential library on the assassination, death, and funeral of Abraham Lincoln includes Ralph Borreson, *When Lincoln Died;* William T. Coggeshall, *The Journeys of Abraham Lincoln . . . From Washington to Springfield;* Dorothy Meserve Kunhardt and Philip B. Kunhardt Jr., *Twenty Days;* Lloyd Lewis, *Myths After Lincoln;* B. F. Morris, *Memorial Record of the Nation's Tribute to Abraham Lincoln;* Carl Sandburg, *Abraham Lincoln: The War Years;* Victor Searcher, *The Farewell to Lincoln;* Edward Steers Jr., *Blood on the Moon: The Assassination of Abraham Lincoln* and *The Lincoln Assassination Encyclopedia;* James L. Swanson, *Manhunt: The 12-Day Chase for Lincoln's Killer;* Scott D. Trostel, *The Lincoln Funeral Train;* and Thomas Reed Turner, *Beware the People Weeping: Public Opinion and the Assassination of Abraham Lincoln.*

Abott, A. Abott, *The Assassination and Death of Abraham Lincoln, President of the United States of America, at Washington, on the 14th of April, 1865* (New York: American News Company, 1865).

Abraham Lincoln: An Exhibition at the Library of Congress in Honor of the 150th Anniversary of His Birth (Washington, D.C.: Library of Congress, 1959).

Allardine, Bruce S., *More Generals in Gray: A Companion Volume to "Generals in Gray."* (Baton Rouge: Louisiana State University Press, 1995).

Allen, Felicity, *Jefferson Davis: Unconquerable Heart* (Columbia: University of Missouri Press, 1999).

Arnold, Isaac N., *Sketch of the Life of Abraham Lincoln* (New York: John B. Bachelder, 1869).

Baker, Jean H., *Mary Todd Lincoln: A Biography* (New York: W. W. Norton, 1987).

Ballard, Michael B., *A Long Shadow: Jefferson Davis and the Final Days of the Confederacy* (Jackson: University Press of Mississippi, 1986).

Bancroft, A. C., *The Life and Death of Jefferson Davis, Ex-President of the Southern Confederacy: Together with Comments of the Press and Funeral Sermons* (New York: J.S. Ogilvie, 1889).

Bartlett, John Russell, *The Literature of the Rebellion: A Catalogue of Books and*

Pamphlets Relating to the Civil War in the United States (Boston: Draper and Halliday, 1866).

Basler, Roy P., *The Lincoln Legend: A Study in Changing Conceptions* (Boston: Houghton Mifflin, 1935).

———, ed., *The Collected Works of Abraham Lincoln,* 8 vols., plus index and supplements (New Brunswick, NJ: Rutgers University Press, 1953).

Bates, David Homer, *Lincoln in the Telegraph Office* (New York: Century Co., 1907).

Bates, Finis L., *Escape and Suicide of John Wilkes Booth* (Memphis, TN: Finis L. Bates, 1907).

Beale, Howard K., ed., *The Diary of Edward Bates, 1859–1866,* vol. 4 of the *Annual Report of the American Historical Association* (Washington, D.C.: Government Printing Office, 1933).

———. *Diary of Gideon Welles,* 3 vols. (New York: W. W. Norton, 1960).

Bell, John, *Confederate Sea Seadog: John Taylor Wood in Exile* (Jefferson, NC: McFarland, 2002).

Benham, William Burton, *Life of Osborn H. Oldroyd: Founder and Collector of Lincoln Mementos* (Washington, D.C.: privately printed, 1927).

Berkin, Carol, *Civil War Wives: The Lives & Times of Angelina Grimké Weld, Virginia Howell Davis & Julia Dent Grant* (New York: Alfred A. Knopf, 2009).

Bernstein, Iver, *The New York City Draft Riots* (New York: Oxford University Press, 1990).

Bingham, John Armor, *Trial of the Conspirators for the Assassination of President Lincoln, s.c. Argument of John A. Bingham, Special Judge Advocate* (Washington, D.C.: Government Printing Office, 1865).

Bishop, Jim, *The Day Lincoln Was Shot* (New York: Harper & Brothers, 1955).

Blackett, R. J. M., *Thomas Morris Chester, Black Civil War Correspondent: His Dispatches from Virginia* (Baton Rouge: Louisiana State University Press, 1989).

Blair, William, *Cities of the Dead: Contesting the Memory of the Civil War in the South 1865–1914* (Chapel Hill: University of North Carolina Press, 2004).

Blake, Mortimer, *Human Depravity John Wilkes Booth: A Sermon Occasioned by the Assassination of President Lincoln, and Delivered in the Winslow Congregational Church, Taunton, Massachusetts on Sunday Evening, April 23, 1865, by the Pastor.* (Champlain: privately printed at the Moorsfield Press, 1925).

Bleser, Carol K. and Lesley J. Gordon, eds., *Intimate Strategies of the Civil War:*

Military Commanders and Their Wives (New York: Oxford University Press, 2001).

———, eds., "The Marriage of Varina Howell and Jefferson Davis: A Portrait of the President and the First Lady of the Confederacy," *Intimate Strategies of the Civil War: Military Commanders and Their Wives* (New York: Oxford University Press, 2001), 3–31.

Blight, David W., *Race and Reunion: The Civil War and American Memory* (Cambridge, MA: Belknap Press, 2001).

Blue, Frederick J., *Salmon P. Chase: A Life in Politics* (Kent, OH: Kent State University Press, 1987).

Bohn, Casimir, *Bohn's Hand-Book of Washington* (Washington, D.C.: Casimir Bohn, 1856).

Boritt, Gabor S., ed., *Why the Confederacy Lost* (New York: Oxford University Press, 1992).

Borreson, Ralph, *When Lincoln Died* (New York: Appleton-Century, 1965).

Boyd, Andrew, *Abraham Lincoln, Foully Assassinated April 14, 1865* (Albany, NY: Joel Munsell, 1868).

———. *Boyd's Washington and Georgetown Directory: 1864* (Washington, D.C.: Hudson Taylor, 1863).

———. *A Memorial Lincoln Bibliography: Being an Account of Books, Eulogies, Sermons, Portraits, Engravings, Medals, etc., Published upon Abraham Lincoln, Sixteenth President of the United States, Assassinated Good Friday, April 14, 1865* (Albany, NY: Andrew Boyd, Directory Publisher, 1870).

Bradley, Mark L., *This Astounding Close: The Road to Bennet Place* (Chapel Hill: University of North Carolina Press, 2000).

Brenner, Walter C., *The Ford Theatre Lincoln Assassination Playbills* (Philadelphia: privately printed, 1937).

Brooks, Noah, *Washington in Lincoln's Time* (New York: Century Co., 1895).

Brooks, Stewart M., *Our Murdered Presidents: The Medical Story* (New York: Frederick Fell, Inc., 1966).

Brown, George William, *Baltimore and the 19th of April, 1861* (Baltimore, MD: N. Murray, 1887).

Browning, Orville Hickman, *The Diary of Orville Hickman Browning,* edited with introduction and notes by Theodore Calvin Pease and James G. Randall (Springfield: Illinois State Historical Library, 1925–33), two vols.

Brubaker, John H., *The Last Capital: Danville, Virginia, and the Final Days of the Confederacy* (Danville, VA: Danville Museum of Fine Arts and History, 1979).

Bruce, George A., *The Capture and Occupation of Richmond* (n.p.: privately printed, 1927).

Bryan, George S., *The Great American Myth: The True Story of Lincoln's Murder* (New York: Carrick & Evans, 1940).

Bryan, Vernanne, *Laura Keene: A British Actress on the American Stage, 1826–1873* (Jefferson, NC: McFarland & Co., 1997).

Buchanan, Lamont, *A Pictorial History of the Confederacy* (New York: Crown Publishers, 1951).

Buckingham, J. E. *Reminiscences and Souvenirs of the Assassination of Abraham Lincoln* (Washington, D.C.: Press of Rufus H. Darby, 1894).

Budiansky, Stephen, *The Bloody Shirt: Terror After Appomattox* (New York: Viking, 2008).

Burlingame, Michael, *Abraham Lincoln: A Life,* 2 vols. (Baltimore, MD: Johns Hopkins University Press, 2008).

Cable, Mary, *The Avenue of the Presidents* (Boston: Houghton Mifflin, 1969).

Cain, Marvin R., *Lincoln's Attorney General: Edward Bates of Missouri* (Columbia: University of Missouri Press, 1965).

Campbell, W. P., *The Escape and Wanderings of J. Wilkes Booth Until Ending of the Trail by Suicide in Oklahoma* (Oklahoma City, OK: privately printed, 1922).

Carpenter, Francis B., *Six Months in the White House with Abraham Lincoln* (New York: Hurd & Houghton, 1866).

Carroll, Gordon, ed., *The Desolate South 1865–1866: A Picture of the Battlefields and of the Devastated Confederacy by John T. Trowbridge* (New York: Duell, Sloan and Pearce, 1956).

Cashin, Joan E., *First Lady of the Confederacy: Varina Davis's Civil War* (Cambridge, MA: Belknap Press, 2006.

Chamlee, Roy Z., *Lincoln's Assassins: A Complete Account of Their Capture, Trial, and Punishment* (Jefferson, NC: McFarland, 1990).

Chase, Salmon P., *Inside Lincoln's Cabinet: The Civil War Diaries of Salmon P. Chase,* ed. David Donald (New York: Longmans, Green, 1954).

Chesson, Michael B., *Richmond After the War 1865–1890* (Richmond: Virginia State Liberty, 1981).

Clark, Allen C., *Abraham Lincoln in the National Capital* (Washington, D.C.: Press of W. F. Roberts Co., 1925).

Clark, James C., *Last Train South: The Flight of the Confederate Government from Richmond* (Jefferson, NC: McFarland & Co., 1984).

Clarke, Asia Booth, *The Unlocked Book: A Memoir of John Wilkes Booth by His Sister* (New York: G. P. Putnam's Sons, 1938).

Clarke, Champ, *The Assassination: Death of a President* (Alexandria, VA: Time-Life Books, 1987).

Coggeshall, William T., *Lincoln Memorial: The Journeys of Abraham Lincoln: From Springfield to Washington, 1861, as President Elect; and from Washington to Springfield, 1865, as President Martyred* (Columbus: Ohio State Journal, 1865).

Cole, Donald B., and John J. McDonough, eds., *Benjamin Brown French: Witness to the Young Republic: A Yankee's Journal, 1828–1870* (Hanover, NH: University Press of New England, 1989).

Coleman, Winston J., Jr., *Last Days, Death and Funeral of Henry Clay* (Lexington, KY: Henry Clay Memorial Foundation, 1951).

Collins, Donald E., *The Death and Resurrection of Jefferson Davis* (Lanham, MD: Rowman & Littlefield, 2005).

Connelly, Thomas L., *The Marble Man: Robert E. Lee and His Image in American Society* (Baton Rouge: Louisiana State University Press, 1977).

Cook, Robert J. *Troubled Commemoration: The American Civil War Centennial, 1961–1965* (Baton Rouge: Louisiana State University Press, 2007).

Cooling, Benjamin Franklin, *Symbol, Sword and Shield: Defending Washington During the Civil War* (Shippensburg, PA: White Mane Publishing Company, 1991).

Cooper, William J., Jr., *Jefferson Davis, American* (New York: Alfred A. Knopf, 2000).

———. *Jefferson Davis and the Civil War Era* (Baton Rouge: Louisiana State University Press, 2008).

———, ed. *Jefferson Davis: The Essential Writings* (New York: Modern Library, 2003).

Coski, John M., *The Confederate Battle Flag: America's Most Embattled Emblem* (Cambridge, MA: Harvard University Press, 2005).

Coulter, E. Merton, *The Confederate States of America 1861–1865* (Baton Rouge: Louisiana State University Press, 1950).

Craughwell, Thomas J., *Stealing Lincoln's Body* (Cambridge, MA: Belknap Press, 2007).

Crist, Lynda Lasswell, et al., eds., *The Papers of Jefferson Davis,* 12 vols. (Baton Rouge: Louisiana State University Press, 1971–2008).

Cullen, Jim, *The Civil War in Popular Culture: A Reusable Past* (Washington,

D.C.: Smithsonian Institution Press, 1995).

Dabney, Virginius, *Richmond: The Story of a City* (Charlottesville: University Press of Virginia, 1990), revised and expanded edition of original 1976 edition.

Dana, Charles A., *Recollections of the Civil War: With the Leaders at Washington and in the Field in the Sixties* (New York: D. Appleton, 1889).

Davis, Burke, *The Long Surrender* (New York: Random House, 1985).

Davis, Jefferson, *The Rise and Fall of the Confederate Government*, 2 vols. (New York: D. Appleton & Co., 1881).

Davis, Varina Howell, *Jefferson Davis, Ex-President of the Confederate States, a Memoir, by His Wife*, 2 vols. (New York: Belford Co., 1890).

Davis, William C., *An Honorable Defeat: The Last Days of the Confederate Government* (New York: Harcourt, 2001).

———. *Jefferson Davis: The Man and His Hour* (New York: HarperCollins, 1991).

———. *Look Away!: A History of the Confederate States of America* (New York: Free Press, 2002).

———. *The Lost Cause: Myths and Realities of the Confederacy* (Lawrence: University Press of Kansas, 1996).

De Chambrun, Marquis Adolphe. *Impressions of Lincoln and the Civil War: A Foreigner's Account.* (New York: Random House, 1952).

Dewitt, David Miller, *The Assassination of Abraham Lincoln and Its Expiation* (New York: Macmillan, 1909).

———. *The Impeachment and Trial of Andrew Johnson* (New York: Macmillan, 1903).

———. *The Judicial Murder of Mary E. Surratt* (Baltimore, MD: J. Murphy, 1895).

Dirck, Brian R., *Lincoln & Davis: Imagining America, 1809–1865* (Lawrence: University Press of Kansas, 2001).

Dixon, Thomas, *The Victim: A Romance of the Real Jefferson Davis* (New York: Grosset & Dunlap, 1914).

Donald, David H., *Lincoln* (New York: Simon & Schuster, 1995).

———. *Lincoln at Home: Two Glimpses of Abraham Lincoln's Family Life* (New York: Simon & Schuster, 2000).

Doster, William E., *Lincoln and Episodes of the Civil War* (New York: G. P. Putnam's Sons, 1915).

Dowdey, Clifford, and Louis H. Manarin, eds., *The Wartime Papers of Robert E. Lee* (Boston: Little, Brown, 1961).

Downes, Alan S., *The Autobiography of Joseph Jefferson* (Cambridge, MA: Belknap Press, 1964).

Eaton, Clement, *A History of the Southern Confederacy* (New York: Macmillan, 1954).

———. *Jefferson Davis* (New York: Free Press, 1977).

Eckert, Edward K., *"Fiction Distorting Fact": Prison Life, Annotated by Jefferson Davis* (Macon, GA: Mercer University Press, 1987).

Edwards, William C., and Edward Steers Jr., *The Lincoln Assassination: The Evidence* (Urbana: University of Illinois Press, 2009).

Eicher, David J., *The Civil War in Books: An Analytical Bibliography* (Urbana: University of Illinois Press, 1997).

Eisenschiml, Otto, *The Case of A. L______, Aged 56* (Chicago: Abraham Lincoln Book Shop, 1943).

———. *In the Shadow of Lincoln's Death* (New York: Wilfred Funk, Inc., 1940).

———. *Why Was Lincoln Murdered?* (Boston: Little, Brown, 1937).

Emerson, Bettie A. C., *Historic Southern Monuments: Representative Memorials of the Heroic Dead of the Southern Confederacy* (New York: Neale Publishing Co., 1911).

Epperson, James F., ed., *The Positive Identification of the Body of John Wilkes Booth, Civil War Naval Chronology* (Washington, D.C.: Government Printing Office, 1971).

Epstein, Daniel Mark, *Lincoln and Whitman: Parallel Lives in Civil War Washington* (New York: Ballantine Books, 2004).

———. *The Lincolns: Portrait of a Marriage* (New York: Ballantine Books, 2008).

Eskew, Garnett Laidlaw, *Willard's of Washington: The Epic of a Capital Caravansary* (New York: Coward-McCann, 1954).

Evans, Eli N., *Judah Benjamin: The Jewish Confederate* (New York: Free Press, 1988).

Evans, W. A., *Mrs. Abraham Lincoln: A Study of Her Personality and Her Influence on Lincoln* (New York: Alfred A. Knopf, 1932).

Fahs, Alice, and Joan Waugh, eds., *The Memory of the Civil War in American Culture* (Chapel Hill: University of North Carolina Press, 2004).

Faust, Drew Gilpin, *The Creation of Confederate Nationalism: Ideology and Identity in the Civil War South* (Baton Rouge: Louisiana State University Press, 1988).

———. *Mothers of Invention: Women of the Slaveholding South in the American Civil War* (Chapel Hill: University of North Carolina Press, 1996).

———. *This Republic of Suffering: Death and the American Civil War* (New York: Alfred A. Knopf, 2008).

Fehrenbacher, Don E., and Virginia Fehrenbacher, *Recollected Words of Abraham Lincoln* (Stanford, CA: Stanford University Press, 1996).

Fellman, Michael, *The Making of Robert E. Lee* (New York: Random House, 2000).

Ferguson, W. J., *I Saw Booth Shoot Lincoln* (Boston: Houghton Mifflin, 1930).

Field, Maunsell B., *Memories of Many Men and of Some Women* (New York: Harper & Brothers, 1874).

Fleischner, Jennifer, *Mrs. Lincoln and Mrs. Keckly* (New York: Broadway Books, 2003).

Flood, Charles Bracelen, *Lee: The Last Years* (Boston: Houghton Mifflin, 1981).

Flower, Frank A., *Edwin McMasters Stanton* (Akron, OH: Saalfield Publishing Co., 1905).

Foster, Gaines M., *Ghosts of the Confederacy: Defeat, the Lost Cause and the Emergence of the New South, 1865–1913* (New York: Oxford University Press, 1987).

Fowler, Robert H., *Album of the Lincoln Murder: Illustrating How It Was Planned, Committed and Avenged* (Harrisburg, PA: Stackpole Books, 1965).

Freeman, Douglas Southall, *The South to Posterity: An Introduction to the Writing of Confederate History* (New York: Charles Scribner's Sons, 1939).

Frey, Herman S., *Jefferson Davis* (Nashville, TN: Frey Enterprises, 1977).

Furgurson, Ernest B., *Ashes of Glory: Richmond at War* (New York: Alfred A. Knopf, 1996).

———. *Freedom Rising: Washington in the Civil War* (New York: Knopf, 2004).

Furtwangler, Albert, *Assassin on Stage: Brutus, Hamlet, and the Death of Lincoln* (Urbana: University of Illinois Press, 1991).

Gallagher, Gary W., *Causes Won, Lost, and Forgotten: How Hollywood & Popular Art Shape What We Know About the Civil War* (Chapel Hill: University of North Carolina Press, 2008).

———. *The Confederate War: How Popular Will, Nationalism, and Military Strategy Could Not Stave Off Defeat* (Cambridge, MA: Harvard University Press, 1997).

Gallagher, Gary W., and Alan T. Nolan, eds., *The Myth of the Lost Cause and Civil War History* (Bloomington: Indiana University Press, 2000).

Gammans, Harold, *Lincoln Names and Epithets* (Boston: Bruce Humphries, 1955).

Garner, Stanton, *The Civil War World of Herman Melville* (Lawrence: University Press of Kansas, 1993).

Garrison, Webb, *The Encyclopedia of Civil War Usage* (Nashville, TN: Cumberland House, 2001).

Gerry, Margarita Spalding, ed., *Through Five Administrations: Reminiscences of Colonel William H. Crook, Body-Guard to President Lincoln* (New York: Harper & Brothers, 1910).

Gildersleeve, Basil L., *The Creed of the Old South 1865–1915* (Baltimore, MD: Johns Hopkins Press, 1915).

Glatthaar, Joseph T., *General Lee's Army: From Victory to Collapse* (New York: Free Press, 2008).

Gobright, Lawrence A., *An Account of Lincoln's Assassination* (New York: n.p. 1869).

———. *Recollections of Men and Things at Washington During Half a Century* (Philadelphia: n.p. 1869).

Good, Timothy S., *We Saw Lincoln Shot: One Hundred Eyewitness Accounts* (Jackson: University Press of Mississippi, 1995).

Goodrich, Thomas, *The Darkest Dawn: Lincoln, Booth, and the Great American Tragedy* (Bloomington: Indiana University Press, 2005).

Gorham, George C., *Life and Public Services of Edwin M. Stanton*, 2 vols. (Boston: Houghton, Mifflin and Company, 1899).

Gray, Clayton, *Conspiracy in Canada* (Montreal: L'Atelier Press, 1957).

Green, James A., *William Henry Harrison: His Life and Times* (Richmond, VA: Garrett and Massie, 1941).

Grieve, Victoria, *Ford's Theatre and the Lincoln Assassination* (Alexandria, VA: Parks & History Association, 2001).

Grimsley, Mark, and Brooks D. Simpson, *The Collapse of the Confederacy* (Lincoln: University of Nebraska Press, 2001).

Gurley, Phineas Densmore, *Faith in God: Dr. Gurley's Sermon at the Funeral of Abraham Lincoln* (Philadelphia: privately printed, 1940).

———. *The Voice of the Rod: A Sermon Preached on Thursday, June 1, 1865, in the New York Avenue Presbyterian Church, Washington, D.C., by the Rev. P. D. Gurley, D.D., Pastor of the Church* (Washington, D.C.: William Ballantyne Bookseller, 1865).

Haco, Dion, *J. Wilkes Booth, the Assassinator of President Lincoln* (New York: T. R. Dawley, 1865).

Haley, William D., ed., *Philp's Washington Described* (New York: Rudd & Carleton, 1861).

Hall, Charles H., *A Mournful Easter: A Discourse Delivered in the Church of the Epiphany, Washington, D.C., on Easter Day, April 19* [sic], *1865, by the Rector, Rev. Charles H. Hall, D.D., Being the Second Day After the Assassination of the President of the United States, and a Similar Attempt upon the Secretary of State, on the Night of Good Friday* (Washington, D.C.: Gideon & Pearson, 1865).

Hall, James O., and Michael Maione, *To Make a Fortune. John Wilkes Booth: Following the Money Trail* (Clinton, MD: Surrat Society, 2003).

Hanchett, William, *The Lincoln Murder Conspiracies* (Urbana: University of Illinois Press, 1983).

Hanna, A. J., *Flight into Oblivion* (Richmond, VA: Johnson Publishing Co., 1938).

Harnden, Henry, *The Capture of Jefferson Davis: A Narrative of the Past Taken by Wisconsin Troops* (Madison, WI: Tracy, Gibbs & Co., 1898).

Harwell, Richard Barksdale, *The Confederate Hundred: A Bibliographic Selection of Confederate Books* (Urbana, IL: Beta Phi Mu, 1964).

———. *In Tall Cotton: The 200 Most Important Books for the Reader, Researcher and Collector* (Austin, TX: Jenkins Publishing Company, 1978).

Harrell, Carolyn L., *When the Bells Tolled for Lincoln: Southern Reaction to the Assassination* (Macon, GA: Mercer University Press, 1997).

Harris, Neil, *Humbug: The Art of P. T. Barnum* (Boston: Little, Brown & Co, 1973).

Harris, William C., *Lincoln's Last Months* (Cambridge, MA: Belknap Press, 2004).

Harrison, Fairfax, ed., *The Harrisons of Skimino* (n.p.: privately printed, 1910).

Harrison, Mrs. Burton, *Recollections Grave and Gay* (New York: Charles Scribner's Sons, 1911).

Hattaway, Herman, and Archer Jones, *How the North Won* (Urbana: University of Illinois Press, 1983).

Hattaway, Herman, and Richard E. Beringer, *Jefferson Davis, Confederate President* (Lawrence: University Press of Kansas, 2002).

Helm, Katherine, *The True Story of Mary, Wife of Lincoln* (New York and London: Harper & Brothers, 1928).

Helwig, Rev. J. B., *The Assassination of President Lincoln: What Was the Religious Faith of Those Engaged in the Conspiracy That Resulted in the Assassination of President Lincoln at Washington, D.C., on Friday Evening, April 14, 1865* (Springfield, OH: A. D. Hosterman & Co., n.d.).

Henriques, Peter R., *The Death of George Washington: He Died as He Lived* (Mount Vernon, VA: Mount Vernon Ladies Association, 2000).

Hill, Tucker, *Victory in Defeat: Jefferson Davis and the Lost Cause* (Richmond, VA: Museum of the Confederacy, 1986).

Holland, J. G., *The Life of Abraham Lincoln* (Springfield, MA: Gurdon Bill, 1866).

Holzer, Harold, Gabor S. Boritt, and Mark E. Neely Jr., *The Lincoln Image: Abraham Lincoln and the Popular Print* (New York: Scribner's, 1984).

Huntington, Richard, and Peter Metcalf, *Celebrations of Death: The Anthropology of Mortuary Ritual* (Cambridge, UK; New York: Cambridge University Press, 1979).

Hylton, J. Dunbar, *The Præsidicide: A Poem* (Philadelphia: Meichel & Plumly, 1868).

In Memoriam (New York: Trent, Filmer & Co., 1865).

Isacsson, Alfred, *The Travels, Arrest and Trial of John H. Surratt* (Middletown, NY: Vestigium Press, 2003).

Janney, Caroline E., *Burying the Dead but Not the Past: Ladies' Memorial Associations and the Lost Cause* (Chapel Hill: University of North Carolina Press, 2008).

Johnson, Andrew, *Impeachment and Trial of Andrew Johnson* (Philadelphia: Barclay & Co., 1868).

Johnson, Byron Berkeley, *Abraham Lincoln and Boston Corbett, with Personal Recollections of Each: John Wilkes Booth and Jefferson Davis, a True Story of Their Capture* (Boston: Lincoln & Smith Press, 1914).

Johnson, Clint, *Pursuit: The Chase, Capture, Persecution & Surprising Release of Confederate President Jefferson Davis* (New York: Citadel Press, 2008).

Johnston, Joseph E., *Narrative of Military Operations During the Civil War.* New York: D. Appleton & Co., 1874.

Jones, J. William, *The Davis Memorial Volume; or our Dead President, Jefferson Davis, and the World's Tribute to His Memory* (Richmond, VA: B. F. Johnson & Co., 1890).

Jones, Katharine M., ed., *Ladies of Richmond, Confederate Capital* (Indianapolis, IN: Bobbs-Merrill, 1962).

Jones, Thomas A., *J. Wilkes Booth: An Account of His Sojourn in Southern*

Maryland After the Assassination of Abraham Lincoln, His Passage Across the Potomac, and His Death in Virginia (Chicago: Laird & Lee, 1893).

Judson, Edward Zane Carroll, *The Parricides; or, the Doom of the Assassins, the Authors of the Nation's Loss, by Ned Buntline* (New York: Hilton & Co., 1865).

Kammen, Michael, *Digging Up the Dead: A History of Notable American Reburials* (Chicago: University of Chicago Press, 2010).

———. *Mystic Chords of Memory: The Transformation of Tradition in American Culture* (New York: Alfred A. Knopf, 1991).

Kauffman, Michael, *American Brutus: John Wilkes Booth and the Lincoln Conspiracies* (New York: Random House, 2004).

Keckley, Elizabeth, *Behind the Scenes, or, Thirty Years a Slave, and Four Years in the White House* (New York: G. W. Carleton & Co., 1868).

Kendall, John S., *The Golden Age of New Orleans Theater* (Baton Rouge: Louisiana State University Press, 1952).

Kimmel, Stanley, *The Mad Booths of Maryland* (Indianapolis, IN: Bobbs-Merrill, 1940).

———. *Mr. Davis's Richmond* (New York: Coward-McCann, 1958).

———. *Mr. Lincoln's Washington* (New York: Bramhall House, 1957).

King, Willard, *Lincoln's Manager David Davis* (Cambridge, MA: Harvard University Press, 1960).

Kunhardt, Dorothy Meserve, and Philip B. Kunhardt Jr., *Twenty Days* (New York: Harper & Row, 1965).

Laderman, Gary, *The Sacred Remains: American Attitudes Toward Death, 1799–1883* (New Haven, CT: Yale University Press, 1996).

Lamon, Dorothy, ed., *Recollections of Abraham Lincoln, 1847–1865, by Ward Hill Lamon* (Chicago: A. C. McClurg and Co., 1895).

Lamon, Ward Hill, *Recollections of Abraham Lincoln 1847–1865,* ed. Dorothy Lamon Teillard (Washington, D.C.: published by the editor, 1911).

Lankford, Nelson, *Richmond Burning: The Last Days of the Confederate Capital* (New York: Viking, 2002).

Lattimer, Dr. John K., *Kennedy and Lincoln: Medical and Ballistic Comparisons of Their Assassinations* (New York: Harcourt Brace Jovanovich, 1980).

Laughlin, Clara E., *The Death of Lincoln: The Story of Booth's Plot, His Deed and the Penalty* (New York: Doubleday, Page, 1909).

Leale, Charles, *Lincoln's Last Hours* (New York: privately printed, 1909).

Lee, Richard M., *Mr. Lincoln's City* (McLean, VA: EPM Publications, 1981).

Leech, Margeret, *Reveille in Washington, 1860–1865* (New York: Harper & Brothers, 1941).

Leonard, Elizabeth D., *Lincoln's Avengers: Justice, Revenge, and Reunion After the Civil War* (New York: W. W. Norton & Company, 2004).

Lewis, Lloyd, *Myths After Lincoln* (New York: Harcourt, Brace and Co., 1920).

The Lincoln Memorial: A Record of the Life, Assassination, and Obsequies of the Martyred President (New York: Bruce & Huntington, 1865).

Loux, Arthur F., *John Wilkes Booth: Day by Day* (privately printed, 1989).

Lowenfels, Walter, ed., *Walt Whitman's Civil War* (New York: Knopf, 1960).

Lubbock, Francis Richard. *Six Decades in Texas or Memoirs of Francis Richard Lubbock, Governor of Texas in War-Time 1861–63,* ed. C. W. Raines (Austin, TX: Ben C. Jones & Co., 1900).

Mahoney, Ella V., *Sketches of Tudor Hall and the Booth Family* (Bel Air, MD: privately printed, 1925).

Mallon, Thomas, *Henry and Clara* (New York: Ticknor & Fields, 1994).

Manarin, Louis H., ed., *Richmond at War: The Minutes of the City Council 1861–1865* (Chapel Hill: University of North Carolina Press, 1966).

———. *Richmond on the James* (Charleston, SC: Arcadia Publishing, 2001).

Marvel, William, *A Place Called Appomattox* (Chapel Hill: University of North Carolina Press, 2000).

Maynard, Nettie Colburn, *Was Abraham Lincoln a Spiritualist? Or, Curious Revelations from the Life of a Trance Medium* (Philadelphia: Rufus C. Hartranft, 1891).

McClure, Stanley W., *Ford's Theatre and the House Where Lincoln Died* (Washington, D.C.: Government Printing Office, 1969).

———. *The Lincoln Museum and the House Where Lincoln Died* (Washington, D.C.: Government Printing Office, 1949).

McCreary, Donna D., *Fashionable First Lady: The Victorian Wardrobe of Mary Lincoln.* (n.p.: Lincoln Presentations, 2007).

McCulloch, Hugh, *Men and Measures of Half a Century* (New York: Charles Scribner's Sons, 1888).

McElroy, Robert, *Jefferson Davis, the Unreal and the Real,* 2 vols. (New York: Harper & Brothers, 1937).

McPherson, James M., "American Victory, American Defeat," *Why the Confederacy Lost,* ed., Gabor S. Boritt (New York: Oxford University Press, 1992), 15–42.

———. *Battle Cry of Freedom: The Civil War Era* (New York: Oxford University Press, 1988).

———. *Drawn with the Sword: Reflections on the American Civil War* (New York: Oxford University Press, 1996).

McPherson, James M., and William J. Cooper Jr., eds., *Writing the Civil War: The Quest to Understand* (Columbia: University of South Carolina Press, 1998).

Meredith, Roy, *Mr. Lincoln's Contemporaries: An Album of Portraits by Mathew Brady* (New York: Charles Scribner's Sons, 1951).

Miers, Earl Schenk, ed., *When the World Ended: The Diary of Emma Le Conte* (New York: Oxford University Press, 1957).

Miller, Ernest C., *John Wilkes Booth: Oilman* (New York: Exposition Press, 1947).

Mills, Cynthia, and Pamela H. Simpson, eds., *Monuments to the Lost Cause: Women, Art, and the Landscapes of Southern Memory* (Knoxville: University of Tennessee Press, 2003).

Mitchell, Mary H., *Hollywood Cemetery: The History of a Southern Shrine* (Richmond: Virginia State Library, 1985).

Mogelever, Jacob, *Death to Traitors: The Story of General Lafayette C. Baker, Lincoln's Forgotten Secret Service Chief* (Garden City, NY: Doubleday & Co., 1960).

Monaghan, Jay, *Lincoln Bibliography, 1839–1939*, 2 vols. (Springfield: Illinois State Historical Library, 1943–45).

Morris, B. F., comp., *Memorial Record of the Nation's Tribute to Abraham Lincoln* (Washington, D.C.: W. H. & O. H. Morrison, 1865).

Morris, Clara, *Life On Stage: My Personal Experiences and Recollections* (New York: McClure, Phillips & Co., 1901).

Mudd, Nettie, ed., *The Life of Dr. Samuel A. Mudd, Containing His Letters from Fort Jefferson, Dry Tortugas Island, Where He Was Imprisoned Four Years for Alleged Complicity in the Assassination of Abraham Lincoln* (New York: Neale Publishing Company, 1906).

Neely, Mark E., Jr., *The Abraham Lincoln Encyclopedia* (New York: McGraw-Hill, 1982).

———. "Abraham Lincoln vs. Jefferson Davis: Comparing Presidential Leadership in the Civil War," *Writing the Civil War: The Quest to Understand*, eds., James M. McPherson and William J. Cooper Jr. (Columbia: University of South Carolina Press, 1998), 96–111.

———. *The Fate of Liberty: Abraham Lincoln and Civil Liberties* (New York: Oxford University Press, 1991).

Neely, Mark E., Jr., Harold Holzer, and Gabor S. Boritt, *The Confederate Image: Prints of the Lost Cause* (Chapel Hill: University of North Carolina Press, 1987).

Nevins, Allan, ed., *Diary of the Civil War 1860–1865: George Templeton Strong* (New York: Macmillan, 1962).

Nevins, Allan, James I. Robertson Jr., and Bell I. Wiley, *Civil War Books: A Critical Bibliography,* 2 vols. (Baton Rouge: Louisiana State University Press, 1967).

Newman, Ralph G., *"In This Sad World of Ours, Sorrow Comes to All": A Timetable for the Lincoln Funeral Train* (Springfield: Civil War Centennial Commission of Illinois, 1965).

———, ed., *Lincoln for the Ages* (Garden City, NY: Doubleday, 1960).

Nicholson, John Page, *Catalogue of Library of Brevet Lieutenant Colonel John Page Nicholson . . . Relating to the War of the Rebellion 1861–1866* (Philadelphia: privately printed, 1914).

Nicolay, John G., *A Short Life of Abraham Lincoln* (New York: Century Co., 1904).

Nicolay, John G., and John Hay, *Abraham Lincoln,* 10 vols. (New York: Century Co., 1890).

Niven, John, *Gideon Welles: Lincoln's Secretary of the Navy* (New York: Oxford University Press, 1973).

———. *John C. Calhoun and the Price of Union* (Baton Rouge: Louisiana State University Press, 1988).

———. *Salmon P. Chase: A Biography* (New York: Oxford University Press, 1995).

Oates, Stephen B., *With Malice Toward None: The Life of Abraham Lincoln* (New York: Harper & Row, 1977).

Oldroyd, Osborn H., *The Assassination of Abraham Lincoln* (Washington, D.C.: Osborn H. Oldroyd, 1901).

———. *The Oldroyd Lincoln Memorial Collection: Located in the House in Which Lincoln Died* (Washington, D.C.: privately printed by Judd and Detweiler, 1903).

Olszewski, George J., *Restoration of Ford's Theatre* (Washington, D.C.: Government Printing Office, 1963).

Ostendorf, Lloyd, *Lincoln's Photographs: A Complete Album* (Dayton, OH: Rockywood Press, 1998).

Osterweis, Rollin G., *The Myth of the Lost Cause 1865–1900* (Hamden, CT: Archon Books, 1973).

———. *Romanticism and Nationalism in the Old South* (New Haven, CT: Yale University Press, 1949).

Ott, Victoria E., *Confederate Daughters: Coming of Age During the Civil War* (Carbondale: Southern Illinois University Press, 2008).

Otto, John Solomon, *Southern Agriculture During the Civil War Era, 1860–1880* (Westport, CT: Greenwood Press, 1994).

Ownsbey, Betty J., *Alias "Paine": Lewis Thornton Powell, the Mystery Man of the Lincoln Conspiracy* (Jefferson, NC: McFarland, 1993).

Paludan, Phillip Shaw, *The Presidency of Abraham Lincoln* (Lawrence: University Press of Kansas, 1994).

Parker, William Harwar, *Recollections of a Naval Officer 1841–1865*, intro and notes by Craig L. Symonds (New York: Charles Scribner's Sons, 1883; repr. Annapolis, MD: Naval Institute Press, 1985).

Parrish, T. Michael, and Robert M. Willingham Jr., *Confederate Imprints: A Bibliography of Southern Publications From Secession to Surrender* (Austin, TX: Jenkins Publishing Co., n.d.).

Parsons, John E., *Henry Deringer's Pocket Pistol* (New York: William Morrow, 1952).

Patrick, Rembert W., *The Fall of Richmond* (Baton Rouge: Louisiana State University Press, 1960).

Petersen, Merrill D., *Lincoln in American Memory* (New York: Oxford University Press, 1994).

Peterson, T. B., *The Trial of the Alleged Assassins and Conspirators at Washington City, D.C., in May and June, 1865* (Philadelphia: T. B. Peterson & Brothers, 1865).

Phillips, Jason, *Diehard Rebels: The Confederate Culture of Invincibility* (Athens: University of Georgia Press, 2007).

Pinsker, Matthew, *Lincoln's Sanctuary: Abraham Lincoln and the Soldiers' Home* (New York: Oxford University Press, 2003).

Pitman, Benn, *The Assassination of President Lincoln and the Trial of the Conspirators* (Cincinnati, OH: Moore, Wilstach & Baldwin, 1865).

Plowden, David, *Lincoln and His America* (New York: Viking Press, 1970).

Plummer, Mark A., *Lincoln's Rail-Splitter: Governor Richard J. Oglesby* (Urbana: University of Illinois Press, 2001).

Poetical Tributes to the Memory of Abraham Lincoln (Philadelphia: J. B. Lippincott & Co., 1865).

Poore, Ben Perley, *The Conspiracy Trial for the Murder of the President,* 3 vols. (Boston: J. E. Tilton and Company, 1865–1866).

———. *Perley's Reminiscences of Sixty Years in the National Metropolis,* 2 vols. Philadelphia: Hubbard Brothers, 1886).

Porcher, Francis Peyre, *Resources of the Southern Fields and Forests, Medical, Economical, and Agricultural* (Charleston, SC: Steam Power Press of Evans & Cogswell, 1863).

Power, John Carroll, *Abraham Lincoln: His Life, Public Services and Great Funeral Cortege* (Springfield, IL: Edwin W. Wilson, 1875).

———. *History of an Attempt to Steal the Body of Abraham Lincoln* (Springfield, IL: H. W. Rokker, 1890).

Pratt, Harry E., *The Personal Finances of Abraham Lincoln* (Springfield, IL: Abraham Lincoln Association, 1943).

Pryor, Elizabeth Brown, *Reading the Man: A Portrait of Robert E. Lee Through His Private Letters* (New York: Viking, 2007).

Putnam, Sallie Brock, *Richmond During the War* (New York: G. W. Carleton, 1867).

Randall, Ruth Painter, *Lincoln's Sons* (Boston: Little, Brown & Co., 1955).

Reagan, John H., *Memoirs, with Special Reference to Secession and the Civil War* (New York: Neale Publishing Co., 1906).

Reck, W. Emerson, *A. Lincoln: His Last 24 Hours* (Jefferson, NC: McFarland, 1987).

Reid, Whitelaw, *After the War: A Tour of the Southern States, 1865–1866,* ed. C. Vann Woodward (New York: Moore, Wilstach & Baldwin, 1866; repr. New York: Harper & Row, 1965).

Reilly, Bernard F., Jr., *American Political Prints 1766–1876. A Catalog of the Collections in the Library of Congress* (Boston: G. K. Hall & Co., 1991).

Remini, Robert V., *Daniel Webster: The Man and His Time* (New York: W. W. Norton & Co., 1997).

———. *Henry Clay: Statesman for the Union* (New York: W. W. Norton, 1993).

Reynolds, David S., *John Brown, Abolitionist* (New York: Knopf, 2005).

Rhodehamel, John, and Louise Taper, *"Right or Wrong, God Judge Me": The Writings of John Wilkes Booth* (Urbana: University of Illinois Press, 1997).

Risvold, Floyd E., ed., *A True History of the Assassination of Abraham Lincoln and of the Conspiracy of 1865, by Louis J. Weichmann, Chief Witness for the*

Government of the United States in the Prosecution of the Conspirators (New York: Knopf, 1975).

Robertson, David, *Booth: A Novel* (New York: Anchor Books, 1998).

Roscoe, Theodore, *The Web of Conspiracy: The Complete Story of the Men Who Murdered Abraham Lincoln* (Englewood Cliffs, NJ: Prentice-Hall, 1959).

Rowland, Dunbar, *Jefferson Davis, Constitutionalist,* 10 vols. (Jackson: Mississippi Department of Archives and History, 1923).

Rowland, Eron, *Varina Howell, Wife of Jefferson Davis,* 2 vols. (New York: Macmillan, 1927 & 1931).

Rubin, Anne Sarah, *A Shattered Nation: The Rise & Fall of the Confederacy, 1861–1868* (Chapel Hill: University of North Carolina Press, 2005).

Ryan, David D., *Four Days in 1865: The Fall of Richmond* (Richmond, VA: Cadmus, 1993).

Sachsman, Davis B., S. Kittrell Rushing, and Debra Reddin van Tuyll, eds., *The Civil War and the Press* (New Brunswick, NJ: Transaction Pub., 2000).

Samples, Gordon, *Lust for Fame: The Stage Career of John Wilkes Booth* (Jefferson, NC: McFarland and Co., 1982).

Sandburg, Carl, *Abraham Lincoln: The War Years,* 4 vols. (New York: Harcourt Brace & Co., 1939).

Searcher, Victor, *The Farewell to Lincoln* (New York: Abingdon Press, 1965).

Semmes, Raphael, *Memoirs of Service Afloat, During the War Between the States* (Baltimore, MD: Kelly, Piet & Co., 1869).

Sermons Preached in Boston on the Death of Abraham Lincoln, Together with the Funeral Services in the East Room of the Executive Mansion at Washington (Boston: J. E. Tilton & Co., 1865).

Seward, Frederick W., *Reminiscences of a War-Time Statesman and Diplomat, 1830– 1915* (New York: G. P. Putnam's Sons, 1916).

Seward, William H., *William H. Seward: An Autobiography from 1801 to 1834, with a Memoir of His Life, and Selections from His Letters, 1831–1872, by Frederick W. Seward,* 3 vols. (New York: Derby and Miller, 1891).

Seymour, Mark Wilson, ed., *The Pursuit of John H. Surratt: Despatches from the Official Record of the Assassination of Abraham Lincoln* (Austin, TX: Civil War Library, 2000).

Shea, John Gilmary, ed., *The Lincoln Memorial: A Record of the Life, Assassination and Obsequies of the Martyred President* (New York: Bruce & Huntington, 1865).

Sheehan-Dean, Aaron, *Why Confederates Fought: Family & Nation in Civil War*

Virginia (Chapel Hill: University of North Carolina Press, 2007).

Shelton, Vaughn, *Mask for Treason: The Lincoln Murder Trial* (Harrisburg, PA: Stackpole Company, 1965).

Sherman, Edwin A., *Lincoln's Death Warrant: Or the Peril of Our Country* (Milwaukee: Wisconsin Patriot, c. 1892).

Sherman, William Tecumseh, *Memoirs of William Tecumseh Sherman, By Himself*, 2 vols. (New York: D. Appleton & Co., 1875).

Shingleton, Royce Gordon, *John Taylor Wood: Sea Ghost of the Confederacy* (Athens: University Press of Georgia, 1979).

Shutes, Milton H., *Lincoln and the Doctors: A Medical Narrative of the Life of Abraham Lincoln* (New York: Pioneer Press, 1933).

Silber, Nina, *The Romance of Reunion: Northerners and the South, 1865–1900* (Chapel Hill: University of North Carolina Press, 1993).

Simms, Henry H., *Life of Robert M. T. Hunter* (Richmond, VA: William Byrd Press, 1935).

Smith, Jean Edward, *Grant* (New York: Simon & Schuster, 2001).

Smoot, Richard Mitchell, *The Unwritten History of the Assassination of Abraham Lincoln* (Clinton, MA: Press of W. J. Coulter, 1908).

Sotos, John G., *The Physical Lincoln* (Mount Vernon, VA: Mt. Vernon Book Systems, 2008).

Speer, Bonnie Stahlman, *The Great Abraham Lincoln Hijack* (Norman, OK: Reliance Press, 1990).

Spencer, William V., *Lincolniana: In Memoriam* (Boston: William V. Spencer, 1865).

Starr, John W., Jr., *Further Light on Lincoln's Last Day* (Harrisburg, PA: privately printed, 1930).

———. *Lincoln's Last Day* (New York: Frederick A. Stokes Co., 1922).

———. *New Light on Lincoln's Last Day* (Harrisburg, PA: privately printed, 1926).

Steers, Edward, Jr., *Blood on the Moon: The Assassination of Abraham Lincoln* (Lexington: University Press of Kentucky, 2001).

———. *The Escape & Capture of John Wilkes Booth* (Gettysburg, PA: Thomas Publications, 1992).

———. *His Name Is Still Mudd: The Case Against Doctor Samuel Alexander Mudd* (Gettysburg, PA: Thomas Publications, 1997).

______. *The Lincoln Assassination Encyclopedia* (New York: William Morrow, 2010).

———, ed., *The Trial: The Assassination of President Lincoln and the Trial of the Conspirators* (Lexington: University Press of Kentucky, 2003).

Stern, Philip van Doren, *The Man Who Killed Lincoln: The Story of John Wilkes Booth and His Part in the Assassination* (New York: Random House, 1939).

Stevens, L. L., *Lives, Crimes, and Confessions of the Assassins* (Troy, NY: Daily Times Steam Printing Establishment, 1865).

Stewart, David O., *Impeached: The Trial of President Andrew Johnson and the Fight for Lincoln's Legacy* (New York: Simon & Schuster, 2009).

Strode, Hudson, *Jefferson Davis: American Patriot 1808–1861* (New York: Harcourt, Brace & Co., 1955).

———. *Jefferson Davis: Confederate President* (New York: Harcourt, Brace & Co., 1959).

———. *Jefferson Davis: Tragic Hero. The Last Twenty-five Years 1864–1889* (New York: Harcourt, Brace & World, 1964).

Surratt, John H., *Life and Extraordinary Adventures of John H. Surratt, the Conspirator* (Philadelphia: Barclay & Co., 1867).

———. *Trial of John H. Surratt in the Criminal Court for the District of Columbia*, 2 vols. (Washington, D.C.: Government Printing Office, 1867).

Swanson, James L., and Daniel R. Weinberg, *Lincoln's Assassins: Their Trial and Execution* (Santa Fe, NM: Arena Editions, 2001).

Taft, Charles Sabin, *Abraham Lincoln's Last Hours: From the Notebook of Charles Sabin Taft, M.D., an Army Surgeon Present at the Assassination, Death and Autopsy* (Chicago: privately printed, 1934).

Taylor, John M., *William Henry Seward: Lincoln's Right Hand* (New York: HarperCollins, 1991).

Taylor, Welford Dunaway, *Our American Cousin: The Play That Changed History* (Washington, D.C.: Beacham Publishing, 1990).

Temple, Wayne C., *Abraham Lincoln: From Skeptic to Prophet* (Mahomet, IL: Mayhaven Publishing, 1995).

The Terrible Tragedy at Washington: Assassination of President Lincoln (Philadelphia: Barclay & Co., 1865).

Thomas, Benjamin P., *Abraham Lincoln* (New York: Knopf, 1952).

Thomas, Benjamin P., and Harold M. Hyman, *Stanton: The Life and Times of Lincoln's Secretary of War* (New York: Knopf, 1962).

Thomas, Emory, *The Confederate Nation: 1861–1865* (New York: Harper & Row, 1979).

Tidwell, William A., *April '65: Confederate Covert Action in the American Civil War* (Kent, OH: Kent State University Press, 1995).

Tidwell, William A., with James O. Hall and David Winfred Gaddy, *Come Retribution: The Confederate Secret Service and the Assassination of Lincoln* (Jackson: University Press of Mississippi, 1988).

Townsend, George Alfred, *The Life, Crime, and Capture of John Wilkes Booth* (New York: Dick & Fitzgerald, 1865).

———. *Rustics in Rebellion: A Yankee Reporter on the Road to Richmond* (Chapel Hill: University of North Carolina Press, 1950).

Townshend, E. D., *Anecdotes of the Civil War* (New York: D. Appleton-Century Co., 1884).

Trindal, Elizabeth Steger, *Mary Surratt: An American Tragedy* (Gretna, LA: Pelican Publishing Company, 1996).

Trostel, Scott D., *The Lincoln Funeral Train* (Fletcher, OH: Cam-Tech Publishing, 2002).

Trudeau, Noah Audre, *Southern Storm: Sherman's March to the Sea* (New York: Harper, 2008).

Turner, Justin G., and Linda Levitt Turner, eds., *Mary Todd Lincoln: Her Life and Letters* (New York: Knopf, 1972).

Turner, Thomas Reed, *The Assassination of Abraham Lincoln* (Malabar, FL: Krieger Publishing Company, 1999).

———. *Beware the People Weeping: Public Opinion and the Assassination of Abraham Lincoln* (Baton Rouge: Louisiana State University Press, 1982).

Valentine, David T., *Obsequies of Abraham Lincoln, in the City of New York* (New York: Edmund Jones & Co., 1866).

Van der Heuvel, Gerry, *Crowns of Thorns and Glory: Mary Todd Lincoln and Varina Howell Davis: The Two First Ladies of the Civil War* (New York: E. P. Dutton, 1988).

Van Deusen, Glyndon G., *William Henry Seward* (New York: Oxford University Press, 1967).

Vowell, Sarah, *Assassination Vacation* (New York: Simon & Schuster, 2005).

Wakelyn, Jon L., *Biographical Dictionary of the Confederacy* (Westport, CT: Greenwood Press, 1977).

Waldrep, Christopher, *Vicksburg's Long Shadow: The Civil War Legacy of Race and Remembrance* (Lanham, MD: Rowman & Littlefield, 2005).

The War of the Rebellion: A Compilation of the Official Records of the Union and Confederate Armies, 128 vols. (Washington, D.C.: Government Printing Office, 1880–1901).

Warner, Ezra J., *Generals in Blue* (Baton Rouge: Louisiana State University Press, 1964).

———. *Generals in Gray* (Baton Rouge: Louisiana State University Press, 1959).

Warren, Robert Penn, *Jefferson Davis Gets His Citizenship Back* (Lexington: University Press of Kentucky, 1980).

Watson, Harry L., and Larry J. Griffin, eds., *Southern Cultures: The Fifteenth Anniversary Reader* (Chapel Hill: University of North Carolina Press, 2008).

Welles, Gideon, *Diary of Gideon Welles: Secretary of the Navy under Lincoln and Johnson,* 3 vols. (Boston: Houghton Mifflin, 1911).

Werstein, Irving, *Abraham Lincoln Versus Jefferson Davis* (New York: Thomas Y. Crowell Company, 1959).

West, Richard S., *Gideon Welles: Lincoln's Navy Department* (Indianapolis, IN: Bobbs-Merrill, 1943).

Weyl, Nathaniel, *Treason: The Story of Disloyalty and Betrayal in American History* (Washington, D.C.: Public Affairs Press, 1950).

Whiteman, Maxwell, *While Lincoln Lay Dying: A Facsimile Reproduction of the First Testimony Taken in Connection with the Assassination of Abraham Lincoln as Recorded by Corporal James Tanner* (Philadelphia: Union League of Philadelphia, 1968).

Wiggins, Sarah Woolfolk, ed., *The Journals of Josiah Gorgas 1857–1878* (Tuscaloosa: University of Alabama Press, 1995).

Williamson, David Brainerd, *Illustrated Life, Services, Martyrdom, and Funeral of Abraham Lincoln* (Philadelphia: T. B. Peterson & Brothers, 1865).

Wilson, Charles Reagan, *Baptized in Blood: The Religion of the Lost Cause 1865–1920* (Athens: University of Georgia Press, 1980).

Wilson, Francis, *John Wilkes Booth: Fact and Fiction of Lincoln's Assassination* (Boston: Houghton Mifflin, 1929).

———. *Joseph Jefferson: Reminiscences of a Fellow Player* (New York: Scribner's, 1905).

Wilson, Rufus Rockwell, *Lincoln Among His Friends: A Sheaf of Intimate Memories* (Caldwell, ID: Caxton Printers, 1942).

Winik, Jay, *April 1865: The Month That Saved America* (New York: HarperCollins, 2001).

Winston, Robert W., *High Stakes and Hair Trigger: The Life of Jefferson Davis* (New York: Henry Holt & Co., 1930).

Winter, William, *Life and Art of Edwin Booth* (New York: Moffat, 1893).

Wise, John Sergeant, *End of an Era* (Boston: Houghton, Mifflin & Co., 1899).

Woods, Rufus, *The Weirdest Story in American History: The Escape of John Wilkes Booth* (Wenachee, WA: privately printed, 1944).

Woodward, C. Vann, *The Burden of Southern History* (Baton Rouge: Louisiana State University Press, 1968, rev. ed.).

Woodworth, Steven E., *Jefferson Davis and His Generals: The Failure of Confederate Command in the West* (Lawrence: University Press of Kansas, 1990).

NOTES

1: "FLITTING SHADOWS"

5 *"A perfect Sunday"* Constance Cary Harrison, *Recollections Grave and Gay* (New York: Scribner's, 1911), 207.

5 *"On Sunday, the 2d of April"* Jefferson Davis, *The Rise and Fall of the Confederate Government* (New York: D. Appleton, 1881), 2:668. Hereafter cited as *Rise and Fall.*

6 *"I see no prospect of doing more than holding our position"* Robert E. Lee, *The Wartime Papers of Robert E. Lee,* ed. Clifford Dowdey (New York: Da Capo Press), 924.

7 *"I happened to sit in the rear"* Harrison, *Recollections Grave and Gay,* 207.

7 *"the people of Richmond had been too long"* Davis, *Rise and Fall,* 2:667.

7 *"Before dismissing the congregation"* Harrison, *Recollections Grave and Gay,* 207.

7 *"On the sidewalk outside the church"* Harrison, *Recollections Grave and Gay,* 207–8.

8 *"About 11:30 a.m. on Sunday, April 2d, a strange agitation"* Clement Sulivane, "The Fall of Richmond: The Evacuation," in *Battles and*

Leaders of the Civil War, eds. Robert Underwood Johnson and Clarence Buel (New York: Thomas Yoseloff, 1956), 4:725.

8 *"I went to my office and assembled the heads"* Davis, *Rise and Fall,* 2:667.

9 *"My own papers"* Davis, *Rise and Fall,* 2:667.

9 *"By this time the report that Richmond was to be evacuated"* Davis, *Rise and Fall,* 2:667.

10 *"He said for the future his headquarters must be in the field"* Varina Howell Davis, *Jefferson Davis, Ex-President of the Confederate States of America: A Memoir by His Wife* (New York: Belford, 1890), 2:575. Hereafter cited as *A Memoir.*

10 *"Very averse to flight"* Varina Davis, *A Memoir,* 2:575.

10 *"I have confidence in your capacity"* Varina Davis, *A Memoir,* 2:575.

10 *"If I live"* Varina Davia, *A Memoir,* 2:575.

10 *"I do not expect to survive the destruction of constitutional liberty"* Varina Davis, *A Memoir,* 2:575.

11 *"All women like bric-a-brac"* Varina Davis, *A Memoir,* 2:576.

11 *"They may be exposed to inconvenience or outrage"* Varina Davis, *A Memoir,* 2:576.

11 *"He showed me how to load, aim, and fire"* Varina Davis, *A Memoir,* 2:576.

11 *"Col. Brown will please order these cartridges"* Lynda Lasswell Crist, ed., *The Papers of Jefferson Davis* (Baton Rouge: Louisiana State University Press, 1971–2008), 11:481. Hereafter cited as *Papers.*

12 *"You can at least, if reduced to the last extremity"* Varina Davis, *A Memoir,* 2:576.

12 *"Leaving the house as it was, and taking only our clothing"* Varina Davis, *A Memoir,* 2:576.

12 *"You cannot remove anything in the shape of food"* Varina Davis, *A Memoir,* 2:576.

12 *"Mr. Davis almost gave way"* Varina Davis, *A Memoir,* 2:577.

13 *"The movement of Gen. Grant to Dinwiddie"* Robert E. Lee, *Lee's Dispatches: Unpublished Letters of General Robert E. Lee, C.S.A., to Jefferson Davis and the War Department of the Confederate States of America, 1862–1865,* ed. Douglas Southall Freeman (1915; New York: G. P. Putnam's Sons, 1957), 358–60.

14 *"The question is often asked of me"* Crist, *Papers,* 11:493.

14 *"My best hope was that Sherman while his army was worn"* Crist, *Papers,* 11:489.

15 *"Arrived here safely very kindly treated"* Crist, *Papers,* 11:491.

16 *"The progress of our arms"* Abraham Lincoln, *The Collected Works of Abraham Lincoln,* ed. Roy P. Basler (New Brunswick: Rutgers University Press, 1952), 8:332–33. Hereafter cited as *Collected Works.*

17 *"During this interview I inquired of the President"* William Tecumseh Sherman, *Memoirs of General William T. Sherman* (New York: Appleton, 1887), 2:326.

18 *"I hope you will stay to see it out"* Lincoln, *Collected Works,* 8:378.

20 *"Dispatches frequently coming in"* Lincoln, *Collected Works,* 8:382.

20 *"All that Sabbath day the trains came and went"* Sulivane, "The Fall of Richmond," 725.

22 *"To move to-night will involve the loss of many valuables"* Crist, *Papers,* 11:499.

22 *"Your telegram recd. I think it will be necessary"* Lee, *Lee's Dispatches,* 375.

22 *"Mrs. Lincoln: At 4:30 p.m. to-day General Grant telegraphs"* Lincoln, *Collected Works,* 8:383

23 *"I think it absolutely necessary that we should abandon our position"* Lee, *Wartime Papers,* 925.

23 *"It is absolutely necessary that we should abandon our position"* Lee, *Wartime Papers,* 926.

24 *"Allow me to tender you, and all with you"* Lincoln, *Collected Works,* 8:383.

24 *"The furniture in the executive mansion it would be well to pack"* Crist, *Papers,* 11:500.

25 *"This was the saddest trip I had ever made"* Francis Richard Lubbock, *Six Decades in Texas; or, Memoirs of Francis Richard Lubbock, Governor of Texas in War Time, 1861–1863,* ed. Cadwell Walton Raines (Austin: B. C. Jones, 1900), 563.

25 *"Richmond would be isolated, and it could not have been defended"* Davis, *Rise and Fall,* 2:668.

26 *"the rabble who stood ready to plunder"* Stephen R. Mallory, "Last Days of the Confederate Government," *McClure's Magazine* 26, no. 2 (Dec. 1900), 102.

27 *"While waiting at the depot"* William Harwar Parker, *Recollections of*

a Naval Officer, 1841–1865, rev. ed., ed. Craig L. Symonds (New York: Charles Scribner's Sons, 1883; Annapolis: Naval Institute Press, 1985), 375. Citations are to the Naval Institute Press edition.

27 *"Not only inside, but on top"* Parker, *Recollections,* 374–75.

28 *"submit to the invading army"* John H. Reagan, *Memoirs, with Special Reference to Secession and the Civil War,* ed. Walter Flavius McCaleb (New York: Neale Publishing Company, 1906), 197.

28 *"left but small opportunity"* Reagan, *Memoirs,* 198.

28 *"Silence reigned over the fugitives"* Mallory, "Last Days," 104.

28 *"The terrible reverses"* Mallory, "Last Days," 104.

29 *"It was near midnight"* Reagan, *Memoirs,* 198.

29 *"another soul to enter"* Parker, *Recollections,* 375.

29 *"The scenes at the depot were a harbinger"* Parker, *Recollections,* 376.

30 *"By nightfall all the flitting shadows"* Harrison, *Recollections,* 208.

30 *"ominous groups of ruffians"* Sulivane, "The Fall of Richmond," 725.

30 *"About 2 o'clock on the morning"* Thomas Thatcher Graves, "The Fall of Richmond: The Occupation," in *Battles and Leaders of the Civil War,* eds. Robert Underwood Johnson and Clarence Buel, 4:726.

30 *"I saw a government on wheels"* John S. Wise, *The End of an Era* (Boston: Houghton, Mifflin & Co., 1899), 415.

31 *"By daylight, on the 3d"* Sulivane, "The Fall of Richmond," 726.

32 *"As we neared the city the fires"* Graves, "The Fall of Richmond," 726–27.

32 *"I looked over at the President's house"* Harrison, *Recollections,* 214.

32 *"A young woman has just passed"* Harrison, *Recollections,* 218.

33 *"His hope and good humor"* Mallory, "Last Days," 104.

33 *"adroit economy"* Mallory, "Last Days," 105.

33 *"They were quiet"* Mallory, "Last Days," 105.

33 *"An earnest, enthusiastic, big-hearted man"* Mallory, "Last Days," 105.

33 *"As the morning advanced"* Mallory, "Last Days," 105.

34 *"This morning Gen. Grant reports"* Lincoln, *Collected Works,* 8:384.

35 *"The news spread like wildfire through Washington"* Noah Brooks, *Washington in Lincoln's Time* (New York: The Century Co., 1895; New York: Georgia University Press, 1971), 218. Citations are to the Georgia University Press edition.

36 *"I congratulate you"* Lincoln, *Collected Works,* 8:384.

37 *"Heads of departments"* Mallory, "Last Days," 103.

38 *"Yours received. Thanks for your caution"* Lincoln, *Collected Works,* 8:385.

39 *"The day of jubilee did not end"* Brooks, *Washington in Lincoln's Time,* 221.

39 *"The ending of the first day of occupation"* Harrison, *Recollections,* 214.

40 *"more or less"* Mallory, "Last Days," 105.

41 *"April 4 and the succeeding four days"* Mallory, "Last Days," 105.

41 *"Some asserted, upon the faith"* Mallory, "Last Days," 105.

41 *"Thus were passed five days"* Mallory, "Last Days," 106.

2: "IN THE DAYS OF OUR YOUTH"

42 *"Thank God that I have lived"* David Dixon Porter, *Incidents and Anecdotes of the Civil War* (New York: D. Appleton and Company, 1885), 294.

42 *"Here we were in a solitary boat"* Porter, *Incidents,* 294.

44 *"Admiral, this brings to mind"* Porter, *Incidents,* 294.

44 *"The street along the river-front"* Porter, *Incidents,* 295.

44 *"There was a small house"* Porter, *Incidents,* 295.

45 *"Don't kneel to me"* Porter, *Incidents,* 295.

45 *"Oh, all ye people clap"* Porter, *Incidents,* 296.

45 *"The crowd immediately became"* Porter, *Incidents,* 297.

46 *"My poor friends, you are free"* Porter, *Incidents,* 298.

46 *"We will pull it down!"* Porter, *Incidents,* 298.

47 *"Is it far"* Graves, "The Fall of Richmond," 726.

47 *"At the Davis House"* Graves, "The Fall of Richmond," 726.

47 *"look of unutterable weariness"* Michael Burlingame, *Abraham Lincoln: A Life* (Baltimore, Md.: Johns Hopkins University Press, 2008), 2:790.

51 *"Abraham Lincoln, his hand and pen"* Lincoln, *Collected Works,* 1:1.

54 *"If slavery is not wrong"* Lincoln, *Collected Works,* 7:282.

57 *"Ann M. Rutledge is"* The copy of *Kirkham's Grammar* inscribed by Lincoln is in the collection of the Library of Congress.

62 *"At length he asked me"* Graves, "The Fall of Richmond," 728.

62 *"quite a small affair compared with"* Porter, *Incidents,* 302.

63 *"news and book agents"* Carte de visite in author's collection.

63 *"President Lincoln replied"* Graves, "The Fall of Richmond," 728.

64 *"Don't drown, Massa Abe"* Burlingame, *Lincoln,* 2:792.

64 *"I walked alone on the street"* Burlingame, *Lincoln,* 2:792.

3: "UNCONQUERABLE HEARTS"

66 *"The baggage cars"* Crist, *Papers,* 11:515.

67 *"Please give me"* Crist, *Papers,* 11:501. Official Records of the War of the Rebellion I, 47, part 3. Hereafter cited as OR.

67 *"Selma has fallen"* Crist, *Papers,* 11:502.

68 *"My Brigade was lost"* Crist, *Papers,* 11:502.

68 *"To the people of the Confederacy"* Jefferson Davis, *Jefferson Davis, Constitutionalist: His Letters, Papers, and Speeches,* ed. Dunbar Rowland (Jackson, MS: Mississippi Department of Archives and History, 1923), 6:529–31. Cited hereafter as *Jefferson Davis.*

71 *"My Dear Wife"* Crist, *Papers,* 11:504.

73 *"We need your personal sanction"* Lincoln, *Collected Works,* 8:387.

73 *"In my letter of yesterday I gave you all of my prospects"* Crist, *Papers,* 11:510.

73 *"I took a long walk"* Eliza Frances Andrews, *The War-time Journal of a Georgia Girl, 1864–1865* (New York: D. Appleton & Co., 1908), 135. Cited hereafter as *War-time Journal.*

74 *"I shall be tonight at Farmville"* Lee, *Wartime Papers,* 931.

74 *"The news of Richmond came upon me"* Crist, *Papers,* 11:514.

75 *"Lieut. Gen. Grant"* Lincoln, *Collected Works,* 8:392.

77 *"The dispute made it all the way"* Lincoln, *Collected Works,* 8:405–6.

77 *"Your dispatch of the 6th"* Crist, *Papers,* 11:526.

79 *"Most people were sleeping soundly"* Brooks, *Washington in Lincoln's Time,* 223.

79 *"Guns are firing"* Gideon Welles, *Diary of Gideon Welles* (Boston: Houghton Mifflin, 1911), 2:278.

81 *"After four years of arduous service"* Lee, *Wartime Papers,* 935.

81 *"fell upon the ears of all"* Mallory, "Last Days," 107.

83 *"We set to work at once"* Burton N. Harrison, "The Capture of Jefferson Davis," *Century Illustrated Monthly Magazine* 27 (Nove. 1883), 131.

83 *"To Mayor J. M. Walker"* Rowland, *Jefferson Davis,* 6:543.

84 *"Of course, recalled Harrison"* Harrison, "Capture," 131.

85 *"Much rain had fallen"* Mallory, "Last Days," 107.

86 *"I remarked on the freshness"* Harrison, "Capture," 131.

86 *"At ten o'clock, Cabinet officers and other chiefs"* Mallory, "Last Days," 107.

86 *"That young lady"* Harrison, "Capture," 132.
87 *"A sharp explosion"* Harrison, "Capture," 132.
88 *"No provision had been made"* Mallory, "Last Days," 107.
88 *"[The owners] of the house"* Harrison, "Capture," 132.
88 *"The people in that part of North Carolina"* Harrison, "Capture," 132.
89 *"Its distinguished hosts"* Mallory, "Last Days," 107.
90 *"The Secty. Of War did not join me at Danville"* Crist, *Papers,* 11:531.
91 *"Mr. President: It is with pain that I announce"* Lee, *Wartime Papers,* 935–36.
91 *"how vast our resources"* Crist, *Papers,* 11:540.
92 *"After reading it, he handed it without comment"* Crist, *Papers,* 11:542, note 2.
92 *"The rumors of a raid on Charlotte"* Crist, *Papers,* 11:540–41.
93 *"The Capitol made a magnificent display"* Benjamin Brown French, *Witness to the Young Republic: A Yankee's Journal, 1828–1870,* eds. Donald B. Cole and John McDonough (Hanover: University Press of New England, 1989), 468.
93 *"Everything was bright and splendid"* John Rhodehamel and Louise Taper, eds., *"Right or Wrong, God Judge Me": The Writings of John Wilkes Booth* (Urbana: University of Illinois Press, 1997), 144.
94 *"Dear Winnie I will come to you if I can"* Crist, *Papers,* 11:541.
94 *"I thank you for the assurance"* Lincoln, *Collected Works,* 8:413.

4: "BORNE BY LOVING HANDS"

100 *"A man running down 10th Street"* Seaton Munroe, "Recollections of Lincoln's Assassination," *North American Review* (Apr. 1896), 424.
101 *"Finding it impossible to go further"* W. Emerson Reck, *A. Lincoln: His Last 24 Hours* (Jefferson, NC: McFarland, 1987), 128.
101 *"I was at Grover's"* Transcript in the collection of the Surratt Society, James O. Hall Research Center.
102 *"We were about getting into bed"* Ralph G. Newman, "The Mystery Occupant's Eyewitness Account of the Death of Abraham Lincoln," *Chicago History* (Spring 1975).
103 *"When we arrived to the street"* Charles A. Leale, *Address Delivered Before the Commandery of the State of New York, Military Order of the Loyal Legion of the United States* (New York: self-published, 1909).
103 *"Where can we take him?"* Reck, *A. Lincoln,* 128.

103 *"I saw a man"* Leale, *Address Delivered.*

103 *"They carried him on out"* Newman, "Mystery Occupant's."

104 *"My balcony being twelve or fourteen feet above"* Reck, *A. Lincoln,* 129.

105 *"Take us to your best room"* Reck, *A. Lincoln,* 128.

105 *"When the president was first laid in bed"* Leale, *Address Delivered.*

106 *"She was perfectly frantic."* Newman, "Mystery Occupant's."

106 *"I went to Mrs. Lincoln"* Leale, *Address Delivered.*

108 *"The first person I met"* Ralph Borreson, *When Lincoln Died* (New York: Appleton-Century, 1965).

109 *"about twenty-five minutes"* Charles Sabin Taft, "Abraham Lincoln's Last Hours," *Century Magazine* 45 (February 1893), 634–36. Charles Sabin Taft, *Abraham Lincoln's Last Hours: From the Note-book of Charles Sabin Taft, M.D., an Army Surgeon Present at the Assassination, Death, and Autopsy* (Chicago: Blackcat Press, 1934).

109 *"I was introduced to Dr. Stone"* Leale, *Address Delivered.*

109 *"It was owing to Dr. Leale's quick judgement"* Charles Sabin Taft, "Abraham Lincoln's Last Hours."

109 *"At about 11p.m. the right eye"* Leale, *Address Delivered.*

111 *"The giant sufferer"* Borreson, *When Lincoln Died,* 41.

112 *"he had last night the usual dream"* Welles, *Diary,* 2:282.

113 *"On a common bedstead"* Transcript in the collection of the Surratt Society.

114 *"Mr. Lincoln is assassinated in the theater"* Storey and Sumner bio.

115 *"Charles Sumner, they have murdered my husband"* Jeremiah Chaplin, J. D. Chaplin, and William Claflin, *The Life of Charles Sumner* (Boston: D. Lothrop & Co., 1874), 417.

116 *"A stroke from Heaven"* Reck, *A. Lincoln,* 138.

117 *"The Hospital Steward"* Leale, *Address Delivered.*

117 *"Mrs. Lincoln accompanied by"* Leale, *Address Delivered.*

117 *"She implored him"* Reck, *A. Lincoln,* 139.

118 *"I awoke and saw that the streetlamps"* French, *Witness,* 469.

118 *"As morning dawned"* Leale, *Address Delivered.*

119 *"Are not the doings of last night dreadful"* French, *Witness,* 469–70.

119 *"I took Mrs. Lincoln by the hand"* French, *Witness,* 470.

119 *"Mrs. Welles was not up"* French, *Witness,* 470.

120 *"Her last visit was most painful"* Reck, *A. Lincoln,* 148.

120 *"pierced every heart"* Reck, *A. Lincoln,* 147.

120 *"Just as the day was struggling"* Transcript in the collection of the Surratt Society.

120 *"As Mrs. Lincoln sat on a chair"* Reck, *A. Lincoln,* 148.

122 *"It was evident to every dispassionate mind"* Stephen R. Mallory, "Last Days of the Confederate Government . . . Last Cabinet Conferences and Negotiations for Johnston's Surrender," *McClure's Magazine*, vol. xvi, December 1901, 242. Hereafter cited as "Last Days," part 2.

122 *"Heavy rains had recently fallen"* Harrison, "Capture," 134.

124 *"immediately after death"* Charles Sabin Taft, "Abraham Lincoln's Last Hours," 636. Charles Sabin Taft, *Abraham Lincoln's Last Hours.*

125 *"1 horsehair covered sofa"* George Olszewski, *House Where Lincoln Died: Furnishing Study* (Washington, D.C.: National Park Service, 1967), 42.

126 *"I stepped to the window and saw the coffin"* Borreson, *When Lincoln Died,* 46

127 *"A dismal rain was falling"* Charles Sabin Taft, "Abraham Lincoln's Last Hours," 636. Charles Sabin Taft, *Abraham Lincoln's Last Hours.*

127 *"Wandering aimlessly up F Street"* Brooks, *Washington in Lincoln's Time,* 231–32.

130 *"Even then I could fancy the relic hunter"* Munroe, "Recollections," 433–34.

130 *"I joined Mr. Petersen's son"* Ferguson.

130 *"make it the center and outstanding part of the large painting"* Reck, *A. Lincoln,* 129.

5: "THE BODY OF THE PRESIDENT EMBALMED!"

131 *"At nine o'clock we took her home"* Surratt Society.

132 *"As I started to go down the stairs"* Borreson, *When Lincoln Died,* 53.

133 *"I went immediately to the room"* French, *Witness,* 470.

134 *"The room . . . contained but little furniture"* Edward Curtis, "Was at the Lincoln Autopsy . . . Dr. Edward Curtis Describes the Scene at the White House—Lincoln's Brain and Physique—Finding of the Bullet—An Account Not Before Printed," *Sun,* April 12, 1903, sec. 4, 4. Cited hereafter as "Lincoln Autopsy."

135 *"Dr. Woodward and I proceeded to open the head"* Curtis, "Lincoln Autopsy."

135 *"During the post-mortem examination"* Charles Sabin Taft, "Abraham

Lincoln's Last Hours," 636. Charles Sabin Taft, *Abraham Lincoln's Last Hours.*

136 *"Silently, in one corner of the room, I prepared the brain"* Curtis, "Lincoln Autopsy."

141 *"On reporting to the President"* Edward D. Townsend, *Anecdotes of the Civil War in the United States* (New York: D. Appleton and Company, 1884), 77.

144 *"[Mrs. Lincoln] was nearly exhausted with grief"* Elizabeth Keckly, *Behind the Scenes, or, Thirty Years a Slave, and Four Years in the White House* (New York: G. W. Carleton & Co., 1868), 189.

144 *"I saw the remains of the President"* French, *Witness,* 471.

144 *"We stood together"* French, *Witness,* 471.

145 *"was looking as natural as life"* Orville Hickman Browning, *The Diary of Orville Hickman Browning,* eds. Theodore Calvin Pease and James G. Randall (Springfield: Illinois State Historical Library, 1925–1933). 2:22.

145 *"It tells a long story of duns and loiterers"* George Alfred Townsend, *The Life, Crime, and Capture of John Wilkes Booth* (New York: Dick and Fitzgerald, 1865), 59.

148 *"Proposed arrangements for the Funeral and disposition of the Remains" Papers of George Harrington,* Missouri Historical Society, St. Louis, Missouri. To avoid repetition, all documents drafted by Harrington, and all correspondence to and from him, come from the same source, his personal papers.

151 *"We agreed . . . to return at 7 to meet"* French, *Witness,* 472.

6: "WE SHALL SEE AND KNOW OUR FRIENDS IN HEAVEN"

160 *"On the night of the seventeenth"* Brooks, *Washington in Lincoln's Time,* 232.

163 *"Well . . . it is only a dream"* Ward Hill Lamon, *Recollections of Abraham Lincoln, 1847–1865, by Ward Hill Lamon,* ed. Dorothy Lamon (Chicago: A. C. McClurg and Company, 1895), 112–14.

164 *" 'Excuse me,' he said, 'but I cannot talk' "* Burlingame, *Lincoln,* 2:177. Burlingame covers the death of Ellsworth in detail.

165 *"My dear Sir and Madam, In the untimely loss"* Lincoln, *Collected Works,* 4:385.

170 *"Well, Nicolay, my boy is gone"* Burlingame, *Lincoln,* 2:298

170 *"When you came to the door here"* Burlingame, *Lincoln,* 2:299.
174 *"He comes to me every night"* Katherine Helm, *The True Story of Mary, Wife of Lincoln* (New York: Harper and Brothers, 1928), 227.
177 *"Dear Fanny"* Lincoln, *Collected Works,* 6:17
181 *"I found one of the sleeves of his shirt"* Surratt Society.
182 *"We discussed . . . whether"* Sherman, *Memoirs,* 2:351.
182 *"Our necessities exclude the idea"* Rowland, *Jefferson Davis,* 6:549.
183 *"During all this march Mr. Davis was singularly equable"* Harrison, "Capture," 136.
186 *"Approach and look at the dead man"* Townsend, *John Wilkes Booth,* 14.
189 *"I am the resurrection and the life"* William Turner Coggeshall, *The Journeys of Abraham Lincoln: From Springfield to Washington, 1861, as President Elect and From Washington to Springfield, 1865, as President Martyred* (Columbus: Ohio State Journal, 1865), 119. Cited hereafter as *Journeys.*
193 *"The cortege passed to the left"* Townsend, *John Wilkes Booth,* 18.

7: "THE CAUSE IS NOT YET DEAD"

194 *"there seemed to be nothing to do"* Harrison, "Capture," 136.
195 *"My friends, I thank you for this evidence of your affection"* Crist, *Papers,* 11:549.
196 *"President Lincoln was assassinated"* Crist, *Papers,* 11:544.
196 *"At Charlotte . . . we received the melancholy news"* Reagan, *Memoirs,* 208.
196 *"conviction of Mr. Lincoln's"* Mallory, "Last Days," part 2, 244.
197 *"fearful news"* Crist, *Papers,* 11:544–46.
197 *"Give me a good force of cavalry"* OR, 47, III, 813–14.
198 *"At night the jets of gas"* Townsend, *John Wilkes Booth,* 18.
199 *"We saw him the last time"* Newman, "Mystery Occupant's."
204 *"Genl. Breckinridge . . . telegraphs to me"* Crist, *Papers,* 11:551.
205 *"Train will start for you at midnight"* OR, 47, III, 814.
205 *"Mr. President: The apprehension I expressed during the winter"* Lee, *Wartime Papers,* 938.
206 *"No other course now seemed open"* Mallory, "Last Days," part 2, 246.
209 *"There was never a moment throughout the whole journey"* Townsend, *Anecdotes,* 224.
212 *"Paroled men and stragglers seized my train"* OR, 47, III, 819.

215 *"The body of this hearse"* Coggeshall, *Journeys,* 144.

215 *"were tastefully arranged evergreens"* Townsend, *Anecdotes,* 224–25.

216 *"No bearers, except the veteran guard"* Townsend, *Anecdotes,* 224.

218 *"A driving rain and the darkness of the evening"* Townsend, *Anecdotes,* 225.

219 *"If you should propose to cross"* OR, 47, III, 829.

219 *"[I] wait for suggestions or directions"* Crist, *Papers,* 11:556, note.

220 *"No mere love of excitement"* Coggeshall, *Journeys,* 149.

220 *"the Square was brilliantly illuminated"* Coggeshall, *Journeys,* 152.

221 *"I have never had a feeling politically"* Lincoln, *Collected Works,* 4:241.

222 *"On the old Independence bell"* Coggeshall, *Journeys,* 153.

223 *"After a person was in line"* Coggeshall, *Journeys,* 156.

224 *"My Dear Winnie / I have asked Mr. Harrison to go in search of you"* Crist, *Papers,* 11:557–60.

227 *"it was fourteen feet long"* Coggeshall, *Journeys,* 172.

228 *"The police, by strenuous exertions"* Coggeshall, *Journeys,* 163.

228 *"The world never witnessed"* Coggeshall, *Journeys,* 181.

228 *"There was no trace of the interior"* Coggeshall, *Journeys,* 164.

229 *"The deportment of the people"* Coggeshall, *Journeys,* 167.

229 *"Captain Parker Snow"* Coggeshall, *Journeys,* 169.

229 *"With practiced fingers"* Coggeshall, *Journeys,* 169–70.

231 *"As a mere pageant"* Coggeshall, *Journeys,* 198.

8: "HE IS NAMED FOR YOU"

232 *"The line of the Hudson River road"* Townsend, *Anecdotes,* 233.

234 The dispute between Townsend and Stanton over the photographing of Lincoln's corpse appears in the Official Records of the War of the Rebellion, 1, 46, 111, at pages 952–67.

238 *"His friends . . . saw the urgent"* Mallory, "Last Days," part 2,246.

238 *"If you think it better"* OR, 47, III, 841.

238 *"There is increasing hazard of desertion"* Crist, *Papers,* 11:566.

241 *"The ladies . . . through"* Coggeshall, *Journeys,* 205.

242 *"The last tribute"* Coggeshall, *Journeys,* 206.

243 *"[T]wo days after"* Mallory, "Last Days," part 2, 246.

243 *"By your advice"* OR, 47, III, 846.

246 *"As the President's remains went farther westward"* Townsend, *Anecdotes,* 235.

246 *"a rare privilege to kiss the coffin"* Coggeshall, *Journeys,* 208.

247 *"You have confidence in yourself"* Lincoln, *Collected Works,* 6:79 .
248 *"After we had joked"* Reagan, *Memoirs,* 210.
248 *"The President of the Confederacy cannot"* Reagan, *Memoirs,* 211.
248 *"his unselfish and patriotic devotion"* Reagan, *Memoirs,* 211.
249 *"Miss Fields, of Wilson Street"* Coggeshall, *Journeys,* 218.
250 *"To a gentleman, a stranger"* Townsend, *Anecdotes,* 236.
250 *"It is surely not the fate"* Crist, *Papers,* 11:569.
251 *"On our way to Abbeville"* Reagan, *Memoirs,* 210.
252 *"dripping like tears on the remains"* Coggeshall, *Journeys,* 219.
253 *"Bonfires and torches were lit"* Coggeshall, *Journeys,* 219.
256 *"The white people seemed to be doing all they could"* Crist, *Papers,* vol. 11, n. 12.
258 *"But he was slain—slain by slavery"* Coggeshall, *Journeys,* 251.
261 *"At midnight the route"* Townsend, *Anecdotes,* 237.
261 *"A succession of arches"* Townsend, *Anecdotes,* 237.
263 *"who got a fresh scab from the arm of a little negro"* Harrison, "Capture," 138.
264 *"A magnificent arch spanned the street"* Townsend, *Anecdotes,* 238.
265 *"nearly every building on Michigan Avenue"* Townsend, *Anecdotes,* 238.
266 *"Captain, I am very sorry to hear that"* Parker, *Recollections,* 391.
266 *"Mr. President, if you remain here you will be captured"* Parker, *Recollections,* 391.
267 *"We witnessed . . . the raids made on the provisions"* Reagan, *Memoirs,* 211.
267 *"When we reached Abbeville"* Crist, *Papers,* vol. 11, n. 7. Reagan, *Memoirs,* 211.
267 *"The escort was here collected"* Mallory, "Last Days," part 2, 246.

9: "COFFIN THAT SLOWLY PASSES"

269 *"Do not try to meet me"* Crist, *Papers,* 11:576.
271 *"As usual, night was forgotten"* Townsend, *Anecdotes,* 239.
273 *"The courier has just delivered yours and I hasten to reply"* Crist, *Papers,* 11:580.
274 *"Thus closed this marvelous exhibition"* Townsend, *Anecdotes,* 242.
276 *"The guard of honor having thus"* Townsend, *Anecdotes,* 242.
277 *"My friends, no one, not in my situation"* Lincoln, *Collected Works,* 4:190.
278 *"[Breckinridge] told me that after he reached Washington"* Reagan, *Memoirs,* 214.

279 *"I inquired where he was going"* Reagan, *Memoirs,* 211.

279 *"We found no federal cavalry"* Reagan, *Memoirs,* 211.

280 *"About noon the town was thrown into the wildest excitement"* Andrews, *Journal,* 175, 181, 189, 190, 201, 206, 212.

280 *"The troops are on the west side"* Crist, *Papers,* 11:583.

283 *"Far more eyes have gazed upon the face"* Coggeshall, *Journeys,* 308.

283 *"Standing, as we do today, by his coffin"* Coggeshall, *Journeys,* 319.

284 *"Evergreen carpeted the stone floor"* Carl Sandburg, *Abraham Lincoln: The War Years,* 4:413.

288 *"I am in such a state of excitement"* Andrews, *Journal,* 204–6.

288 *"It is with deep regret"* Crist, *Papers,* 11:584. Rowland, *Jefferson Davis,* 6:586–87.

288 *"After some delay at Washington"* Reagan, *Memoirs,* 212.

289 *"The President left town about ten o'clock"* Andrews, *Journal,* 206.

289 *"The talk now is"* Andrews, *Journal,* 217.

289 *"This, I suppose, is the end"* Andrews, *Journal,* 217.

289 *"Twenty days after the terrible night"* Coggeshall, *Journeys,* 325.

294 *"Mr. Lincoln, on his way from Springfield to Washington"* Townsend, *Anecdotes,* 243.

301 *"Fully realizing that so large a party"* Lubbock, *Six Decades,* 571.

302 *"we halted on a small stream near Irwinville"* Lubbock, *Six Decades,* 571.

302 *"We had all now agreed"* Harrison, "Capture," 142.

302 *"The President notified us to be ready"* Reagan, *Memoirs,* 218.

303 *"After getting that promise from the President"* Harrison, "Capture," 142.

303 "Time wore on" Lubbock, *Six Decades,* 571.

10: "BY GOD, YOU ARE THE MEN WE ARE LOOKING FOR"

304 *"From thence we proceeded to a blind woods"* OR, 49, I, 532.

305 *"Impressing a negro as a guide"* OR, 49, I, 532.

305 *"[J]ust as the earliest dawn appeared"* OR, 49, I, 536.

307 *"Colonel, do you hear the firing?"* Harrison, "Capture," 142.

307 *"As soon as one of them came within range"* Harrison, "Capture," 142.

307 *"At this moment"* W. T. Walthall, "The True Story of the Capture of Jefferson Davis," *Southern Historical Society Papers* 5, no. 3 (Mar. 1878).

307 *"We sprang immediately to our feet"* Lubbock, *Six Decades,* 571.

308 *"When this firing occurred the troops in our front"* Reagan, *Memoirs,* 219.

308 *"What does that mean? Have you any men"* Harrison, "Capture," 142.

309 *"The Federal cavalry are upon us"* Reagan, *Memoirs,* 220.

310 *"Knowing he would be recognized"* Chester Bradley, "Was Jefferson Davis Disguised as a Woman When Captured?" *Journal of Mississippi History* 36 (Aug. 1974), 243–68.

311 *"As I started, my wife thoughtfully threw over my head"* Varina Davis, *A Memoir,* 2:701–2.

311 *"in a short time they were in possession of very nearly everything"* Lubbock, *Six Decades,* 572.

312 *"I emptied the contents of my haversack"* Harrison, "Capture," 144.

312 *"This is a bad business"* Walthall, "True Story," 14.

314 *"The hardest to bear of all the humiliations"* Andrews, *Journal,* 238.

315 *"As soon as the firing ceased I returned to camp"* OR, 49, I, 536.

316 *"I had been astonished to discover"* Harrison, "Capture," 144.

316 *"The man who a few days before"* Lubbock, *Six Decades,* 572.

316 *"[S]he bore up with womanly fortitude"* Lubbock, *Six Decades,* 573.

320 *"We have not got your saddle bags"* Reagan, *Memoirs,* 221.

322 *"When we reached Macon"* Reagan, *Memoirs,* 221.

323 *"As one of the means of making the Confederate cause odious"* Reagan, *Memoirs,* 221.

323 *"When I came up from breakfast"* French, *Witness,* 477.

324 *"Intelligence was received this morning"* Welles, *Diary,* 2:306.

324 *"I am sitting in the President's Office"* Townsend, *John Wilkes Booth,* 57–58.

326 *"I am glad to sit in his chair"* Townsend, *John Wilkes Booth,* 62.

328 *"Barnum is a shrewd businessman"* Strong, *Diary,* 3:598.

330 *"ample provision being made for the families"* OR, 49, I, 516.

11: "LIVING IN A TOMB"

333 *"They have him in his prison house"* Lincoln, *Collected Works,* 2:403–7.

335 *"Mrs. Mary Lincoln left the City on Monday evening"* French, *Witness,* 479.

336 *"[T]he great review of the returning armies"* Welles, *Diary,* 2:310.

337 *"I put a gilded eagle over the front door"* French, *Witness,* 478.

346 *"I hate the Yankees more and more"* Andrews, *Journal,* 371.

349 *"I am now permitted to write you"* Crist, *Papers,* 12:13.

349 *"With regret and apprehension I have heard"* Crist, *Papers,* 12:44.

351 *"Last Christmas we had a home"* Crist, *Papers,* 12:80.

351 *"I hope that you will not think me a rude little girl"* Crist, *Papers,* 12:114.

352 *"It is true that I have not made [Jefferson Davis]"* OR, 914.

354 *"The prison life by Dr. Craven"* Crist, *Papers,* 12:153.

355 *"I urged that the welfare of the whole country"* Reagan, *Memoirs,* 231.

356 *"poor Davis . . . wasted and careworn"* Crist, *Papers,* 12:210.

357 *"Last Friday [June 1], Hollywood was glorified with flowers"* Crist, *Papers,* 12:214.

12: "THE SHADOW OF THE CONFEDERACY"

365 *"I did not like the man"* Strode, *Tragic Years,* 459–62.

366 *"I have been compelled to prove General Sherman"* Strode, *Tragic Years,* 473.

374 *"Mr. Chairman and Fellow Citizens"* Rowland, *Jefferson Davis,* 10:47.

375 "Permit me to cordially congratulate you" Rowland, *Jefferson Davis,* 10:72.

376 *"The package containing all of our correspondence"* Crist, *Papers,* 1:348.

377 *"Dreams my dear Sarah we will agree"* Crist, *Papers,* 1:345.

377 *"The shadow of the Confederacy"* Varina Howell Davis to Constance Cary Harrison, transcript in the collection of the author, courtesy of the *Papers of Jefferson Davis.*

INDEX

Page numbers in *italics* refer to illustrations.

Abbeville, Ga., 298, 299, 305, 330
Abbeville, S.C., 219, 250–51, 262–63, 266–67, 273–74, 280
abolitionists, xi, 53, 356
Abraham Lincoln Presidential Library and Museum, 392–93
Adams, John, 53
Adams, John Quincy, 167, 180
Agriculture Department, U.S., Farm Security Administration of, 399–400
Aiken's Landing, 17
Alabama, 77, 366, 377, 383
Albany, N.Y., 199, 234, 235, 237, 241
Alexander, Dr., 138, 186, 286
Alexander, John, 285
Alexandria, Va., 163–65, 192, 208, 331, 382
American Museum, 345
American Revolution, 220–21
Anderson, Finley, 126
Anderson, Joseph Reid, 62
Anderson, S.C., 245
Andersonville Prison, 344, 350
Andrews, Eliza Frances, 73–74, 280, 288–89, 313–14, 346–47
Andrews, Garnett, 73, 288
Annapolis, Md., 214
Antietam, battle of, 112
Antietam, Md., 16, 156
Appomattox Court House, Va., 77–78, 82, 94, 147, 206, 361
Appomattox River, 6, 21–22
Arkansas, 123
Arlington National Cemetery, 396
Army, U.S., 53, 102, 140–41, 148, 192, 237, 354
  Company D of the Seventy-fourth Regiment of, 242
  First Wisconsin Cavalry of, 256, 298, 305–6, 313–14, 330
  Fourth Michigan Cavalry of, 256, 298, 304–7, 310, 313–16, 321, 322, 330
  frontier wars of, 361
  Lincoln's assassination and, 108
  Lincoln's funeral and, 150
  Northern Department of Ohio in, 247
  Second Cavalry Division, 297
  Sixteenth New York Cavalry of, 314
  Twelfth Veteran Reserve Corps of, 208
  see also Union Army

Army of Georgia (Union), 336
Army of Northern Virginia (Confederate), 5, 13–14, 24, 26, 41, 67, 69, 74, 76
  surrender of, 78, 79–80, 81–82, 91, 95, 102, 195, 197, 205–6, 224
Army of Tennessee (Confederate), 67
Army of the Potomac (Union), 176, 329–30, 336, 337
Army of the Tennessee (Union), 336
Army of the West (Union), 329–30
Arnold, Isaac N., 151
Asheville, N.C., 245
Ashmun, George, 158–59
Associated Press, 151
Atlanta, Ga., 3, 76, 279, 323, 371, 372, 383
Atlanta Constitution, 367, 370–71
Auburn, N.Y., 391
Augur, C. C., 116, 126, 150–51
Augusta, Ga., 245, 262

Bahamas, 122, 273
Baker, C., 126
Baker, Edward D., 167–68, 180
Ball's Bluff, battle of, 167
Baltimore, Md., 98, 199, 203, 212–17, 218, 239, 346
Baltimore and Ohio Railroad, 208
Barnes, Joseph K., 108–9, 117, 133–34, 188, 354
Barnum, P. T., 327–29, 344–45
Barringer, Victor C., 122
Bates, Edward, 170
Bates, Lewis F., 194–97
Beauregard, P. G. T., 67, 87, 204
Beauvoir, 362–64, *362*, 366–67, 371, 373, 375–76, 377, 382, *403*
  Hurricane Katrina and, 402–3
Bedell, Grace, 246
Beecher, Henry Ward, 239–40
Ben-Hur (Wallace), 240
Benjamin, Judah, 8, 32–33, 41, 86, 247, 273, 274
  departure of, 278–79, 317
Benton, Thomas Hart, 52
Bersch, Carl, 104–5, 130
Bible, xii, 52, 60, 292
Biloxi, Miss., 362
Black Hawk, 58
Black Hawk War, 57–58
blacks:
  in captured Richmond, 26, 39, 44–46, 64
  Davis's location reported to Union troops by, 256, 302
  Davis's views on, 53–54, 60–61
  liberation of, 26, 32, 44–47
  Lincoln honored by, 44–46, 64, 158, 217, 261, 340
  Lincoln's views on, 53–54, 61
  New York draft riots and lynching of, 34
  racism directed towards, 398
  Reconstruction and, 361
  voting rights for, 90
  white supremacy and, 53–54, 60–61
Blair, Montgomery, 170
Bonham, Milledge L., 94
Booth, John Wilkes, xi, 34, 93, 96–97, 129, 130, 367
  allegorical lithograph of, 136, *137*
  capture and death of, 238, 241, 243, 244, 287, 314, 317, 342
  Confederate connection of, 121
  conspirators of, 147, 196, 214, 215, 293
  images of, 240
  Lincoln's assassination by, 100–101, 106–7, 110, 136, 192, 276, 325
  manhunt for, 111, 147, 204, 244, 268, 296–98, 317, 400
  tried and executed co-conspirators of, 319, 340–41, 342–44, *343*, 347, 356
  wax figures of, 344, 392–93
Boston, Mass., 99
Boutell, Henry, 306
Boyd, Andrew, 125
Boyd, William H., 125
Bradford, A. W., 214, 217
Bradford, David, 72
Brady, Mathew, 189, 192, 205, 239
Bragg, Braxton, 14, 204
Breckinridge, John C., 76, 196, 204–5, 211–12, 247, 278–82, 288, 317
  1860 presidential candidacy of, 53
  Richmond evacuation and, 5–6, 9, 23–24, 26–27, 41, 72
Brierfield, 72, 248, 377

Broad River, S.C., 247, 250
Brooks, Daniel, 265
Brooks, Noah, 35, 39, 79, 127, 160, 166
Brooks, Preston, 115
Brooks Brothers, 139
Brough, John, 247
Brown, Charles, 138, 186, 281–82, 286
Brown, John, xi-xii, xiii, 356
Brown, Simeon B., 245
Browning, Orville Hickman, 133, 145, 153, 171, 200
Brown's Ferry, 298
Buchanan, James, 6, 220
Buena Vista, battle of, 52, 309
Buffalo, N.Y., 199, 241–43
Bull Run, battle of, 112
Bureau of Military Justice, 296, 320
Burke, Francis, 95, 114
Burnside, Ambrose, 176
Burr, Frank A., 367, 371
Burt, Armistead, 266
Bush, George W., 400
Bush, Laura, 400

Cadwalader, George, 217
California, 167, 309
Cameron, Simon, 141, 158–59
Campbell, Givhan, 301
Campbell, John Archibald, 62
Canada, 243
Capitol, U.S., 49, 93, 119, 133, 143, 152, 336–37, 350
  Davis statue in, 400, 401
  funeral procession to, 149–50, 158, 190–93, 209, 286, 330, 382, 386, 400
  Lincoln's lying in state at, 140, 148, 150, 185, 197–200, 207–9, 215, 286, 391
Carolina Life Insurance Company, 359
Carroll, William T., family vault of, 171, 173
Cary, Constance, 5, 6–7, 32, 39–40, 312
  marriage to Burton Harrison by, 6, 377, 390
Cattell, Harry P., 138
Chamberlain, Joshua Lawrence, 336–37
Chancellorsville, battle of, 247, 341
Chapin, C. W., 187
Charleston, S.C., 3, 182
Charlotte, N.C., 181, 243, 245, 256–57, 296
  Davis in, 198, 204–5, 206, 212, 219, 238
  Davis unwelcome in, 194–96, 223, 252, 280
  Varina in, 10, 15, 66, 74, 83, 92
Chase, Salmon P., 170
Chatham Artillery, 371
Chattahoochee River, 279
Chattanooga, Tenn., 68, 279
Chester, S.C., 92
Chestnut, Mary, 175–76
Chicago, Ill., 84, 136, 143, 145, 152–53, 155, 156, 199, 235, 261, 263–65, *264*, 270–71, 281, 286, 290, 293
Chicago Board of Trade, 265
Chicago Historical Society, 125
Chicago Tribune, 263–64, 265, 270–71, 338
City Point, Va., 16–17, 18–19, 22, 36, 73, 74–76, 210
Civil War, U.S., 1, 14, 36–37, 60, 126, 136, 141, 166–67, 277, 361, 374, 394
  beginning of, 370
  cranial surgery at time of, 110
  Davis's determined continuation of, 200
  death toll in, xiv, 3, 17, 47, 76, 294, 366, 369, 388
  embalming and, 147–48, 235
  ending of, xi
  Lincoln's funeral train as symbol of cost of, 213
  150th anniversary of, 401
Clark, Micajah H., 9, 288
Clarke, William, 105–6, 133, 287
Clay, Clement C., 25, 268–69, 314
Clay, Virginia, 368
Cleary, William, 268–69
Cleveland, Ohio, 199, 247, 248–50, 252–54, *253*
Clover Station, 30
Cobb, Howell, 67–68, 77
Coffin, Charles C., 47
Cokesbury, Ga., 280
Cokesbury, S.C., 265
Cold Harbor, 246
Columbian Expedition of 1893, 125
Columbus, Ga., 67, 77, 257

Columbus, Ohio, 199, 253–55, 258–59, 284
Comanche nation, 123
Confederate archives, 28
Confederate History Month, 401
Confederate Memorial Hall, 382
Confederate ordnance department, 11
Confederate railroad system, 27, 36, 66, 86, 89, 122, 204–5, 211–12
Confederate States of America, xi, 30, 36, 123, 276, 283–84, 288, 364, 380
- armies of, 176, 197, 273, 357
- armies of, after Lee's surrender, 147, 182, 195, 206–7, 279, 280–81, 318, 329
- Booth's connection to, 121
- controversy surrounding, 401–2
- Davis elected president of, 50–51
- Davis's determination to continue, 82–84, 95
- ending of, 288–89
- as Lost Cause, 30, 84, 370, 373, 381, 388, 397, 402, 403
- modern response to racial politics of, 401–2
- Northern perception of, xiv
- postwar policy towards hierarchy of, 17–18
- resilience of, 38
- Richmond as capital of, 3; *see also* Danville, Va.
- waning enthusiasm for, 194

Confederate treasury, 9, 28, 29–30, 212, 266, 274, 278, 289
- and gold rumors, 244–45, 251, 256, 313–15, 347–48

Congressional Cemetery, 167, 180
Connecticut, 187
Constitution, U.S., 48, 49, 143, 154
- slavery and, 53

Cooper, Samuel, 37
Cooper Union, 156, 340
Copperheads, 34, 226
corpse photography, 235
- of Davis, 379, *379*
- of Lincoln, 23–41, *230,* 234–37, 248, 292, 379

Crane, Charles H., 109, 117, 133, 188
cranial surgery, 110, 354
Craven, John J., 341–42, 349–50, 352, 354
Crook, W. H., 16
Crowell, J. H., 126
Croxton, John Thomas, 257
Cuba, 273, 279, 317
Currier & Ives, 99–100, 241
- Richmond burning print of, 39, *40*

Curtin, Andrew, 200–201, 217
Curtis, Joseph, 133–38, 354
Custer, George Armstrong, 361

Dana, Charles A., 239
Dan River, 28
Danville, Va., 15, 219
- evacuation from, 82–84, 89
- evacuation to, 10, 12, 28–30, 33, 383
- as new Confederate capital, 8–9, 36–38, 40–41, 66–68, 73, 76, 77, 81–82, 88, 90, 95
- Richmond refugees in, 38, 41

Danville Railroad, 6, 13
David, King, 292
Davis, Charles Henry, 236, 237
Davis, George, 8
Davis, Jefferson, *4,* 66, 75, 77, 118, 127, 212, *360, 362, 377*
- Abbeville departed by, 273–74
- appearance of, 54–55
- arrival in Charlotte by, 194–96
- arrival in Danville by, 36–38, 383
- attacks on, 7, 60, 62
- cabinet of, 8–9, 27, 81–82, 83, 88–90, 95, 122, 182, 273, 288, 308
- capture of, *310,* 312, 314–18, 320–29, 352, 398–99, *399*
- charisma of, 55
- Charlotte speech of, 195
- children of, 10, 12, 59, 72, 171, 263, 301, 315, 316, 317, 366
- Confederate gold rumor and, 244–45, 251, 256, 314, 347–48, 365
- corpse photograph of, 379, *379*
- correspondence by, 14–15, 22, 71–72, 73, 90, 93, 95, 223–25, 238, 250, 273–74, 288, 348–49, 350
- criticism of, 283–84
- cross-dressing escape rumors about, 310–11, *310,* 322–23, *323,* 324, 326–29, *328,* 344, 346–47, *346*

in Danville evacuation, 82–89
in Danville exile, 40–41, 67–68, 76, 77
"Danville Farewell" of, 83–84
Danville Proclamation of, 68–70, 91
death mask of, 379–80
death pageant for, 378–85, *379, 380, 384, 385,* 397
demonstrations against, 276
determination to continue Confederacy by, 71, 182–83, 199–200, 206–7, 219
eclipse of, 48, 397–98, 400–402
education of, 48, 54, 60
eightieth birthday of, 374–76
fate of, 293
final illness and death of, 377–78, *378*
financial insecurity of, 359, 361–62
at Fort Monroe, 330, 331, 333–35, *334, 335,* 337–40, 341–42, 349–53, 355
funeral train of, *380,* 382–83, 397
in Greensboro, 87–92
Greenville speech of, 91–92
health of, 354, 366, 372–73
honor and principle of, 58, 61–62, 248, 316–17, 322
humor of, 223
inauguration of, 170, 374, 377
indifference to luxuries of, 55
Johnson's reward offer for, 268–69, 313–14
and Johnston's surrender, 238, 365
Joseph's death and, 174–76
journey of, xi, xiii, xiv, 2, 147, 181–83
journey south of, 243–45, 247–48, 250–52, 256–58, 265–67, 268–70, 278–81, 289, 296, 298–303, *300*
as keeper of Confederate memory, 359–77, 388
Kentucky birth of, 51
Lee's surrender and, 81–82, 90–91, 92
as legend, 397
Lincoln's assassination and, 121–22, 181–82, 196–97, 198, 199–200, 204, 219, 223
Lincoln's murder attributed to, 147, 204, 276, 287, 296, 314, 319–20, 326–27, 340–41, 342–44, *343*
marriages of, *see* Davis, Varina Howell; Taylor, Sarah Knox
military background of, 48, 49, 52, 58, 123, 309, 312, 402
military intelligence and, 37, 67
in Montgomery, 366–71
monuments for, 386–87, 397–400, *399*
New Orleans funeral of, 382–83, *384*
news of Union threat to Richmond and, 5–9, 20, 21–24
Northern opinion about, xiv, 340–41
optimism of, 183
in Pierce administration, 49, 53
plantation of, 48–49, 72, 248, 357, 377
post-release conduct of, 357–58, 359–77
presidential prospects of, 48, 50, 53
as president of the Confederacy, 50–51, 369, 380, 397
prison Christmas of, 351
refusal to flee the country by, 122, 182, 266–67, 273, 279
reward for, 268–69, 313–14, 320
in Richmond evacuation, 21–33, 279, 296, 317, 383
Richmond reburial of, 383–86, 390, 397–98
satires and mockery of, *343,* 344–47, *344*
secession and, 50–51, 355, 356, 363
Sherman criticized by, 365–66
slavery and, 49, 60–61, 357, 367, 401
speaking tours of, 371–74, 397
strength of, 54–55
Texas as possible goal for, 197
toughness of, 122–23
Union pursuit of, 296–302, *297,* 304–18, 368
in U.S. Senate, 49–50, 369
Varina reunited with, 298–303, 309–14
in Washington, Ga., 278–80, 288–89
wax figures of, 344–45
Davis, Jefferson, Jr., 12, 175, 299
Davis, Joseph E. (brother), 354, 355
Davis, Joseph Evan (son), 2, 56, 174–76, 357, 385
Davis, Joseph R. (nephew), 68
Davis, Mary, 364
Davis, Samuel Emory, 71, 72, 171, 351, 383–85

Davis, Varina Anne "Winnie," 263, 364,
Davis, Varina Howell, 50, 74, 123, 175, 333, 348, 356, 357, 366, 369, 373
 better treatment for Davis sought by, 352–53
 capture of, 315–17, 322
 in Charlotte, 66, 74, 83
 in Chester, S.C., 92
 correspondence between Davis and, 15, 71–72, 73, 74, 93–94, 95, 197, 219, 223–25, 238, 250–51, 262–63, 269–70, 281, 348–49, 351
 Davis reunited with, 298–303, 309–14
 Davis's courtship of, 58–59, 369
 Davis's death and, 377–78, 379, 382
 Davis's permanent gravesite and, 382, 390
 evacuation from Richmond of, 10–12, 15
 freeing of, 330–31
 life after Davis for, 390
 Wilde and, 364–65
Davis Bend, 56
Declaration of Independence, 54, 221–22
Democratic Party, 50, 53
Dennison, William, 207
Detroit, Mich., 298
District of Columbia, 90, 98, 140
Dix, John Adams, 199, 229, 236–37, 239–40, 246
"Dixie," 84
Dixon, Elizabeth, 108, 113, 117, 119, 120, 131–32
Dixon, James, 108, 119
Dorsey, Sarah, 362
Douglas, Stephen A., 48, 49–50, 53
Douglass, Frederick, 340, 392
Drake, Mrs., 246
Dred Scott decision, 334
Dr. Harker's patent medicine company, 376
Dublin, Ga., 298–99
Duke, Angier Biddle, 396
Dunkirk, N.Y., 246

Early, Jubal, 376, 379
Eckert, Major, 235
Edward H. Jones (engine), 210
Edwards, Elizabeth Todd, 389
Effie Afton case, 156
elections, U.S.:
 of 1860, 50, 51, 53, 98, 246, 292
 of 1864, 76, 292
Ellsworth, Elmer, 163–67, 168, 382
Ellsworth's Zouaves, 163–64, 166–67
Emancipation Proclamation, 261
Ennis, George O., 53
Euclid, Ohio, 247
Europe, 122, 317

Fairfax, Mrs., 381
Ferguson, James, 96
Ferguson, William J., 130
Fessenden, William P., 141–42, 151
Field, Maunsell, 93, 108, 151, 158
Fillmore, Millard, 242
fires, 99–100
Flannery, Lot, 342
Flint River, 297
Florida, 12, 250–51, 257, 267, 273, 279, 299, 317
Foot, Solomon, 144, 151
Forbes, Charles, 95
Ford's Theatre, xi, 94–97, 99–101, 103, 106–7, 109–11, 116, 119, 127, 130, 144, 181, 240, 287, 392
Ford's Theatre Society, 400
Fort Crawford, Wisc., 58
Fort Monroe, Va., 329–31, 333–34, *335*, 337–40, 341–42, 349–53, 355
Fortress Monroe, Va., 16
Fox, Gustavus, 164
Frailey, Commander, 330
France, 309
Francis, George, 101–2, 103–6, 113, 199
Francis, Huldah, 101–2, 103, 105, 113, 199
Franklin, Sir John, 229
Fredericksburg, battle of, 341
Fredericksburg, Va., 176–77
Freemasons, 218, 261
French, Benjamin Brown, 93, 117–18, 133
 on Davis's capture, 323–24
 diary of, 118, 139
 on Grand Review, 337
 Lincoln's assassination and, 117, 119
 Lincoln's funeral and, 133, 144, 150–51, 152, 210–11, 286, 391

on Mary Lincoln's departure, 335
mourning arrangements made by, 119, 133, 140, 337

Gardner, Alexander, 188–89, 239, 241
Garfield, James, 361
Garrett's Farm, 367
Garrison, William Lloyd, 340
Garrison's Landing, N.Y., 232
Gary, Martin Witherspoon, 265
Georgetown, 171, 173, 192, 208
Georgia, 67, 250, 252, 257, 296, 309, 313, 318
Confederacy supported in, 280
Georgia Historical Society, 371
Georgia State Fair, 373
Gettysburg, battle of, 112, 121
Gettysburg, Pa., 16, 156
national cemetery at, 140, 181
Gist's Ferry, 250
Gobright, Lawrence A., 151
Gordon, John B., 367
Gorgas, Josiah, 11
Grady, Henry, 371
Grand Review, 329–30, 331, 333, 336–37
Grant, Julia, 96
Grant, Ulysses S., 36, 41, 75, 77, 92, 94, 96, 207, 245, 246, 248, 289
advance on Richmond by, 13–14, 16–17, 20, 24
Davis's surrender and, 330–31
Grand Review ordered by, 329–30, 331
Lee's surrender to, 77–79
Petersburg captured by, 22, 24, 34, 38
Graves, Thomas Thatcher, 30, 32, 47, 62–63
Great Britain, 132
Great Central Fair, 177
Great Western Railroad, 277
Greeley, Horace, 356
Greene, Nathaniel, 371
Greensboro, N.C., 15, 67, 85, 252
Cabinet Car in, 89
as Confederate capital, 82, 84, 87–92, 95, 121–22, 280
evacuation of, 122
Greenville, Ala., 382
Grover's Theatre, 96, 101, 114
Gulf of Mexico, 299
Gunther, Charles F., 125
Gurley, Phineas D., 123–24, 172–73, 189–90, 207, 284
Gurney, Jeremiah, 292
Gurney, Thomas, *230*, 239–41, 379

Hall, Rev. Mr., 188
Halleck, Henry W., 115, 150–51, 244, 330
Hampton, Wade, 197, 207, 219, 238, 243–44
Hampton Roads, 330
Hampton Roads peace conference, 47
"Hang Him on the Sour Apple Tree," 348
Harding, Warren G., 398
Harnden, Henry, 256, 298
Harpers Ferry, Va., xi
Harper's Weekly, 50, 350
Harrington, George, 141–43, *142*, 148–52, 156–59, 184–85, 188, 287, 315, 396
Harris, Clara, 96, 106–8, 113
Harris, J. H., 398
Harrisburg, Pa., 199, 217, 218, 219, 239
Harrison, Burton, 183, 274, 355, 361, 390
in Charlotte, 194, 223
in departure from Greensboro, 122
in evacuation to Greensboro, 83, 84–88
and final capture, 307–9, 311–12, 315–16, 317, 320–22
marriage of Constance Cary to, 6
Varina Davis assisted by, 10, 66, 82–83, 224, 262–63, 298–99, 302–3
Harrison, William Henry, 142, 188
Hathaway, Lieutenant, 315, 321
Hawes, G. W., 230
Hawkinsville, Ga., 299
Hay, John, 181, 324
Hayes, Margaret Davis, 387, 390
Helm, Emilie Todd, 174
Herndon, William, 57
Herold, David, 341
Hill, Ben, 371
Hollywood Cemetery, Richmond, 2, 357, 383–85, 398
Hooker, Joseph, 247
House of Representatives, U.S., 149, 151, 170
Howell, Jefferson Davis, 315
Howell, Maggie, 94, 315–16
Hudson, N.Y., 232, 233
Hudson River, 225, 232, 235, 394

Hudson River Railroad, 231
Hurricane Katrina, 402–3

Illinois, 51, 53, 57, 95, 98, 143, 148, 151, 152, 154, 155, 167, 168, 171, 207, 209, 293, 295
Illinois Central railroad, 156
Independence Hall, 157, 220–23, 225
Indiana, 51, 155, 247, 259–61, 293
Indianapolis, Ind., 199, 261
Indiana State Journal, 259–61
Industrial Revolution, 361
Irwinville, Ga., 298, 300, 302, 304, 317–18, 321

Jackson, Andrew, 123
Jackson, James W., 164
Jackson, Miss., 248
Jackson, Thomas J. "Stonewall," 72
James River, 6, 13, 17, 22
Jefferson, Thomas, 38, 53, 100, 381
Jefferson Davis (Varina Howell Davis), 390
Jefferson Davis Monument Association, 386
Jersey City, N.J., 225
Johnson, Andrew, 196, 276, 337, 341
  hostility to South by, 121, 199–200, 204
  and Mary Lincoln's refusal to vacate the White House, 287, 324
  reward for Davis offered by, 268–69, 313–14
  Varina's petitions to, 352, 353
Johnson, Lyndon B., 398
Johnston, Albert Sidney, 9, 312
Johnston, Joseph E., 14, 67, 87, 90, 91–92, 182, 243–44, 296
  Confederate gold rumor and, 244, 365
  surrender by, 212, 238, 244, 256, 365
Johnston, William Preston, 9, 25, 33, 299, 302
  and capture with Davis, 307, 312–13, 315, 321, 322
John Sylvester, 355
Jones, Jim, 306–7
Jones, J. William, 379
Justice Department, U.S., 157

Keckly, Elizabeth, 59, 108, 143–44, 146–47
Keene, Laura, 94, 130
Kennedy, Jacqueline, 396
Kennedy, John F., funeral of, 396
Kentucky, 51, 152, 247, 400
King, Albert F. Africanus, 106–7, 108
Kinney, Constance, 119
Kinney, Mary, 119
Knoxville, Tenn., 68

Lake Erie, 246, 249
Lamon, Ward Hill, 90, 140, 152, 161–63
Lancaster, Pa., 219–20
"Last Ditch Polka," 348
Leale, Charles A., 102–3, 105–10, 116–18, 120–21, 133
Lee, Mary Anna Randolph Custis, 5, 21, 37, 80
Lee, Robert E., 26, 29, 37, 40–41, 67, 73, 76, 123, 273–74, 296, 386
  correspondence by, 6, 13, 20, 21, 22–24, 74, 90–91, 205–6
  on Davis, 62
  Davis urged to surrender by, 205–6
  death of, 359–61
  General Order No. 9 of, 81, 84
  letter on Davis's release, 357
  mausoleum of, 365
  popularity of, 7–8, 19, 69, 74, 347
  retreat from Petersburg by, 5–6, 8
  return to Richmond by, 80, 205
  surrender by, 77–82, 86, 90–92, 95, 102, 147, 195, 197, 205–6, 224
  surrender inevitable to, 76–77, 205–6
  threat of defeat to, 15, 19, 24, 75
  and threat to Richmond, 5–7, 13–15, 19–24
Lee, Robert E., Jr., 92
Leesburg, Va., 167
Lexington, Ky., 59
Lexington, Va., 365
Libby Prison, 46, 125
Liberty Bell, 221–22
Library of Congress, 398, 400
Lincoln, Abraham, 3, 27, 35, *43*, 77, 289, 329
  appearance of, 54–55, 160
  assassination of, xi, xii, xiii, xiv, 57, 100–102, 111, 114–16, 121–22, 123, 129, 140, 147, 181–82, 183, 196, 204, 207, 214, 219, 239, 240, 242,

254, 268, 276, 289, 293, 296, 314, 318, 326–27, 347, 350, 354, 356, 367, 374–75, 388, 389
autopsy of, 134–38, *134, 135,* 145, 243, 354
Baker and, 167–68
in Black Hawk War, 57–58
blood relics of, 181, 395–96
burial place of, 139, 143, 145, 152–53
cabinet of, 94, 111–12, 147, 149, 170
cartes de visite in honor of, xii-xiii, *xiii*
changing views on blacks held by, 53–54, 90
charisma of, 55
coffin of, 160–61, 191–92, 209, 216–17, 218, 221–22, 252–53, 254, 264–65, 275, 281–82, 284, 286, 293, 392, 393
Confederate surrender and, 78–79
correspondence by, 20, 22, 24, 34, 38, 75, 165–66, 177
deathbed of, 105–21, *112,* 123–25, 128–30, *128*
death of, 121, 123, 125, *132,* 160, 183, 354, 378, 382, 385
death pageant of, 293–96, 388, 392
early romance of, *see* Rutledge, Ann M.
East Room viewing and funeral in, 146, 148–50, 152, 158–59, 160–63, 181, 183–91, *187,* 197–98, 199, 249, 277, 281, 285
in 1860 election, 51, 53, 98, 246, 292
in 1864 election, 76, 292
Ellsworth and, 163–67, 168
embalming of, 138–39, 144, 145, 147–48, 155, 186, 252
in field with Union Army, 15–20, 24, 33–34, 36, 38–39, 73, 74–75
first inaugural address of, 6, 287
Ford's Theatre outing of, 94–96
funeral of, 133
funeral expenses of, 284–86
funeral pageant for, xiii, xiv
funeral procession of, 149–50, 158, 190–93, 209, 286, 330, 382, 386, 400
funeral train of, 154–58, 187, 207–18, *207, 211,* 219–23, 225–31, 232–37, 241–43, 245–47, 248–50, 252–56, 258–62, *260,* 263–65, 270–73, 274–78, *275,* 281, 286, 289–96, 324, 331, 382, 383, 388, 391, 393–95
Gettysburg address of, 140, 181
hearses of, 191–92, *191,* 208–9, *221,* 226–27, *226,* 252, 255, 264–65, 286, 290–91, 391
"House Divided" speech of, 281
in House of Representatives, 52, 98, 156, 167
humor of, 44, 55, 64, 94
inaugural train of, 27, 140, 153–54, 156, 213–14, 242, 277–78, 292
inauguration of, 167, 191
indifference to luxuries of, 55, 138–39, 290–91
Kentucky birth of, 51
lack of vengefulness towards Davis and Confederates by, 17–18, 63–64, 71, 78, 80, 182, 196, 200, 296
law office of, *272,* 273
as lawyer, 51–53, 57, 139, 152, 156, 163, 270, 271
life mask of, 380
lying in state of, 140, 148, 150, 185, 197–200, 207–9, 215, 286, 391
McCullough and, 176–77
manual labor of, 54, 146
marriage of, *see* Lincoln, Mary Todd
modesty of, 45
mourning of, as national phenomenon, 123–24, 185, 388
mourning ribbons for, *190, 213,* 396
national day of mourning for, 340
open casket for, 155
Philadelphia speech of, 221–22
poems of loss and death fascinating to, 168
postwar policies and plans of, 90, 94
prefiguring dreams of, 111–12, 161–63
railroads and waterways supported by, 155–56
Richmond visit of, 38–39, 42–47, 62–65, 68, 73, 210, 356
as saint and martyr, 108, 148, 198, 293, 293–95, 388
second inaugural address of, xii, 16, 18, 143, 193, 283

Lincoln, Abraham *(cont'd)*
second inauguration of, 188–89, 208
self-education of, 51–52, 60
in Senate, 48, 53
skeptical but accepting nature of, 61
slavery opposed by, xii, 52–54, 60–61, 333–34
smallpox of, 146
souvenirs of, 129–30, *395*, 396
Springfield farewell speech of, 277–78
strength of, 54–55
theater loved by, 96, 181, 325–26
threats to, 36, 60, 213–14
two hundredth birthday celebrations for, 400
and U.S. Arsenal tragedy, 177–81, 342
wax figures of, 344, 392
White House private passage of, 145–46
Willie's death and, 1–2, 168–74
as writer and storyteller, 64–65
Lincoln, Abraham, remains of, 186–87, 216–17, 252, 291, 379–80, 392
cottage industry surrounding, 401
felt absence at Grand Review of, 336–37
ghost train legend and, 393–94
photo of, *230*, 234–37, 239–41, 248, 292, 379
plot to kidnap, 361
and transportation to Peterson house, 102–5
transported to White House, 126–27, 131–32, 207
undertakers traveling with, 210, 229–30, 235, 252, 281–82
in White House Guest Room, 132–39, 143–46, 160
Lincoln, Eddie, 161, 167, 168, 174, 389
Lincoln, Ill., 271
Lincoln, Mary Todd, 16, 19–20, 22, 56, 118, 165, 167, 188, 236
death of, 389
in delayed departure from White House, 286, 287, 324, 335, 389
difficult nature of, 59–60, 113, 155
extravagance of, 59, 161
at Ford's Theatre, 94–96
and Lincoln's burial place, 139, 143, 145, 152–53, 271
Lincoln's courtship of, 59
Lincoln's deathbed and grief of, 106, 107–8, 113–14, 117, 119–21, 124, 125
and Lincoln's prophetic dream, 161–63
lock of hair requested by, 135
mental instability of, 200, 287, 335, 389
mourning for Lincoln by, 131–33, 143–44, 146–47, 155, 183, 200–201, 209–10, 281, 285, 335, *389*
Richmond toured by, 90
and Willie's death, 173–74, 177
Lincoln, Robert:
in Civil War, 16, 34, 94
on funeral train, 209
at Lincoln's deathbed, 114, 119, 124
in Lincoln's funeral procession, 192
Lincoln, Robert Todd, 96, 139, 146, 152–53, 168, 335, 389–90
Lincoln, Sarah Bush Johnson, 51, 254
Lincoln, Tad, 16, 44, 46, 96, 168–69, 173, 174, 335
death of, 389
and father's assassination, 114, 117, 146–47, 183, 209–10, 324
in Lincoln's funeral procession, 192
Lincoln's relationship with, 19, 209–10, 389
Lincoln, Thomas, 51, 146
Lincoln, William Wallace "Willie," 1–2, 60, 114, 138, 144, 146, 161, 168–73, 176–77, 186, 389
coffin of, on Lincoln's funeral train, 208–9, 255–56, 259, 281–82, 284
Lincoln-Douglas debates of 1858, 48, 49–50, 53, 270
Lincoln Memorial, 396, 397–98
Little Bighorn, battle of, 361
Little Falls, N.Y., 241–42
Lockwood, H. H., 215
Locust Grove, La., 56
Longstreet, James, 197
Louisiana, 56, 247
Louisville, Ky., 157
Lubbock, Francis R., 9, 25, 33, 122, 301–3, 307–8, 311–12, 315, 316, 321, 322
Lynchburg, Va., 74

McCallum, Daniel C., 202, 277
McClellan, George B., 76, 141, 167
McCook, General, 258
McCulloch, Hugh, 120, 142
McCullough, Fanny, 176–77
McElfresh, Sallie, 180
Macon, Ga., 262, 298, 320–23, 372
Madison, James, 53, 381
malaria, 56
Mallory, Stephen, 206–7, 238, 243, 247, 266–67, 273
 in Danville evacuation, 81–82, 85
 in Greensboro, 88–90, 122
 on Lincoln's assassination, 196–97
 post resigned by, 270, 288
 in Richmond evacuation, 9, 28, 29, 37, 38, 40–41
Manhunt: The 12-Day Chase for Lincoln's Killer (Swanson), xi
Manifest Destiny, 52
Marine Corps, U.S., 76, 192
Maryland, 121, 155, 213, 296
Massachusetts, 187
Maurin, Major, 315
Mayo, Joseph, 24–25, 72
Meade, George Gordon, 16–17
Meigs, Montgomery, 140–41, 211
Melville, Herman, 293
Metairie Cemetery, 382
Mexican War, 9, 49, 52, 78, 123, 141, 309, 402
Mexico, 52, 80, 122, 123, 197, 224, 317
 lancers from, 309
Michigan City, Ind., 261
Miles, Nelson A., 341, 349–50, 352, 354
Miller, John K., 245
Minnigerode, Charles, 5, 7, 350
Mint, U.S., 400
Minty, R. H. G., 257, 297
Mississippi, 48, 50, 72, 123, 299, 352, 357, 376, 400
 Gulf Coast of, 362–63, 366, 402–3
 secession of, 50
Mississippi City, Miss., 373–74
Mississippi Rifles, 49, 52, 309
Mississippi River, 14, 56, 80, 156, 197, 206, 219, 224, 225, 248, 250–51, 256–57, 266, 267, 279, 317–18, 329, 365
Missouri Compromise, 52
Mitchell, William S., 158
Mobile, Ala., 3, 382
Mobile Register, 363–64
Monroe, James, 53
Monterrey, battle of, 52
Montgomery, Ala., 67, 77, 170, 366–71, 382–83
Montreal, 121
Moody, Captain, 303, 315
Morris, General, 214
Morristown, N.J., 225
Morton, Emily Jessie, 351
Mount Vernon, 188
Mumler (spirit photographer), 389
Munroe, Seaton, 100, 129–30
Museum of Funeral Customs, 393
Museum of the Confederacy, 397, 401–2

Napoleon I, Emperor of France, 309
Nashville, Tenn., 123
Nassau, 182
Natchez, Miss., 123
Natchez Trace, 123
National Park Service, 130
National Portrait Gallery, 130
National Republican, 168
Native Americans, 300, 354
Naval Academy, U.S., 9
Navy, U.S., 44, 192
Navy Department, Confederate, 29
New England, 187
New Jersey, 155, 292
New Orleans, La., 3, 76, 156, 225, 363, 364, 376, 377, 380, 397
 Davis's funeral in, 382–83, *384*
New Orleans *Picayune,* 364
New Paris, Ohio, 259
New Salem, Ill., 55, 56–57, 156, 173
Newseum, 400
New York, N.Y., 99, 156, 157, 165, 199, 203, 234, 246, 276, 281–82, 290, 291, 328, 340, 354, 391, 394–95
 Civil War draft riots in, 226
 Davis hung in effigy in, 344–45
 funeral procession in, 225–31, *226, 230,* 232
 Lincoln corpse photograph in, *230,* 234–37, 239–41, 292, 379

New York, N.Y. *(cont'd)*
Southern sympathizers in, 34, 226, 390
Varina in, 390
New York Central Railroad, 393
New York Herald, 164, 228–29
New York State, 155, 245, 292
New York Times, 80, 116, 190, 231, 239, 253, 345, 368, 371, 372
New York Tribune, 241, 326–27
New York World, 367
Nichols, George W., 151
Nichols, William A., 150, 152, 158–59
Nicolay, John G., 170
North:
Confederacy as seen by, xiv
Davis as seen by, xiv
Davis's admiration for, 49
mixed feelings about Davis in, 340–41, 355, 356–57
national day of mourning for Lincoln in, 340
North Carolina, 14, 37, 41, 66, 194, 257, 280, 297, 365
Confederate army in, 87
North East, Pa., 246
Notson, H. M., 133

Oak Hill Cemetery, 171, 173, 208, 383
Oak Ridge Cemetery, 282
Ocmulgee River, 297
O'Connor, Charles, 356
Ogeechee River, 289
Oglesby, Richard J., 151
Ohio, 123, 155, 292
Ohio River, 123
Old Arsenal, U.S. Army Penitentiary at, 287, 342
Old Capitol Prison, 350
O'Melia, Mary, 24–25, 71
Oregon, 167
Our American Cousin, 94, 96

Palmer, General, 313
Papers of Jefferson Davis, 401
Parker, William, 27, 29–30, 266–67, 269, 273
Parks, W. P., 376
Pelouze, Louis H., 126
Pennsylvania, 155, 292
Penrose, Charles B., 16, 44
Pensacola, Fla., 247
Peoria Speech, Lincoln's, 52–53
Petersburg, Va., 205
Confederate withdrawal from, 5–6, 8, 13, 34
Grant's capture of, 22, 24, 34, 38
Petersen, Anna, 98, 124
Petersen, Anne, 124
Petersen, Julia, 124
Petersen, William, 98–99, 124, 128
boarding house of, 95–97, 98–121, *104*, *112*, 123–230, *128*, 131–32, 143–44, 146, 151, 158, 161, 181, 188, 192, 199, 207, 286, 287
Philadelphia, Pa., 99, 157, 177, 199, 203, 219–23, *221*, 225, 239, 281
Philadelphia Inquirer, 220, 223
Pickett's Charge, 194, 373
Piedmont Railroad, 67
Pierce, Franklin, 49, 118, 355
Piqua, Ohio, 259
Pittsburgh, Pa., 201
Pleasant Valley, Ohio, 259
Polk, James K., 52
Porter, David Dixon, 17, 42, 44–47, 62–63
Port Hudson, La., 156
Port Royal, Va., 238
Post Office, U.S., 371, 400
Post Office Department, Confederate, 289
Potomac River, 163, 164
Powell, Lewis, 121, 147, 341
press, Northern, 289–92, 315
Prison Life of Jefferson Davis (Craven), 354
Pritchard, Benjamin D., 256, 298, 304–6, 308–9, 314–15, 320–21
Proctor, Redfield, 380–81
Providence, R.I., 340
Purington, Lieutenant, 305–6

racism, 398
Radford, William, 330
Raleigh, N.C., *385*
Rathbone, Henry, 96, 106–7
Raymond, Henry J., 239–40

Raynor, Mary Virginia, 242
Reagan, John, 196, 247–48, 251, 266–67, 278–79, 288–89, 302, 355
  and capture with Davis, 308, 312, 314, 320–23
  in Richmond evacuation, 9, 27–28
Reconstruction, 358, 361
Republican National Convention of 1860, 141, 143, 152, 270
Revolutionary War, American, 371
Richmond, Ind., 259
Richmond, Va., 2, 5, 76, 86, 123, 140, 176, 331, 346
  burning of, 28–29, 30–32, 39, *40*, 42, 77, 331, 332
  as Confederate capital, 3, 82, 83, 125, 170, 176, 198, 274, 332, 351, 356
  Davis memorial for, 386–87
  Davis's reburial in, 383–86, 390, 397–98
  Davis's return to and court appearance in, 355–56
  evacuation of, 9–13, 20–29, 34, 67, 68, 85, 87, 88, 89, 95, 147, 183, 279, 296, 299, 317, 383
  fall of, 8, 15, 26, 34–39, *35*, 68–70, 73–74, 77, 80, 82, 102
  food shortages in, 12, 31–32
  Lee's return to, 80, 205
  liberated blacks in, 26, 39, 44–46, 64
  Lincoln in, 38–39, 42–47, 62–65, 68, 73, 210, 356
  looting and riot in, 26, 29–30, 31–32, 39–40, 332
  monuments for Davis in, 398
  peace negotiations in, 62, 73
  severe rainstorm in, 331–32
  theater fire in, 100
  as tourist destination, 90, 93
  Union threat to, 5–8, 13–15, 18–26, 31, 205
Richmond and Danville Railroad, 26–27
Richmond Times, 331–32, 338–40
Richview, Ill., 376
Rise and Fall of the Confederate Government (Davis), 363, 366
River Queen, 16–17, 19, 24, 42–43, 75–76
Robert E. Lee mausoleum, 365
Rocketts, 44, 356
Rucker, D. H., 126, 207
Rutledge, Ann M., 56–57, 161, 168, 173

Safford, Henry, 100–103, 105
St. John's Church, 192
St. Louis, Mo., 58
St. Paul's Episcopal Church, 5, 7–8, 20, 176
Salisbury, N.C., 181
Saluda River, 256
Sandburg, Carl, 284
Sandersville, Ga., 298
Sandford, Moses, 101–2, 181
Sands, Frank, 210, 229–30, 281–82
Sands & Harvey, 286
Sangamon River, 156
Sanitary Commission, U.S., 178
Sanitary Fair, 345
Saunders, George N., 268–69
Savannah, Ga., 3, 76, 323, 329, 371–73
Savannah River, 278, 280–81
"Savior of Our Country, The," 255–56
Scaife's Ferry, 250
Schofield, John M., 14
Scott, Winfield, 141, 191
secession, 50–51, 363
  legal status of, 355, 356
Selden & Co., 63
Selma, Ala., 67, 77
Semple, James A., 89
Senate, U.S., 115, 149, 151, 167, 170
  Davis in, 49–50, 369
  Lincoln in, 48, 53
Seven Pines, battle of, 341
Seward, Frances Adeline Miller, 115, 391
Seward, Frederick, 110, 115, 147
Seward, William H., 19, 73, 78, 170
  attack on, 110–11, 114–15, 119, 121, 144, 147, 196, 268, 276, 287, 391
Shakespeare, William, 34, 52, 60, 174, 226
Sheridan, Philip H., 28, 75
Sherman, William Tecumseh, 14, 17–18, 182, 212, 289, 329
  Davis conspiracy theory of, 365–66
  Johnston's surrender to, 238, 256, 365
Shiloh, battle of, 9, 312
Shubrick, William B., 151

Simpson, Matthew, 184, 189, 282–84
Sing Sing, N.Y., 232, *233*
slavery, 100, 258–59, 356, 357, 367, 382, 388, 401
  Brown's raid and opposition to, xi-xii
  in Confederacy, 3
  Davis and, 49, 60–61, 357, 367, 401
  expansion of, 52, 54
  Lincoln's opposition to, xii, 52–54, 60–61, 333–34
  U.S. Constitution and, 53
smallpox, 262–63
Smith, Anna Davis, 56
Smith, Gerrit, 356
Smith, Kirby, 14
Smithsonian Institution, 49, 124
Smithsonian National Museum of American History, 400
Snow, Parker, 229
Soldiers' Home, 93
South, 276
  agricultural empire of the, 2
  Brown as seen in, xii
  Davis as spokesman for, 50
  feeling of superiority in, 322
  impact of Lincoln's death on, 196
  Johnson and, 121, 196, 199–200
  postwar, 361
  reconstruction of, 92
  Union army as seen by, xiv, 322
South Carolina, 238, 245, 250, 252, 257, 268, 297
  secession by, 98–99
Southern Express Company, 195
Southern Historical Society, 361
Southern Historical Society Papers, 363
Speed, James, 157, 207
Speed, Phillip, 157
Spotswood Hotel, 356
Springfield, Ill., 27, 59, 143, 145, 153, 155, 161, 163, 199, 201–2, 213, 247, 293, 295, 331, 391
  Abraham Lincoln Presidential Library in, 392–93
  Lincoln's burial in, 271–78, *275, 283,* 287, 389
Springfield (Mass.) *Republican,* 340
Stanton, Edwin M., 79, 92, 130, 136, 170, 180, 313
  arrests ordered by, 240
  Booth manhunt and, 111, 147, 204, 244, 268, 296
  on Davis capture, 330–32
  Davis's captivity and, 339, 341, 352, 354
  Davis suspected of assassination involvement by, 276, 287, 297, 320, 326
  and fall of Richmond, 17, 18, 20, 34–36
  Grand Review and, 331
  Lincoln corpse photograph and, 234–37, 239–41, 248
  Lincoln's corpse and, 138–40, 379–80
  Lincoln's funeral and, 140–42, 184–85, 189
  Lincoln's funeral train and, 154–58, 199, 200–204, 207–9, 217, 219, 225, 246, 274, 277
  at Petersen house, 113, 120–21
  post-funeral planning by, 143, 151, 153
  on Seward attack and Lincoln assassination, 110–11
  women's clothes rumor propagated by, 324, 327–29, 344
State Department, U.S., 110
states' rights, 363
Stephens, Alexander, 47
Stevens, Thaddeus, 220
Stevenson, Job E., 258–59, 284
Stone, Robert King, 109, 117, 133, 135–36, 169
Stoneman, George, 88
  Confederate gold rumor and, 244–45
Strong, George Templeton, 328
Stuart, J. E. B., 2
Sulivane, Clement, 8, 20, 31
Sumner, Charles, 114–15
Supreme Court, U.S., 52, 62, 171
Surratt, Mary, 341, 344
Sutherlin, William T., 37–38
Swan, Otis D., 157
Swancey's Ferry, S.C., 256
Swanson, Claude A., 387
Syracuse, N.Y., 242

Taft, Charles Sabin, 106–7, 108–9, 120, 124, 126–27, 133, 135
Taft, William Howard, 398
Taltavul's Star Saloon, 96, 102
Tanner, James, 126
Taylor, Sarah Knox, 55–56, 58, 175, 357, 366, 376–77
Taylor, Zachary, 9, 50, 55, 142, 188
telegraph, 18–19, 20
Texas, 14, 33, 80, 247
  as possible new Confederate center, 197, 219, 279, 299, 317
T. Gurney & Son, 239–40
Thomas, D. C., 126
Thomas, George, Confederate gold rumor and, 244–45
Thomas, Lorenzo, 141
Thomas, William B., 157
Thompson, Jacob, 268–69
Tifton, Ga., 398–99, *399*
Tillson, Davis, 245
torpedo general, 85–87
Townsend, Edward D., 141, *203*, 352
  Lincoln corpse photograph and, *230*, 234–38, 248
  Lincoln's funeral train and, 201–3, 209, 213–19, 225, 232–33, 243, 245–46, 248–50, 253, 255, 261–62, 271, 274–77, 294
Townsend, George Alfred, 145–46, 185–87, 192–93, 198
  on cleaning out of Lincoln's office, 324–26
Treasury Department, U.S., 93, 108, 141–42, 150, 151, 156, 158, 187, 192, 287, 336–37
Tredegar Iron Works, 11, 401
Trenholm, George, 9, 32, 33, 38
Tucker, Beverly, 268–69, 314
Tuscarora, 330
Tyger River, 250
Tyler, E. B., 214

Ulke, Henry, 128–30
Ulke, Julius, 128–30
Union Army, 82, 87, 92, 182, 213–14, 275, 289, 356
  approach to Richmond by, 5, 13–14, 18–21, 26, 30, 31
  cavalry of, 88, 194, 244–45, 278–79, 299–302, 304–18
  Davis honored by, 321–22
  Grand Review of, 329–30, 331, 333, 336–37
  Lincoln's popularity with, 75–76
  prisoners from, 46, 344
  in Richmond, 44
  as seen by South, xiv
  as threat to Davis, 200
  see also Army, U.S.
Union League Association, 220
Union League Club, 157
Union Pacific Railroad, 391
United Confederate Veterans, 386
United Daughters of the Confederacy, 386, 399
United States Military Railroad, 27, 208–9, 277, 331, 382
Urbana, Ohio, 259
U.S. Army Medical Museum, 136, 137, 395
U.S. Arsenal, tragedy at, 177–81, 342
Usher, John P., 117

Valentine, Edward, 386
Vanderbilt, Cornelius, 356
Vicksburg, battle of, 112
Vicksburg, Miss., 3, 76, 123, 156, 248
Vignodi, Professor, 344
Vincent, Thomas, 116
Virginia, 40–41, 95, 121, 194, 206, 297, 317, 401
  aristocracy of, 3
  as Confederacy's principal state, 37, 70, 82

Walker, J. M., 83
Wallace, Lew, 215, 240
War Department, Confederate, 288
War Department, U.S., 18, 35, 101, 150, 152, 154, 181, 190, 235, 239–41, 249, 296, 297, 330, 341, 344, 350, 352, 380
War of 1812, 132, 141
Washington, D.C., 1–2, 16, 18, 19, 27, 48, 73, 75, 77, 78, 90, 100, 119, 121, 123, 126, 130, 141, 152–53, 155, 167, 173,

Washington, D.C. *(cont'd)*
178, 199, 203, 209, 213, 214, 234, 237, 239, 242, 271, 273, 275, 276, 277, 281, 287, 295, 296, 331, 358, 361, 371, 383, 391, 393
antebellum, 352
boardinghouse culture in, 98
Davis's capture and, 318–19, 321, 323–24, 326
Davis's death and, 380
fall of Richmond celebrated in, 34–36, 39, 102
Grand Review in, 329–30, 331, 333, 336–37
Lee's surrender celebrated in, 79–80, 81, 84, 92–93, 102
Lincoln's funeral ceremonies planned in, 141, 143, 148–52
reaction to Lincoln's assassination in, 116
Richmond's nearness to, 3
Washington, Ga., 73, 92, 273, 278–80, 288–89
Washington, George, 53, 143, 170, 188, 221, 265, 278, 293, 382
Washington and Lee University, 386
Washington Daily Morning Chronicle, 180, 323–24
Washington Evening Star, 34–35, 124, 170, 171–72, 173, 178–79, 180
Washington Navy Yard, 94, 156, 164
Washington Post, 381
wax figures:
of Booth, 344, 392–93
of Davis, 344–45
of Lincoln, 344, 392
Webster, Daniel, 52
Weitzel, Godfrey, 47, 62, 63
Welles, Gideon, 75, 79–80, 92, 108, 110–12, 119, 138, 170, 207, 237, 324, 330–31, 336
Welles, Mary Jane, 108, 119–20, 133
Western Railroad Corporation, 187
Westfield, N.Y., 246
West Point, U.S. Military Academy at, 48, 54, 55, 232
Wheeler, Joseph, 238
White House, 3, 15, 16, 18, 60, 78, 84, 93, 96, 103, 110, 111, 114, 119, 130, 131, 140, 151, 153, 156, 161, 164, 167, 215, 337, 400
East Room viewing and funeral in, 146, 148–50, 152, 158–59, 160–63, 181, 183–91, *187,* 197–98, 199, 249, 277, 281, 285
Lincoln's corpse in Guest Room of, 132–39, 143–46, 160
Lincoln's corpse transported to, 126–27, 131–32, 207
Mary Lincoln's delayed departure from, 286, 287, 324, 335, 389
other funerals in, 165, 170–71, 173
in War of 1812, 132
White House of the Confederacy, 2, 3, 5, 15, 60, 170, 174, 397, 401–2
evacuation of, 9–10, 12–13, 22, 24–25
Lincoln at, 47, 62–64, 66, 73
photographs sold of, 63
Union seizure of, 31, 32
Whitman, Walt, xiv, 34, 124, 262, 294–95
Wilcox's Mills, Ga., 204
Wilde, Oscar, 363–65, *364*
William P. Clyde, 330
Willis, Lee H., 376
Wilson, Henry, 164
Wilson, James, 256–58, 296–98, 313, 320–22, 330
Winslow, Edward F., 257
Wirt, William, 52
Wirz, Henry, 344, 350, 356
Wisconsin, 57
Wise, John S., 30
Wofford, William T., 67–68
Wood, John Taylor, 9, 25, 33, 88, 92, 122, 302, 310, 312–13, 318
Woodward, Janvier J., 133–36, 138, 354
Wormley, James, 158
wrestling, 55

Yates, Richard, 151
York, Pa., 218
Yorkville, S.C., 243